THE CITY CULTURES READER

The new millennium is marked by the fact that, for the first time, the majority of the human inhabitants of the Earth will dwell in cities. This is a collection of 60 texts on the cultural aspects, frameworks and perceptions of cities, reflecting the concerns for diversity and difference, rather than conventional notions of universality.

The City Cultures Reader is organized into twelve sections, drawing on material from cultural studies, architectural criticism and theory, cultural and human geography, urban design and critical theory. Topics covered include city form, city cultures in relation to politics and economics, everyday life, memory and identity, cultural representation, ecology, social justice, resistance, utopian and dystopian visions, and possible futures.

This exciting collection of readings draws out the comparisons and contrasts by the juxtaposition of texts in each section, and adds a critical and informative dimension through the general and section introductions.

Malcolm Miles is Reader in Art and Design at the University of Plymouth, **Tim Hall** is Lecturer in Human Geography at Cheltenham and Gloucester College of Higher Education, and **Iain Borden** is Director of Architectural History and Theory, and Reader in Architecture and Urban Culture at The Bartlett, University College London.

THE CITY CULTURES READER

edited by

Malcolm Miles, Tim Hall
and
Iain Borden

London and New York

First published 2000
by Routledge
11 New Fetter Lane, London EC4P 4EE

Simultaneously published in the USA and Canada
by Routledge
29 West 35th Street, New York, NY 10001

Routledge is an imprint of the Taylor & Francis Group

© 2000 Selection and editorial matter Malcolm Miles, Tim Hall and Iain Borden

Typeset in Sabon by Solius (Bristol) Ltd, Bristol

Printed and bound in Great Britain by
TJ International Ltd, Padstow, Cornwall

British Library Cataloguing in Publication Data
A catalogue record for this book is available from the British Library

Library of Congress Cataloging in Publication Data
The city cultures reader / edited by Malcolm Miles, Tim Hall and Iain Borden
    p.   cm.
    1. Cities and towns.   2. City and town life.   3. Culture.   I. Miles, Malcolm.
II. Hall, Tim.   III. Borden, Iain.
HT151.C5822   2000   99-16477
307.76–dc21

ISBN 0–415–20733–9 (hbk)
ISBN 0–415–20734–7 (pbk)

CONTENTS

III   The Culture Industries

IV   Culture and Technologies

2   LIVING URBAN CULTURE

V   Everyday Life

VI   Memory, Imagination and Identity

VII   Representations of the City

ACKNOWLEDGEMENTS

The editors are particularly grateful to Susie Medley, who undertook the task of tracing rights holders and seeking permissions from them; and to Sarah Lloyd who originally commissioned the Reader, and Sarah Carty who saw through its later stages at Routledge.

The editors are grateful to the following publishers and authors for permissions to reprint material in this collection. Considerable effort has been made to trace and contact copyright holders in order to gain replies prior to publication. The editors and publisher apologize for any oversights or omissions. If notified, the publisher will endeavour to remedy these at the earliest opportunity.

Henry N. Abrams Inc.: Susanna Torre (1996) 'Claiming the Public Space: The Mothers of Plaza de Mayo' in D. Agrest, P. Conway and L.K. Weisman (eds) *The Sex of Architecture*, pp. 241–50.

The Architectural Association: Christian Norberg-Schulz (1976) 'The Phenomenon of Place', *AA Quarterly*, 8(4): 3–10.

The Architectural Association and Bernard Tschumi Architects: Bernard Tschumi (1994) 'Spaces and Events' from *Architecture and Disjunction*, pp. 139–49.

A & U Publishing Co: Herman Hertzberger (1991) 'The Public Realm' (first section), *Architecture and Urbanism*, April: 12–45.

Artforum: Patricia Phillips (1988) 'Out of Order: The Public Art Machine', *Artforum*, Dec.: 92–6.

Beacon Press: Lewis Mumford (1956) 'The Monastery and the Clock' from *The Human Prospect*, extract from section 1, pp. 3–9.

Marshall Berman: 'Robert Moses: The Expressway World' from *All That Is Solid Melts Into Air* (Simon & Schuster, 1982) pp. 290–312.

Black Rose Books: Murray Bookchin (1986) 'Introduction' from *The Limits of the City*, pp. 6–10.

Blackwell: David Harvey (1989) 'From Managerialism to Entrepreneurialism: The Transformation in Urban Governance in Late Capitalism', *Geografiska Annaler*, 71B: 4–17; Henri Lefebvre [1974] (1991) *The Production of Space*, pp. 38–9 and 78–9; N.J. Thrift (1995) 'A Hyperactive World' in R.J. Johnston, P.J. Taylor and M.J. Watts (eds) *Geographies of Global Change: Remapping the World in the Late Twentieth Century*.

Marion Boyars: Ivan Illich (1986) 'The Dirt of Cities', 'The Aura of Cities', 'The Smell of the Dead', 'Utopia of an Odorless City' from H_2O *and the Waters of Forgetfulness*, pp. 47–54.

The Center for Visionary Leadership: Corinne McLaughlin and Gordon Davidson (1985) '*ARCOSANTI*: Urban "Arcology" – Ecologically Sound Architecture' from *Builders of the Dawn*, pp. 251–5.

Harvard University Press: Siegfried Kracauer [1963] (1995) 'The Hotel Lobby' from *The Mass Ornament*, pp. 173–85.

Viola Klawan: photographs in Heinz Paetzold, 'The Philosophical Notion of the City', in H. Paetzold (ed.) *City Life* (Jan van Eyck Akademie, 1997) pp. 15–37.

Macmillan: Sharon Zukin (1996) 'Space and Symbols in an Age of Decline' in Anthony King (ed.) *Re-presenting the City*, pp. 43-59.

MIT Press: Indra Kagis McEwen (1993) 'Between Movement and Fixity: The Place for Order' from *Socrates' Ancestor*, pp. 79-89; Anne Pendleton-Jullian (1996) extract from *The Road That is Not a Road – The Open City, Ritoque, Chile*, pp. 23–5; Aldo Rossi (1982) 'Typological Questions' and 'The Collective Memory' from *The Architecture of the City*, pp. 35, 40–1, 130–1.

The New Press: Rosalyn Deutsche (1991) 'Alternative Space' in B. Wallis (ed.) *If You Lived Here*, pp. 45–66.

Office for Metropolitan Architecture: Rem Koolhaas (1995) 'Whatever Happened to Urbanism?' and 'Singapore Songlines' from *S, M, L, XL*, pp. 961–71 and 1075–85.

Heinz Paetzold and Jan van Eyck Akademie, Maastricht: Heinz Paetzold (1997) 'The Philosophical Notion of the City', in H. Paetzold (ed.) *City Life*, pp. 15–37.

Penguin Putnam Inc.: Gaston Bachelard (1969) 'The House. From Cellar to Garret. The Significance of the Hut' from *The Poetics of Space*, pp. 10–12.

Pine Forge Press: Saskia Sassen (1994) 'The New Inequalities within Cities' from *Cities in a World Economy*, pp. 99–117.

Pluto Press: Kenneth Frampton (1985) 'Towards a Critical Regionalism' in H. Foster (ed.) *Postmodern Culture*, pp. 20–3.

Princeton Architectural Press: Lebbeus Woods (1995) 'Everyday War' in P. Lang (ed.) *Mortal City*, pp. 46–53.

Random House: Roland Barthes [1970] (1982) 'Center-City, Empty Center' and 'No Address' from *Empire of Signs*, pp. 30–6; Jane Jacobs (1961) 'The Uses of Sidewalks: Contact' from *The Death and Life of Great American Cities*, pp. 55–7.

Reed Educational Publishing: Jonathan Charley (1995) 'Industrialization and the City: Work, Speed-up, Urbanization', in I. Borden and D. Dunster (eds) *Architecture and the Sites of History: Interpretations of Buildings and Cities*, pp. 344–57.

Resurgence: Hassan Fathy (1984) 'Palaces of Mud', *Resurgence*, 103, March/April: 16–17.

Sage: Sarah Buie (1996) 'Market as Mandala: The Erotic Space of Commerce', *Organization*, 3(2): 225–32.

Estate of Raphael Samuel: 'Theme Parks – Why Not?' (1995).

Ken-ichi Sasaki and Jan van Eyck Akademie, Maastricht: (1997) 'For Whom is City Design? Tactility versus Visuality' in H. Paetzold (ed.) *City Life*.

Joost Smiers: 'European Cities: First Sow, then Reap' from *Rough Weather* (1997) previously published only in Dutch and French.

Souvenir Press: Judith Shklar (1973) 'The Political Theory of Utopia: From Melancholy to Nostalgia' in F.E. Manuel (ed.) *Utopias and Utopian Thought*, pp. 101–15.

Suhrkamp Verlag: Ernst Bloch [1928] (1991) 'Ludwigshafen–Mannheim' from *Heritage of Our Times*, Cambridge, Polity, pp. 191–4; Siegfried Kracauer [1963] (1995) 'The Hotel Lobby' from *The Mass Ornament*, pp. 173–85.

Thames & Hudson: Victor Papanek (1995) 'Designing for a Safer Future' from *The Green Imperative*, pp. 29–39.

University of California Press and Andre Deutsche: Samir al-Khalil (1991) 'The Kitsch of Baghdad' from *The Monument: Art, Vulgarity and Responsibility in Iraq*, pp. 19–32.

University of California Press: Elizabeth Wilson (1991) 'World cities' from *The Sphinx in the City*, pp. 121–34.

Verso: Theodor W. Adorno and Max Horkheimer [1944] (1997) 'The Culture Industry: Enlightenment as Mass Deception' from *Dialectic of Enlightenment*, pp. 126–7; George McKay (1996) 'Eco-Rads on the Road' from *Senseless Acts of Beauty*, pp. 151–4; Jeremy Seabrook (1996) 'People of the City I' from *In the Cities of the South*, pp. 74–85.

J. Wiley & Sons: Archigram (1972) 'Instant City' in P. Cook (ed.) *Archigram* Studio Vista, pp. 86–101; William McDonough (1997) 'The Hannover Principles' in C. Jencks and K. Kropf (eds) *Theories and Manifestos of Contemporary Architecture*, Academy Editions, p. 160; William Mitchell (1996) 'Soft Cities', *Architectural Design*, special issue 'Architects in Cyberspace', pp. 84–7.

Patrick Wright: 'Abysmal Heights' from *A Journey Through Ruins: A Keyhole Portrait of British Postwar Life and Culture* (Flamingo, 1993) pp. 98-109.

Zed Books: Farha Ghannam (1997) 'Reimagining the Global: Relocation and Local Identities in Cairo' in A. Öncü and P. Weyland (eds) *Space, Culture and Power*, pp. 119–39; R. Patel (1997) 'Urban Violence – An Overview' in J. Beall (ed.) *A City for All*, pp. 96–103.

In addition, there are the following texts previously published by Routledge:

Jon Bird (1993) 'Dystopia on the Thames' in Bird *et al.* (eds) *Mapping the Futures*, pp. 120–35.

I. Taylor, K. Evans and P. Fraser (1996) 'Shop 'til you Drop' from *A Tale of Two Cities: Global Change, Local Feeling and Everyday Life in the North of England*, pp. 141–60.

S. Graham and S. Marvin (1996) 'The Social and Cultural Life of the City' from *Telecommunications and the City: Electronic Spaces, Urban Places*, pp. 209–13.

Dolores Hayden (1996) 'The Power of Place: Urban Landscapes as Public History', in I. Borden, J. Kerr, A. Pivaro and J. Rendell (eds) *Strangely Familiar*, pp. 47–51.

Michael Hough (1995) 'The Contradiction of Values' from *Cities and Natural Process*, pp. 6–10.

Jane M. Jacobs (1996) 'Conclusion' from *Edge of Empire*, pp. 157–63.

Frank Mort (1996) 'Boulevards for the Fashionable and Famous' from *Cultures of Consumption*, pp. 157–63.

Edward Robbins (1996) 'Thinking Space/ Seeing Space: Thamesmead Revisited', *Urban Design International*, 1(3): 283–91.

David Sibley (1995) 'Border Crossings' from *Geographies of Exclusion*, pp. 32–48.

Neil Smith (1996) 'Class Struggle on Avenue B' from *The New Urban Frontier*, pp. 1–12.

Edward Soja (1996) 'The Stimulus of a Little Confusion: On Spuistraat, Amsterdam' in I. Borden, J. Kerr, A. Pivaro and J. Rendell (eds) *Strangely Familiar*, pp. 27–31.

INTRODUCTION

THE CITY

The city is that entity to which we choose to give a name: Berlin or Paris, London or Edinburgh, Lima or Bogota, Nairobi or Johannesburg, New York or Chicago, Seoul or Taipei, Budapest or Moscow. Yet this schema of naming is too easy, for in using a singular word (however convenient that might be) we rush too quickly over the bewildering complexity that a city, many cities, might be. To cite but a few elements of that complexity . . .

A city is a place, a bounded entity marked by a particular core and density. It is located at a given geographic position, spreading outward until it ends at the outermost suburb. And yet a city is far more than this. Where is the centre to be located? Indeed, are there not multiple centres? And who is to say where the city ends, for when does the suburb terminate and the countryside begin? Indeed, there are many arguments for seeing the countryside and the city as simply parts of an overall urban space, for are not the fields and agricultural plains just as manufactured as is the city proper?

A city is a set of objects. In it we find the significant spaces that constitute urban life. It is the place of airports and docks, houses and hospitals, schools and colleges, factories and offices, theatres and stadia, power stations and television masts. Yet different cities offer different constellations, different arrangements – gridded and irregular, cored and multi-cored, compact and dispersed, low-rise and high-rise, circular and linear, undulating and flat, islands and frontiers. A city is, then, a set of practices. It is the place where things happen and people act. It is the place of making and consuming, driving and walking, teaching and learning, jostling and sleeping. It is a place where doing occurs. A city is not a singular text, nor indeed a text at all. It would be the worst kind of illusion to read the city only as objects, for it is a living, social entity. Yet we cannot just do things without spaces and objects, nor can we do everything in terms that are visible.

The city is invisible. It is the place of exchanges and flows, where money, ideas and data are transferred. And it is also the place where people fall in love, where desires and emotions pervade throughout the streets and interiors, filling its spaces with a powerful force indiscernible to clipboard documentarians and cashflow accountants. And these flows also leak out of the city, out into networks that connect city to city, building to building, person to person, across vast distances and instant times.

A city is a set of beliefs. To live here, one must have faith in a huge range of impossible illusions. One has to truly believe that paper is valuable (money), that cars will not invade pavements (traffic behaviour), that strangers are not dangerous (otherness), that governments and institutions are acting for the general good (politics). The city is also where one has faith in oneself, in one's ability to transcend, revel in or reproduce the city. The city is the place which at once threatens and exhilarates, and the metropolitan dweller must be comfortable with that fact. It is the place in which dwellers have a right to exist, and a right to experience.

The city is an understanding of itself. To make it work, to make it operate, to make it liveable, all manner of ideologies, schema, concepts and images are required. Mathematical models, city maps, sign systems, poetic descriptions, painterly and photographic representations, architectural drawings – all these things and many more are the codes by which we consciously struggle to comprehend the city.

The city is urban professions. Planners consider the economic, social and spatial operations of the city at an enormous range of scales. On the one hand, they rule on single buildings, considering their context in zoning plans and streetscapes. On the other hand, they consider the whole city in regional and global contexts, seeking to position cities in economic and political patterns. Architects and urban designers, for their part, deliberate on a similar range of scales, from the single room to the master-planning of city quarters and even, less frequently, the layout of entire new cities. And then there are the many other cultural designers and producers at work in the city: graphic designers, fashion designers, film-makers, university lecturers, artists and musicians, writers and performers . . . all of whom depend on the city for their livelihood, while, conversely, the city depends on them for its.

The city is a place of the spectacular. It is where major historic events take place, grand architecture is constructed, big decisions are taken, great art is produced. It provides some, if not all, of the most significant cultural moments and artefacts by which our different societies express and mark their significance and achievements. Yet the city is also the place of routine events, ordinary buildings, everyday decisions, useful crafts. It is the place where we live our lives in the most banal yet important of ways. It is where all spectacles must find their ultimate relevance, not in themselves but for how they resonate with and are reproduced through everyday lives.

The city is temporal. It has been there before, it is now, and it will become. It is a palimpsest of buildings, memories, ideas and traditions (historical time). It also moves at different paces and rhythms, at once fast and slow, braking and accelerating (differential time). And it repeats certain cycles, such as those of seasons, commuting, shopping patterns or the sequences of traffic lights (cyclical time). It is the time of individual perceptions, from the stillness of boredom and the fast pace of excitement (personal time). And, of course, it is the time of the future, of the city yet to come.

The city is historical. It is forever changing, always different according to different conditions, different people, different ages. It is never exactly the same, never predictable. Yet neither is it a free space, where people can do just what they like. It is a place of opportunity, but under particular circumstances. It is what Henri Lefebvre has called a 'possibilities machine'.

CITY CULTURE

Cities, then, are sites of constant flux, their built form mediated by successive acts of destruction and creation. People's experience of the city is affected by social factors such as gender, class and ethnicity. For different groups in society at different times, the city is a different place. Urban experience is also mediated by representations of the city, as in television news, documentaries and dramas, and by the life of its streets. The city is an event, a performance in which the roles of actor and spectator are interchangeable. Like any performance, the city is a product of culture. Culture, in this sense, means a process of intellectual development through which ideas are formed and changed; it also means the way of life produced by such ideas and the forms, such as art, architecture, film and fashion, in which they take shape. But if the city is a product of culture, it is also where most culture is made and received. The urban character of this culture may be revealed or concealed by its form.

To write of the city as an event reclaims one of its two, reciprocal aspects, because conventionally the city of form, conceptualized in plans, tends to dominate. This city is shaped by intentions, of which power, wealth and technology are the instruments. In a democracy, elected representatives play a role in the processes of regulation. The articulation of plans tends to use a visual language, as in cartography, and this itself privileges static form over mutable experience. But if the plan constructs a stage, the drama is played out by a city's inhabitants. And it may be a matter of improvisation as much as the learning of the script provided by the plan. Although the two aspects overlap, it is possible to write of a dominant or conceptualized city of plans, and an experiential city of everyday life. Not only buildings, but also everyday acts, including shopping, going to the cinema, waiting for a bus and skateboarding, comprise a city's culture.

The city plan is, in any case, determined by social, economic and political factors. These factors are cultural not natural. They change in history and may privilege certain interests. How city form is encountered by different publics, too, is mediated through cultural values and allegiances. The perception of urban development zones, such as Canary Wharf in London or Battery Park City in New York, depends on the specific public to which the observer belongs. The same shining towers can be seen as monuments to corporate utopia, or signs of marginalization and a whole range of things between. Similarly, informal settlements on the edges of cities in the non-affluent world are seen either as threats to civic order, or as a beginning of community. There is no single or coherent category which contains the diversity of urban settlement at the beginning of the twenty-first century. Instead, it is diversity itself – of publics and modes of settlement – which characterizes life in cities today.

Part of that diversity is a widening gap between zones of affluence and those of deprivation, often in close geographical proximity, echoed in the different degrees of access and power of groups within society. For people who have a critical awareness of society and environment, the determination of the city – the concept to which it aspires and the relation of the concept to actuality – is a site for intervention. Appropriations of and resistances to the dominant city are evident in graffiti, fly-posting and squatting. These may denote the existence of sub-cultures, or what has been called, problematically, an 'underclass'. But the divisiveness of contemporary city life indicates a failure of the utopian vision of the city implicit in nineteenth-century notions of progress and eighteenth-century idealism. The inadequacy of modernist efforts to engineer a new society by design, as in the garden city or the progressive social housing schemes of the inter-war and post-war periods, followed an uncritical acceptance of such notions translated as spatial prescriptions rather than social processes, of form privileged over the act of occupation.

Cities are produced, then, according to cultural values. They are also represented in cultural products, and the common image of the city in the cultural representations of affluent societies today is dystopia. Cities are seen as sites of crime, social division and a collapse of value. The fragmentation of cities through mono-functional zoning, and by the edges of awkwardness financial institutions withdraw in search of richer pickings elsewhere, as well as remote rural areas, ageing residential suburbs and large areas of the less developed world.

It is worth remembering that, despite the hype, the cultural lives of the majority of the world's urban populations are, and will continue to be, lived out in the slow-world, rather than the fast-world. This is not to say, however, that these lives will not be profoundly touched by cultures produced in the centres of the fast-world. City cultures of the future seem ever more likely to involve an amalgam of the increasing penetration of cultural products generated in the media 'factories' of the fast-world, mediated through and consumed within the growing spaces of the slow-world. However, this is not to suggest any global hegemony flowing out across the world from a few world-cities. The subversions and transformations of fast-world cultures in the music, cinema, art, literature and everyday lives of the slow-world suggest the futures of cities and their cultures are replete with possibilities and challenges.

THE STRUCTURE AND PROCESS OF THE BOOK

The Reader is structured in twelve sections, grouped in three parts. These parts follow a movement from things, such as urban form or industry, to experience, such as everyday life and the construction of memories, to the negotiation of present conditions and creation of possible futures. Another way to put this is that there is a movement from the received to the interpreted to the envisioned. Brief introductions to each part are given below, and each section begins with an introduction related to the selection of texts in it, and, in some cases, others included in the list of further reading below.

Part 1 is entitled Shaping Urban Culture, denoting how urban forms and the cultures of which they are part are determined. The built city, as a set of objects in space, is only one aspect of the complex entity called a city. Another is the meaning and value attached to its forms by people. Both built form and people's experiences of it are conditioned by politics, economics and culture. Culture itself is mediated, partly by being produced in urban environments, and through the workings of the culture industries. Both built form and the culture with which it has a reciprocal relationship are conditioned by, and sites of development for, technologies which in turn shape ideas.

The four sections in Part 1 are:

I.   Forms and spaces of the city
II. Cultures and urban economics and politics
III. The culture industries
IV. Culture and technologies

Part 2, Living Urban Culture, concerns experience, and, again, mediation. Here issues such as memory and representation are specifically addressed. The first section concerns the processes of daily living. To focus on everyday life, rather than the grander themes of a dominant history and high culture, is to assert an end of the privileging of that supposed greatness and goodness which characterizes high culture. Once it is accepted that looking at commonplace objects in shop windows reveals insights into a society's value structures, history is no longer a narrative of grandeur. From this it follows that representations can be decoded, and identities constructed, that neither is natural, both cultural. What ecology adds to this is a realization that the elements in a system of life are interdependent. Not only are there no more or less important elements, but the survival of all depends on the sustainability of an equilibrium involving all.

The four sections in Part 2 are:

V. Everyday life
VI. Memory, imagination and identities
VII. Representations of the city
VIII. Cultures and ecology

Part 3, Negotiating Urban Cultures, approaches the issues of urban futures and roles of urban dwellers. Questions of the emotional ownership of space and the empowerment of urban publics are crucial to an understanding of urban sustainability. Today, assumptions cannot be made about the continuation of hierarchies, whether economic, political or cultural. These are matters of negotiation, in which professionals and non-professionals may have equal claims. Perhaps a key to a possible future for cities is an equation of the knowledge of dwellers on dwelling with that of designers on design.

The four sections in Part 3 are:

IX. Social justice
X. Cultures of resistance and transgression
XI. Utopia and dystopia
XII. Possible futures

Each section is multi-disciplinary in scope. The balance of well-known writers and those emerging into recognition varies, according to the material available, but the intention of the book as a whole is to introduce writing which may be unfamiliar to people outside the discipline in which it is produced, alongside some more familiar texts. Each section will appeal, it is hoped, to readers from any of the disciplines of the built environment – architecture, art, cultural geography, cultural studies, planning, urban sociology and urban design, amongst others. Even the first section, on urban form, is not composed entirely of architectural writing.

In many cases, it is difficult to categorize a text, and several could have fitted equally well into two or three of the sections. The reader is therefore encouraged to cross-reference between sections, and to regard the divisions as matters of convenience rather than the bones of a new theory of urbanism.

The selection of readings from a large, rapidly expanding and complex literature reflects both intentions and constraints. The intentions are to produce an anthology which is multi-disciplinary throughout, and to focus on the recent rather than historical literature. Although a few theoretical texts from the earlier part of the century are included, by Adorno and Horkheimer, Bloch and Kracauer, there is no attempt to give a comprehensive coverage of, say, the literature of urban planning. This is already done in *The City Reader*, edited by LeGates and Stout (Routledge, 1996, second edition 2000). The aim here is to reflect current debate, and enable those who use this collection to contribute to that debate, aware of the diversity of positions of which it is composed.

Amongst the constraints are the costs of permissions and the appropriateness with which parts of a larger work can be represented outside the whole. To put it simply: some texts were too expensive, their original publishers having a voracious appetite for money; others are too difficult to treat by extract. In a few cases it proved impossible to contact the holders of rights. There was also the constraint of size, governed by the total volume of text which could be included in a book remaining affordable in an academic market.

Throughout, the co-editors have worked as a team, and the selections for each section have been made by the team together. Drafting the introductions has been shared, each editor working on an agreed set, then receiving comments from the others before making suitable revisions. This way, the three fields within which the editors currently teach and carry out research – cultural theory, architecture and cultural geography – inform the whole work.

In the majority of cases the editors have removed reference citations from the texts included in the Reader. This is intended to enhance clarity and ease of reading. Not wishing to create the impression that the authors collected here have made unattributed arguments, we refer the reader to the original publications for full lists of references cited in the texts. A number of important cited references are included in the suggestions for further reading below or in the section introductions. In many cases recent publications have superseded those cited by the authors in the texts collected here. We have tried to make reference to these where relevant.

SUGGESTIONS FOR FURTHER READING

The following titles cover a wide range of approaches to city cultures. For the most part, they are recent publications. Others are included which continue to have a seminal influence on the literature. Where titles are included from which extracts have been reproduced in this Reader, it is to draw attention to the full contents of the book.

Marc Augé, *Non-Places: Introduction to an Anthropology of Supermodernity*, translated by J. Howe, London, Verso, 1995

Stephen Barker, *Fragments of the European City*, London, Reaktion, 1995

Walter Benjamin, *Illuminations*, London, Cape, 1970

Walter Benjamin, *One-Way Street and Other Writings*, London, Verso, 1979

Marshall Berman, *All That is Solid Melts Into Air*, New York, Simon & Schuster, 1982

Jon Bird *et al.* (eds) *Mapping the Futures: Local Futures, Global Change*, London, Routledge, 1993

Iain Borden, Joe Kerr, Alicia Pivaro and Jane Rendell (eds) *Strangely Familiar: Narratives of Architecture in the City*, London, Routledge, 1996

Iain Borden, Joe Kerr, Jane Rendell with Alicia Pivaro, *The Unknown City: Contesting Architecture and Social Space*, Cambridge, Mass., MIT, 2000

Victor Burgin, *Some Cities*, London, Reaktion, 1996

Michel de Certeau, *The Practice of Everyday Life*, Berkeley, Calif., University of California Press, 1984

Le Corbusier, *The City of Tomorrow and its Planning*, London, The Architectural Press, 1947

Mike Davis, *City of Quartz*, London, Verso, 1991

Mike Douglass and John Friedmann, *Cities for Citizens*, Chichester, Wiley, 1998

Ruth Finnegan, *Tales of the City: A Study of Narrative and Urban Life*, Cambridge, Cambridge University Press, 1998

David Frisby and Mike Featherstone (eds) *Simmel on Culture*, London, Sage, 1997

Joel Garreau, *Edge City: Life on the New Urban Frontier*, New York, Doubleday, 1991

Stephen Graham and Simon Marvin, *Telecommunications and the City: Electronic Spaces, Urban Places*, London, Routledge, 1996

Tim Hall, *Urban Geography*, London, Routledge, 1998

Tim Hall and Phil Hubbard (eds) *The Entrepreneurial City*, Chichester, Wiley, 1998

David Harvey, *The Urban Experience*, Oxford, Blackwell, 1989

bell hooks, *Yearning: Race, Gender and Cultural Politics*, London, Turnaround Press, 1989

Brian D. Jacobs, *Fractured Cities: Capitalism, Community and Empowerment in Britain and America*, London, Routledge, 1992

Fredric Jameson, *Postmodernism, Or, the Cultural Logic of Late Capitalism*, London, Verso, 1991

Gilbert M. Joseph and Mark D. Szuchman (eds) *I Saw a City Invincible: Urban Portraits of Latin America*, Wilmington, Del., Jaguar Books, 1996

Gerry Kearns and Chris Philo (eds) *Selling Places: The City as Cultural Capital Past and Present*, Oxford, Pergammon, 1993

Ray Kiely and Phil Marfleet (eds) *Globalisation and the Third World*, London, Routledge, 1998

Anthony King (ed.) *Re-Presenting the City: Ethnicity, Capital and Culture in the 21st Century*, London, Macmillan

Ken Knabb (ed.) *Situationist International Anthology*, Berkeley, Calif., Bureau of Public Secrets, 1981

Rem Koolhaus and Bruce Mau, *S, M, L, XL: Office for Metropolitan Architecture*, Jennifer Sigler (ed.), New York, Monacelli, 1995

Neil Leach (ed.) *Rethinking Architecture: A Reader in Cultural Theory*, London, Routledge, 1997

Henri Lefebvre, *Critique of Everyday Life*, London, Verso, 1991

Henri Lefebvre, *Writings on Cities*, translated and edited by E. Kofman and E. Lebas, Oxford, Blackwell, 1996

Doreen Massey, *Space, Place and Gender*, Cambridge, Polity, 1994

Andy Merrifield and Erik Swyngedouw (eds), *The Urbanisation of Injustice*, London, Lawrence & Wishart, 1995

Malcolm Miles, *Art, Space & the City*, London, Routledge, 1997

W.J.T. Mitchell (ed.) *Art and the Public Sphere*, Chicago, University of Chicago Press, 1992

Frank Mort, *Cultures of Consumption: Masculinities and Social Space in Late Twentieth-Century Britain*, London, Routledge, 1996

Ayse Öncü and Petra Weyland (eds) *Space, Culture and Power: New Identities in Globalizing Cities*, London, Zed Books, 1997

Michael Pacione (ed.) *Britain's Cities*, London, Routledge, 1997

Jane Rendell, Barbara Penner and Iain Borden (eds) *Gender, Space, Architecture: An Interdisciplinary Introduction*, London, Routledge, 1999

George Robertson *et al.* (eds) *Travellers' Tales: Narratives of Home and Displacement*, London, Routledge, 1994

Leonie Sandercock, *Towards Cosmopolis: Planning for Multicultural Cities*, Chichester, Wiley, 1998

Richard Sennett, *Flesh and Stone*, London, Faber & Faber, 1994

John Rennie Short, *The Urban Order: An Introduction to Cities, Culture and Power*, Oxford, Blackwell, 1996

Michael Sorkin (ed.) *Variations on a Theme Park: The New American City and the End of Public Space*, New York, Noonday Press, 1992

Manfredo Tafuri, *Architecture and Utopia: Design and Capitalist Development*, Cambridge, Mass., MIT, 1976

Ian Taylor, Karen Evans and Penny Fraser, *A Tale of Two Cities: Global Change, Local Feeling and Everyday Life in the North of England*, London, Routledge, 1996

Brian Wallis (ed.) *If You Lived Here*, Seattle, Bay Press, 1991

Sophie Watson and Katherine Gibson (eds) *Postmodern Cities and Spaces*, Oxford, Blackwell, 1995

Sallie Westwood and John Williams (eds) *Imagining Cities*, London, Routledge, 1997

Patrick Wright, *A Journey Through Ruins: A Keyhole Portrait of British Postwar Life and Culture*, London, Flamingo, 1993

Sharon Zukin, *The Culture of Cities*, Oxford, Blackwell, 1996

PART 1

Shaping Urban Culture

SECTION I

FORMS AND SPACES OF THE CITY

INTRODUCTION

Questions of the forms and spaces of the city are at the heart of academic enquiry across a number of disciplines from architecture and planning to geography, sociology, cultural and media studies. This section aims to convey something of the richness of questions of city form and space. The extracts gathered here are diverse in the perspectives they represent, the purposes behind their writing and their emphasis. These extracts are concerned with questions such as the centrality of city form in social organization, the meanings of urban spaces and their roles in everyday life and feeling, and for whom urban spaces are designed.

The archetypal modern city was that imagined in the plans and manifestos of a number of highly influential modern architects. The most famous and influential of these was Charles Edouard Jeanneret (1887–1965) a Swiss architect and planner better known by the pseudonym Le Corbusier, which he adopted after moving to Paris. Le Corbusier laid a blueprint for the imagination and planning of the modern city in a series of publications. The most fully realized and complete of these was the polemic *The City of To-morrow and Its Planning* (1929 [1947], The Architectural Press). Le Corbusier was most closely associated with an architectural conception (the modern movement) and an approach to town planning that placed a premium on the guiding principles of rationality, order and efficiency based on a strongly centralized plan. Critiques of the application of Le Corbusier's ideas can be found in James Holston's interesting account of Brasília, *The Modernist City: An Anthropological Critique of Brasília*, (1989, University of Chicago Press), and John Gold's excellent account of the application of modernist ideas by British planning departments, *The Experience of Modernism* (1998, Routledge).

Le Corbusier's approach underpinned a major transformation of the post-war urban landscape across the world as his ideas, or at least parts of them, were adopted by a number of planners and architects. In a few cases whole cities were built on Le Corbusian lines, the most famous example being Brasília. However, his ideas are familiar to most cities in the form of the high-rise housing projects of many cities that other planners thought answered the need to cheaply house growing post-war populations. Quantitatively they may have done, qualitatively many were unmitigated disasters. Le Corbusier has become one of the most derided figures in the history of town planning (see, for example, Peter Hall's critique in *Cities of Tomorrow* (1988, Basil Blackwell). Criticisms of Le Corbusier usually involve two distinct elements: criticisms of the flawed nature of his vision and of the failure of the application of his ideas by others. A guide to one such space is offered by Edward Robbins' (1996) look at the Thamesmead estate in London in 'Thinking Space/Seeing Space: Thamesmead Revisited'.

In many ways the nemesis of Le Corbusier's, and the modernist, vision was Jane Jacobs, a neighbourhood activist and journalist rather than a professional planner, architect or academic. However, her non-professional status did not prevent her work, most notably *The Death and Life of Great American Cities* (1961, Random House), an extract from which begins this section, having a profound impact on urban planning, perceptions of the inner city and patterns of urban living.

Jacobs's key thesis stemmed from her love of certain neighbourhoods such as Greenwich Village in New York and the Italian North End in Boston. She argued that city form and spaces should provide more than just narrowly defined optimal solutions to certain problems of the distribution of people, resources and traffic. Such beliefs, she argued, had dominated modern town planning with disastrous results. Her ideal city was very different from that of Le Corbusier and his followers. Many of what she regards as important qualities in city spaces and neighbourhoods are outlined in the extract from *The Death and Life of Great American Cities*, 'The Uses of Sidewalks: Contact' in this section. Uppermost of these qualities are safety and the feeling of well-being that follows this, the qualities she considered to be absent from the sterile streets and extensive parks of modern planned cities. This extract discusses how safety is generated in busy, crowded streets like those of Boston's North End. While appearing slight in itself, this observation must be taken as part of a number brought together in *The Death and Life* that question the fundamental principles of modern town planning.

Jacobs's book represented a turning point in the way cities were viewed. *The Death and Life* had many knock-on effects in a number of spheres. As well as being part of the growing awareness that was informing the planning profession of the failures of modern town planning, it was hugely influential in shaping public opinion. Partly as a result of *The Death and Life*, movements such as community architecture and the advocacy of grass-roots approaches to urban planning emerged in the 1960s and 1970s. Similarly, *The Death and Life* can be seen as one of the precursors to the rediscovery of the inner city by middle-class professionals. Although Jacobs's is most remembered for *The Death and Life* she has produced other books of note including *The Economy of Cities* (1969) and *Cities and the Wealth of Nations* (1984).

The further troubled progress of over-planned modernization is explored by Rem Koolhaas (1995) in 'Singapore Songlines: Portrait of a Potemkin Metropolis . . . or Thirty Years of Tabula Rasa'. Koolhaas explores the erasures, cultural and physical, involved in the modernization and development of Singapore since the 1960s, the brainchild of one man, Lee Kuan Yew, its first post-colonial president. The disruptions that Koolhaas explores in Singapore serve as a salient warning of the dangers of modernization and erasure through planning.

The grid street-plan is often seen as a characteristic sign of modern city planning, taken to indicate the operations of a market economy, dividing the site of a city such as New York into equal lots to sell at equal prices. But its origins go much further back in history, as investigated by Indra Kagis McEwen in the extract from *Socrates' Ancestor*. McEwen (who begins from her own re-translation of the *Anaximander Fragment* taken as a source previously by Heidegger) links the grid of Greek colonial cities with the way form conveys a sense of order which reflects a cosmos. She further relates this to material culture, as in weaving. The cloth which is finely made has its threads perpendicular to each other. Otherwise, it is as if it does not 'appear' at all. And so for the city.

Elsewhere in this chapter the meaning of the city and its spaces is considered. Sarah Buie (1996), in 'Market as Mandala: The Erotic Space of Commerce', offers an alternative reading of the acts and the spaces of commerce in the city, pointing out the erotics of interaction associated with markets and commercial acts of exchange in a number of cultures. In a conclusion that echoes some of the concerns highlighted by Koolhaas, Buie bemoans the increasing homogeniety of such spaces of commerce in modern western society. It is instructive to contrast Buie's conclusions with the complexities of the shopping act and its spaces in Taylor, Evans and Fraser's extract in Section V and with work on the multiple meanings of shopping and consumption, for example D. Miller, P. Jackson, N. Thrift, B. Holbrook and M. Rowlands *Shopping, Place and Identity* (1998, Routledge).

Finally, the dominance of visuality in the design of the city is challenged by Ken-ichi Sasaki (1997) in 'For Whom is City Design? Tactility versus Visuality'. Ken-ichi Sasaki challenges the dominance (or the tryanny) of the assumption, within the disciplines that produce the city, that the city is a thing to be looked at. We hope this collection will demonstrate that the urban experience is richer than this in a plethora of ways. Among these other experiences of the city is the tactile experience of urban spaces, something long ignored, or at the very least underappreciated by planners, urban designers and architects. This multi-sensual experience of the city is beginning to be highlighted by others interested in the contribution of senses other than the visual to the experience of place. Recent contributions include Paul Rodaway's *Sensuous Geographies: Body, Sense and Place* (1994, Routledge), which draws on the legacy of humanistic geography, and Rob Imrie's discussion of the experiences of place, space and the city by disabled groups, *Disability and the City* (1996, Paul Chapman).

1 JANE JACOBS

'The Uses of Sidewalks: Contact'

from *The Death and Life of Great American Cities* (1961)

Reformers have long observed city people loitering on busy corners, hanging around in candy stores and bars and drinking soda pop on stoops, and have passed a judgment, the gist of which is: "This is deplorable! If these people had decent homes and a more private or bosky outdoor place, they wouldn't be on the street!"

This judgment represents a profound misunderstanding of cities. It makes no more sense than to drop in at a testimonial banquet in a hotel and conclude that if these people had wives who could cook, they would give their parties at home.

The point of both the testimonial banquet and the social life of city sidewalks is precisely that they are public. They bring together people who do not know each other in an intimate, private social fashion and in most cases do not care to know each other in that fashion.

Nobody can keep open house in a great city. Nobody wants to. And yet if interesting, useful and significant contacts among the people of cities are confined to acquaintanceships suitable for private life, the city becomes stultified. Cities are full of people with whom, from your viewpoint, or mine, or any other individual's, a certain degree of contact is useful or enjoyable; but you do not want them in your hair. And they do not want you in theirs either.

In speaking about city sidewalk safety, I mentioned how necessary it is that there should be, in the brains behind the eyes on the street, an almost unconscious assumption of general street support when the chips are down – when a citizen has to chose, for instance, whether he will take responsibility, or abdicate it, in combating barbarism or protecting strangers. There is a short word for this assumption of support: trust. The trust of a city street is formed over time from many, many little public sidewalk contacts. It grows out of people stopping by at the bar for a beer, getting advice from the grocer and giving advice to the newsstand man, comparing opinions with other customers at the bakery and nodding hello to the two boys drinking pop on the stoop, eyeing the girls while waiting to be called for dinner, admonishing the children, hearing about a job from the hardware man and borrowing a dollar from the druggist, admiring the new babies and sympathizing over the way a coat faded. Customs vary: in some neighborhoods people compare notes on their dogs; in others they compare notes on their landlords.

Most of it is ostensibly utterly trivial but the sum is not trivial at all. The sum of such casual, public contact at a local level – most of it fortuitous, most of it associated with errands, all of it metered by the person concerned and not thrust upon him by anyone – is a feeling for the public identity of people, a web of public respect and trust, and a resource in time of personal or neighborhood need. The absence of this trust is a disaster to a city street. Its cultivation cannot be institutionalized. And above all, *it implies no private commitments*.

I have seen a striking difference between presence and absence of casual public trust on two sides of the same wide street in East Harlem, composed of residents of roughly the same incomes and same races. On the old-city side, which was full of public places and the sidewalk loitering so deplored by Utopian minders of other people's leisure, the children were being kept well in hand. On the project side of the street across the way, the children, who had a fire hydrant open beside their play area, were behaving destructively, drenching the open windows of houses with water, squirting it on adults who ignorantly walked on the project side of the street, throwing it into the windows of cars as they went by. Nobody dared to stop them. These were anonymous children, and the identities behind them were an unknown. What if you scolded or stopped them? Who would back you up over there in the blind-eyed Turf? Would you get, instead, revenge? Better to keep out of it. Impersonal city streets make anonymous people, and this is not a matter of esthetic quality nor of a mystical emotional effect in architectural scale. It is a matter of what kinds of tangible enterprises sidewalks have, and therefore of how people use the sidewalks in practical, everyday life.

The casual public sidewalk life of cities ties directly into other types of public life, of which I shall mention one as illustrative, although there is no end to their variety.

2 INDRA KAGIS McEWEN

'Between Movement and Fixity: The Place for Order'

from *Socrates' Ancestor* (1993)

THE *POLIS*

In his book *La naissance de la cité grecque: cultes, espace et société VIII–VII siècles avant J.-C.*, a work based on recent archaeological research, François de Polignac presents evidence for the formative role of what he calls *la cité cultuelle*, or ritual city, in the emergence of the Greek *polis*. Athens, he stresses, was not, as is usually assumed, the paradigm but rather the exception, being, with its centralized structure focused on the Acropolis, something of a unique survivor from the Mycenaean period and its palace cultures. The typical cases were the other Greek cities – among them the cities of Asia Minor, and also, especially, the colonial cities of Sicily, Magna Graecia, and the Black Sea area – which only emerged in the eighth and seventh centuries, after the so-called "dark age" (twelfth to ninth centuries) that followed the Dorian invasion and the subsequent collapse of Mycenaean civilization.

Irad Malkin, in his study of religion and Greek colonization, is vociferous in the defense of the rationally or functionally planned, as opposed to the emergent, or what I would call "made," city, citing the case of Greek colonial foundations as evidence. He also suggests that "colonialization contributed just as much towards the rise of the *polis* as it was dependent on this rise for its own existence." I cannot agree that the colonial founders were urban planners, as Malkin's argument suggests. The regularity of street layouts Malkin brings to bear as evidence seems to me to have much more to do with the notion of allowing *kosmos* to appear through their rhythm than with planning in the modern sense of the word. However, his argument for a reciprocal relationship between the development of *polis* and colonial city is convincing. Furthermore such reciprocity has important implications with respect to the role navigation and ships played in the making and thinking of the new *poleis*. If the colonial foundations influenced the emergent *poleis* as much as vice versa, the fact that Greek colonists were all, necessarily, sailors before they became settlers becomes very significant. Between the *metropolis*, or mother city, and the new foundation, the city existed as a ship.

Weaving the City

What is noteworthy about the new *poleis*, in contrast to the old Mycenaean cities, is the presence of sanctuaries, which had never existed in the Mycenaean civilization, apart from the hearth of the quasi-divine king-father within the Mycenaean palaces themselves. There are tombs at Mycenae and Knossos, but no temples. François de Polignac carefully maps the archaeological traces of the eighth- and seventh-century sanctuaries and shows how they fall into three categories: urban sanctuaries, within the inhabited urban area itself; suburban sanctuaries, placed at the limit of, or at a short distance from, the habitat; and extraurban sanctuaries, which were not part of daily ritual, since they were located some six to twelve kilometers from the town at the very limit of the city's territory (*chōra*). Many of the most celebrated sanctuaries of the Greek world are indeed located on nonurban sites, and in view of this, it is impossible to maintain that the city grew up around the temple, which is the conclusion drawn if Athens, mistakenly, is taken as paradigmatic.

Rather, it would appear to be possible to extrapolate from de Polignac's argument the notion of a *polis* allowed to appear as a surface woven by the activity of its inhabitants: the sequential building of sanctuaries over a period of time, which at times stretched over decades, and the subsequent ritual processions from center to urban limit to territorial limit and back again, in what can be seen as a kind of Ariadne's dance, magnified to cover a territory that was not called *choros* but *chōra*.

In the *Iliad chōrē*, which is the Ionic form of *chōra*, is a scant space (*oligē chōrē*) between, such as that between a horse and a chariot, or the one in which the corpse of Patroclus is dragged to and fro after Hector has slain him, or the narrow rim of shoreline left for the Achaeans to fight in. The verb *choreo* is used in the military sense of giving ground before the enemy.

Chōros, the masculine form of *chōra* or *chōrē*, in general denotes a space that is somewhat more defined than the feminine *chōra*. In one notable passage Hector and Odysseus "measured out a [masculine] space [*chōron diemetreon*]" for the single combat between Paris and Menelaus, who then "took their stand near together in the measured space [*diametrētōi eni chōrōi*]." Relevant in this context is the so-called Pyrrhic dance that was part of every Spartan soldier's military training.

In the *Odyssey*, where the word, more often than not, appears in its masculine, *chōros*, form, the tendency is for it to mean place as location, but also land, country, or territory. According to the Aristotelian definition, *chōra*, translated by Loeb as "room," is similar to place (*topos*), which Aristotle says is a "surface-continent [*epipedon periechon*] that encompasses its content in the manner of a vessel." *Chōra* is different from void (*kenon* or *chaos*, which Aristotle equates and says do not exist).

Now it is, of course, very dangerous to read backward from Aristotle into the archaic period. However, Aristotle's discussion of *topos* and *chōra* does suggest a possible guess as to how the *chōra* of the *polis* may have been understood in earlier times as a territory made to appear through a continual remaking, or reweaving of its encompassing surface, just as the world itself was made to appear when the colonists' ships plied the seas.

In the same chapter of his *Physics*, Aristotle notes that Plato too identified *topos* and *chōra* in the *Timaeus*. Plato says that after the "first kind," which is "self-identical form [*eidos*], ungenerated and indestructible," and the "second kind," the object perceptible by sense "becoming in a place [*topōi*] and out of it again perishing," is a "third kind":

> ever-existing place [*chōra*] which admits not of destruction and provides room for all things that have birth, itself being apprehensible by a kind of bastard reasoning by the aid of non-sensation, barely an object of belief; for ... it is somehow necessary that all that exists should exist in some spot [*en tini topōi*] and occupying some place [*ketachon chōron tina*] and that that which is neither on earth nor anywhere in the heaven is nothing.

Plato's *chōra*, the receptacle of Becoming, is eternal, and indestructible, but the *chōra* of the nascent, archaic *polis* was not. The archaic *polis* was an uncertain place that needed to be anchored at the strategic points of center, middle ground, and outer limit by the new sanctuaries. It was not a vessel with a fixed form, but, like the appearing surface of a woven cloth – of all the traces of material culture one of the most perishable – had continually to be mended or made to reappear.

And why, to return to the question that has long vexed historians of urban form, were the streets of colonial foundations regularly spaced? Why did they intersect at right angles? I suggested earlier, a little vaguely, that their regularity had to do with *kosmos*, but if we think of the city in terms of weaving, as I believe the early Greeks did, the intention made manifest in orthogonal street layouts becomes quite precise. Weaving, says H. Ling Roth, "consists of the interlacing *at right angles* [my italics] by one series of filaments or threads, known as the weft ... of another series, known as the warp, both being in the same plane." *Harmonia*, close fitting, can be a feature of the tightly woven cloth only: a textile with a loose weave is not, so to speak, "harmonious." It does not, properly speaking, appear at all. And one cannot produce a "harmonious," tightly woven fabric if warp and weft threads are not regularly spaced and are not at right angles to one another, perfectly orthogonal. Nor, for that matter, could an ancient Greek shipwright build a watertight boat if the tenons were not at right angles to the planks. In a textile, skewed or unequally distributed threads produce a loosely woven fabric full of holes, just as in a boat skewed tenons make for a leaky vessel. This was the last thing the people who founded new Greek cities in strange, often hostile lands would have wanted.

In the mid-fifth century, the Piraeus was given an orthogonal "plan" whose traces today have all but disappeared. One supposes, however, that the configuration of this plan must have been at least generically similar to the plan of Selinous, the seventh-century colonial foundation in Sicily ... even though Hippodamian planning is usually considered to be based on a square grid.

Now the Piraeus was, as it still remains, the port of Athens, and ports, generally speaking, are populated by shifty characters of every description. Being places where people continually come and go, they are not known for having a "harmonious" or coherent urban fabric. The habitués of ports do not form closely knit communities. The Piraeus of the mid-fifth century was no exception. Practically waterless, with steep, barren hills, it also had an unhealthy climate due to its proximity to the Halipedon Marsh. By the Periclean period it was full of foreigners, who practiced strange, extravagant Great Mother cults. As the largest and most populous deme of Athens, it had also become a hotbed of radical democracy. Everything about the Piraeus lacked *harmonia*: its climate, its topography, its population. Thus when, at Pericles' request, Hippodamus of Miletus "cut up" the Piraeus and imposed an orthogonal grid of streets on its very difficult existing terrain, the intention would have been to make the Piraeus "harmonious," the way a well-built boat or a tightly woven cloth was understood by the early Greeks to be "harmonious" – even as in earlier centuries the founders of colonial cities had sought to ensure the *harmonia* of their new foundations by laying out regularly spaced streets that intersected at right angles.

The Piraeus was something like a colonial foundation in more ways than just this one. Hippodamus' "cutting up" took place a generation after Themistocles had had, as Thucydides puts it, "the audacity to suggest that the Athenians should attach themselves to the sea [*tēs thalassēs anthektea*]." Themistocles, by building the Piraeus as the Athenian port and

laying the foundations of Athens' sea power, made Athens a *metropolis*, a mother city. Graphically umbilical long walls, also built at this time, linked the "child" to its mother and underscored the maternal connection. And just as the parental relation between mother cities and their colonial foundations became reciprocal to become manifest in the rise of the *polis*, so did the roles of Athens and the Piraeus become reversed, for by sustaining the city with imported goods the port became the nurturing mother of the city that had given it birth. As Plutarch, who calls Athens' sea empire the mother of democracy, says, "Themistocles did not, as Aristophanes the comic poet says, 'knead the Piraeus onto the city,' nay, he fastened the city to the Piraeus and the land to the sea."

It has been argued, chiefly by Joseph Rykwert, that orthogonal street layouts had to do with orientation: with squaring the position of the citizen with the geometric configuration of the cosmos. This certainly seems to be so. But in early Greece, where craftsmen were known as *dēmiourgoi*, and before either the cosmos or city streets became geometrical, the experience of weavers had already led them to the discovery that the *kosmos* of a tightly woven cloth depended on equally spacing warp and weft threads and interlacing them at right angles to one another.

When Irad Malkin discusses the choosing of sites for the new sanctuaries in colonial cities, he refers, as evidence for the colonial founders' functional approach, to a certain passage from Aristotle's *Politics*. The Barker translation of this passage reads as follows:

> This site [*topos*] should be on an eminence conspicuous [*epiphaneia*] enough for men to look up and see goodness [*aretē*] enthroned [*ikanōs*] and strong enough [*erymnoterōs*] to command [*geitniōnta*] the adjacent quarters [*merē*] of the city.

"This conspicuousness," says Malkin, "seems straightforward and signifies prominence and impressiveness ... as a criterion for choosing the site of a sacred area." A close look at the original Greek reveals nothing so straightforward. First of all, in the Greek there is not one word about looking up, or about the *topos* being "on an eminence": these are the translator's interpolations. Secondly, *epiphaneia*, a word already discussed at some length, has only a secondary relationship to conspicuousness. *Epiphaneia* has to do with appearing, visibility being the evidence for existence: *epiphaneia* is visible surface, and testifies to coming-to-light. That *epiphaneia* should be read as "prominence and impressiveness" is also an interpolation. Thirdly, *aretē*, goodness, is *ikanōs*, which is not to say "enthroned" but reached, fulfilled, or attained in a becoming or appropriate way. And, finally, *geitniōnta* is a participle of the verb *geinomai*: to be born.

Thus, a legitimate alternative reading of the passage might be: "The place should be such as to have *epiphaneia* so as to see goodness fulfilled and strengthened, so that the regions of the city might come to be." If a city is to be woven, *hyphainein* (brought to light), so that it may appear (have *epiphaneia*), the sites for the temples whose building is integral to that city's appearing must also, themselves, have *epiphaneia*. And if Aristotle's view on the choosing of temple locations can be taken to bear traces of how these sites were in fact chosen in the early colonial cities, it is a view which, when read with an effort to think Greek rather than functional modern, entirely sustains what I have been attempting to articulate about the weaving of the city.

The colonial founder's authority came from the Delphic oracle, which is to say from the god Phoebus Apollo, "and Phoebus it is that men follow when they measure [*diemetrēsanto*] cities; for Phoebus evermore delights in the founding of cities, and Phoebus himself weaves [*hyphainei*] their foundations." Hephaestus, as the hymn cited earlier suggests, taught men *erga*, the crafts that made possible the existence of human communities. The hymn also mentions Athena: "With bright-eyed [*glaukōpis*] Athena he taught men glorious crafts." The craft specific to Hephaestus, god of the forge, was metalworking. The craft specific to Athena, goddess of the city, was weaving. Athena, patroness of all cities, but chiefly of course of Athens, taught people how to weave. If it was Apollo who masterminded colonial expeditions and wove the foundations of cities, it was Athena, bright-eyed patroness, also, of weavers, who taught people how to make these cities visible.

3 REM KOOLHAAS

'Singapore Songlines: Portrait of a Potemkin Metropolis . . . or Thirty Years of Tabula Rasa'

from *Rem Koolhaas and Bruce Mau, S, M, L, XL* (1995)

PROMETHEAN HANGOVER: THE NEXT LAP

From one single, teeming Chinatown, Singapore has become a city *with* a Chinatown. It seems completed.

But as a (former) theater of the tabula rasa, Singapore now has the tenuous quality of a freeze-frame, of an arrested movement that can be set in motion again at any time on its way to yet another configuration; it is a city perpetually morphed to the next state.

The curse of the tabula rasa is that, once applied, it proves not only previous occupancies expendable, but also each *future* occupancy provisional too, ultimately temporary. That makes the claim to finality – the illusion on which even the most mediocre architecture is based – impossible. It makes Architecture impossible.

The anxiety induced by the precarious status of Singapore's reality is exacerbated by the absence of a geometric stability. Its courage to erase has not inspired a new conceptual frame – *guiding concept*? – a definitive prognosis of the island's status, an autonomous identity independent of infill, such as the Manhattan grid. Singapore's proliferating geometry is strained beyond its breaking point when it has to organize the coexistence of the strictly orthogonal super-blocks of average modernity that comprise the vast majority of its built substance. Singapore's "planning" – the mere sum of presences – is formless, like a batik pattern. It emerges surprisingly, seemingly from nowhere, and can be canceled and erased equally abruptly. The city is an imperfect collage: all foreground, no background.

Maybe this lack of geometry is typically Asian; Tokyo is the eternal example. But what does that make the present, almost worldwide condition? Is Paris encircled by an Asian ring? Is Piranesi's Roman Forum Chinese? Or is our tolerance for the imperfection of "other" cultures, "other" standards a camouflaged form of post-colonial condescension?

The resistance of these assembled buildings to forming a recognizable ensemble creates, Asian or not, a condition where the exterior – the classic domain of the urban – appears residual, leftover, overcharged with commercial effluence from hermetic interiors, hyper-densities of trivial commandments, public art, the reconstructed tropicality of landscaping.

As a manifesto of the quantitative, Singapore reveals a cruel contradiction: huge increases in matter, the overall effect increasingly unreal. The sinister quality of the windows – black glass, sometimes purple – creates, as in a model-railroad landscape, an additional degree of abstraction that makes it impossible to guess whether the buildings are empty or teeming with transplanted Confucian life . . .

In spite of its colossal substance, Singapore is doomed to remain a Potemkin metropolis.

That is not a local problem. We can *make* things, but not necessarily make them real. Singapore represents the point where the volume of the new overwhelms the volume of the old, has become too big to be animated by it, has not yet developed its own vitality. Mathematically, the third millennium will be an experiment in this form of soullessness (unless we wake up from our 30-year sleep of self-hatred).

After its monumental achievement, Singapore now suffers a Promethean hangover. A sense of anticlimax is palpable. The "finished" Barthian state is grasping for new themes, new metaphors, new signs to superimpose on its luxurious substance. From external enemies, the attention has shifted to internal demons, of which doubt is so far the most unusual.

Lee resigned in 1990 but remains prominently in the background as an éminence grise. His successor, Goh Chok Tong, must assure the transition from a hyper-efficient garrison state to a more relaxed version of Sparta.

It is a period of transition, revision, marginal adjustments, "New Orientations"; after urbanization comes "leisurization." "Singaporeans now aspire to the finer things in life – to the arts, culture, and sports . . ."

The recent creation of a Ministry for Information and the Arts is indicative. As Yeo, its minister, warns, "It may seem odd, but we have to pursue the subject of fun very seriously if we want to stay competitive in the 21st century . . ."

Singapore is a *city without qualities* (maybe that is an ultimate form of deconstruction, and even of freedom). But its evolution – its songline – continues: from enlightened postwar UN triumvirate, first manifestation of belated CIAM apotheosis, overheated metabolist metropolis, now dominated by a kind of Confucian postmodernism in which the brutal early housing slabs are rehabilitated with symmetrical ornament.

In the eighties, the global consumer frenzy perverted Singapore's image to one of repulsive caricature: an entire city perceived as shopping center, an orgy of Eurasian vulgarity, a city stripped of the last vestiges of authenticity and dignity. But even in a terminal project such as Nge Ann City, the elements of former ideological life are present, latent under the sheen of garish postmodernity (granite, brass, brick) which, in the new rhetoric, is based not only on Asian life but on the resurrection of Asian aesthetics: the Chinese Wall, pagodas, the Forbidden City, etc. Under the forms and decorations it is still a stunning urban machine, with its lavish parking decks on the 11th floor, the diversity of its atriums, the surprising richness of its cellular department stores, mixing Nike with Chanel, Timberland with Thai food: Turbo-Metabolism.

History, especially colonial history, is rehabilitated, paradoxically because it is the only one recognizable *as* history: the Raffles Hotel, painstakingly restored in the front, is cloned in the back to accommodate a shopping-center extension that far exceeds the original in volume.

Paul Rudolph reemerges from limbo. Somewhere in the city one of his American prototypes – it started its conceptual life in the sixties as a stack of mobile homes hoisted in a steel skeleton – stands realized in concrete.

In 1981 he had been part of the Beach Road experiment – presumably unknowingly. For a developer, and without contact with his Singaporean colleagues, the American designs a metabolic project: a rotated concrete tower next to a deformed bulge of a podium, one of the first manifestations of the independent atrium. Thirteen years later, it too stands realized, but in aluminum, the rotation of the tower replaced by indentation, a metallic corncob, its "American atrium" more hollow than its Asian counterparts.

Singapore's center will be hyper-dense; a massive invasion of stark, undetailed forms crowds the city model on the top floor of the planning office. On newly reclaimed land, the last center pieces are being fitted with contextual masterpieces: a "Botta," a posthumous "Stirling." But how can buildings be sympathetic to their environment if there *is* no environment?

Various anxieties (repressed? imported?) come gingerly to the surface, most insidiously about the disappearance of history. "There is a call to preserve and explore our rich cultural heritage . . ."

Goh has identified his reign as the Next Lap (it supersedes Vision 1999). At his November 1990 swearing-in he proclaims, "Singapore can do well only if her good sons and daughters are prepared to dedicate themselves to help others. I shall rally them to serve the country. For if they do not come forward, what future will we have? I therefore call on my fellow citizens to join me, to run the next lap together . . ."

But the name alone betrays an inbuilt fatigue, like a marathon run around a track. Goh's Next Lap is like an invitation to join him on a treadmill.

Mostly, the Next Lap represents further work on Singapore's identity. "Our vision is . . . an island with an increased sense of 'island-ness' – more beaches, marinas, resorts, and possibly entertainment parks as well as better access to an attractive coastline and a city that embraces the waterline more closely as a signal of its island heritage. Singapore will be cloaked in greenery, both manicured by man and protected tracts of natural growth and with waterbodies woven into the landscape." Altogether, Singapore is poised to evolve "Towards a Tropical City of Excellence."

In this climate of relative reconsideration, if not contemplation, nature itself is a prime candidate for rehabilitation, sometimes retroactively. "All of our efforts are marked by the desire to balance development with nature ... Sometimes, as elsewhere around the world, we have tended to over-develop a few. In some such cases, there is a need to roll back time, remove the buildings and rehabilitate the old vegetation." Almost ominously, it even seems as if nature will be the next project of development, throwing the mechanics of the tabula rasa into a paradoxical reverse gear: after development, Eden.

Already in 1963, Lee Kuan Yew "personally launched a tree-planting campaign" as prophylactic compensation for the urban renewal programs that were to be initiated. "Active tree planting was carried out for all roads, vacant plots, and new development sites."

Parallel to the intensification of urban renewal, a "garden city" campaign was started in 1967, "a beautification programme that aims to clothe the republic in a green mantle resplendent with the colors of nature ..."

Now the state is about to complete a "park network," an ambitious web implemented through a "park connector system" that will convert Singapore into a "total playground."

Worldwide, landscape is becoming the new ideological medium, more popular, more versatile, easier to implement than architecture, capable of conveying the same signifiers but more subtly, more subliminally; it is two-dimensional rather than three-dimensional, more economical, more accommodating, infinitely more susceptible to intentional inscriptions.

The irony of Singapore's climate is that its tropical heat and humidity are at the same time the perfect alibi for a full-scale retreat into interior, generalized, non-specific, air-conditioned comfort – *and* the sole surviving element of authenticity, the only thing that makes Singapore tropical, still. With indoors turned into a shopping Eden, outdoors becomes a Potemkin nature – a plantation of tropical emblems, palms, shrubs, which the very tropicality of the weather makes ornamental.

The "tropical" in "tropical excellence" is a trap, a conceptual dead end where the metaphorical and the literal wrestle each other to a standoff: while all of Singapore's architecture is on a flight *away* from the heat, their ensemble is supposed to be its apotheosis.

The only tropical authenticity left is a kind of accelerated decay, a Conradian rot: it is the resistance to *that* tropicality that explains Singapore's uptightness. "It corresponds to a deep primordial fear of being swallowed up by the jungle, a fate that can only be avoided by being ever more perfect, ever more disciplined, always the best ..."

Finally, in a move beyond the reach of irony, the island is now being outfitted with a perimeter beach. "By the year x, through reclamation and replanning, the amount of accessible shoreline is almost doubled, while the inaccessible areas are correspondingly reduced. There are ample opportunities for us to create beaches, promenades, marinas, resorts, etc."

Singapore now becomes a willed idyll – "like in May '68," the former chief planner, Liu Thai Ker, whispers to me. It is a subtle revision. Not "under the pavement, beach," but "after the pavement, beach."

POSTSCRIPT: METASTASIS

As it stands, the Singapore model – sum, as we have seen, of a series of systematic transsubstantiations which make it, in effect, one of the most ideological of all urban conditions – is now poised to metastasize across Asia. The sparkle of its organization, the glamour of its successful uprooting, the success of its human transformation, the laundering of its past, its manipulation of vernacular cultures present an irresistible model for those facing the task of imagining – and building – new urban conditions for the even more countless millions. More and more, Singapore claims itself a laboratory for China, a role that could lift its present moroseness.

The sums are stark: "Eighty percent of China's population is still rural," argues Liu Thai Ker, former head of the URA, now in private practice. "The mere shift of one fourth of them to the city over the next 20 years – an implausibly low figure – would imply a doubling of all their urban substance."

It is unlikely that the deconstructivist model, or any of the other respectable contemporary propositions (what are they anyway?), has a great attraction in these circumstances. Singapore represents the exact dosage of "authority, instrumentality, and vision" necessary to appeal. In numerous architectural offices in Singapore, whose names few of us have ever heard, China's future is being prepared. In these countless new cities the skyscraper is the only surviving typology. After the iconoclasm of communism there will be a second, more efficient Ludditism, helping the Chinese toward the "desired land": market economy – but minus the decadence, the democracy, the messiness, the disorder, the cruelty of the West.

Projecting outward from Singapore, an asymmetrical epicenter, there will be new Singapores across the entire mainland. Its model will be the stamp of China's modernization.

Two billion people can't be wrong.

EXIT

Singapore mantra: don't forget to confirm your return flight.

4 SARAH BUIE

'Market as Mandala: The Erotic Space of Commerce'

from *Organization* (1996)

[. . .]

THE NATURE OF MARKETS

As described by one British geographer, Brian Berry, the market is 'a vital, equilibrating interface between the geography of production and the geography of consumption – the intermeshing of those geographies'. It was the original space for what Adam Smith called 'a certain propensity in human nature ... the propensity to truck, to barter and exchange one thing for another'. What is its nature, and what is its aesthetic?

It is often a matter of scale. Markets are, first and foremost, places of personal exchange. They are only a step or two removed, in the lineage of economic history, from the worlds of gift exchange and reciprocity or barter, when trade was about much more than the actual goods transacted. In traditional marketplaces and bazaars, people make their transactions face-to-face, buyers and sellers often know each other and they may develop longstanding trading relationships. As people, they are embedded in a larger network of social relationships, including family and friends, which engenders loyalty as well as obligations, accountability as well as the steady prosperity a loyal clientele can provide.

In addition, in traditional markets sellers are often closely connected to their product or goods; they may have grown it, or made it, or supervised its making. So often there is a sense of pride and ownership in the seller, as well as standards for and an expertise about the product and its relative value. These factors certainly serve as the foundation for increased trust on the part of buyers; pride in making can also motivate the seller to make exceptional displays of the goods in question. Colorful, tactile, fragrant produce and wares, organized with care and abundantly displayed, create a wide range of sensory pleasures and heighten the satisfaction of making a transaction. The economic exchange is coupled with a sensual one as well.

In marketplaces and bazaars, bargaining is a highly developed practice among both buyers and sellers, requiring skill, perseverance and steely nerves. Yet it is also an art and a dance, and both parties are respectful of each other's intelligence, wile and stamina. If the bargaining transaction is successful, both the buyer and seller will leave fulfilled and happy – the seller will have received a fair price, and the buyer will feel that he has done well, made an interesting transaction. There is also often an artistry and ritual to the measuring of goods for purchase, such that the buyer will feel that he or she is receiving an ampleness that is in fact not there; measuring devices with small bases cause the produce to spill out over the top, there are generous sweeping gestures made of the topping off of the package with something additional not paid for. All of this is part of the play, part of the dance of exchange. And sometimes, with more expensive merchandise, showing what's for sale can also become an extraordinary ritual, with the customer seated on little stools, served tea or soft drinks, and regaled with a display of finely woven saris, or every possible black flowered cotton print currently available.

Like the mandala, marketplaces are radiating centers. First, they are places of gathering, in which virtually all aspects of life are in play, united in a shared space. Beyond satisfying their first function as a place of the exchange of goods, they are centers of social life, of communication, of political and judicial activity, of cultural and religious events. They are places for the exchange of news, information and gossip ... Friends and family meet for casual socializing, or to complete some aspect of a circumcision ritual or a funeral observance. After making the rounds for vegetables, oil, fish and fabric, individuals may make a *puja* or devotion to his or her deity at a shrine in the middle of the road or in a market wall. In the rich melee of social intercourse the market provides, love affairs are begun, marriages are arranged. Disputes and debts are created and settled, and the market day often ends with a siege of drinking, dancing and fighting.

The market schedule is the central factor in the public life of a district or region; marketers either plan their activities around the market-day cycle, or markets are scheduled to coincide with public events and holidays, religious festivals and fairs, the best times to easily assemble a large collection of buyers and sellers. And they are physically central, often developing at a crossroads, or on a trade route where a stronghold or temple exists; in medieval European towns the markets often formed along the wall of a cathedral; the Agora in Athens and the Forum in Rome are both examples of central markets that were arenas for almost all public life. Whether in large cities or small villages, the centrality of the marketplace is essential for effective redistribution of goods in a region. And, when combined with the architecture of religion and politics (as is so often the case), it creates a strong sense of place, a focus for the life of a village or metropolis; the aesthetic character of any particular settlement has been shaped to some degree by the marketplace, its centrality, its demands, its scale, its rhythms.

Markets are perhaps most interesting as a space of opportunity. They serve their participants as a container of many choices, offering them first a variety and multiplicity of goods, but also a sense of potential, adventure, happenstance encounters at the same time that they provide grounded contact with the familiar. A critical mass of participants brings with it diverse possibilities; the rhythmic movement of goods between hands is mirrored in the variety of other kinds of exchange that become possible

as well. As in the mandala, the dualities and dramas of life, gaining and losing, meeting and parting, virtue and vice, the sacred and the mundane, exist in a dynamic interaction within the market container, and a heightened sense of liveliness and possibility results.

The space of opportunity is economic as well. As economic institutions, the earliest markets on records were begun by accomplished Egyptian farmers, and it was hard to distinguish between the buyers and the sellers; everyone was a producer, and there was great parity in the process of exchange. Intrinsically, markets stand as places of mutual opportunity, where no one person's or group's interests prevail, where a diverse group of buyers and sellers compete with relative equality. On a local level this system, among persons embedded in a shared cultural and material world, should be self-generating and self-sustaining. It does not require a great deal of regulation (primarily that of a fair set of weights and the protection of some simple security), nor any artificial devices to stimulate or restrict it.

THE MARKET PRINCIPLE VS THE MARKET PLACE

However, as the needs and desires of a group extend beyond what is locally available, special mechanisms (that of money in the first instance) and regulations must develop in order for the system to function. Slowly this process tends to evolve away from the egalitarian into one dominated by official controls outside and more privileged controls inside. Taxes, import regulations, the imposition of pricing structures – the pressure on small producers becomes too great. Ultimately, the market principle destroys the marketplace ... And as we live increasingly in a world culture, traditional markets are gradually less important for social, political and religious transactions as well. Videos and CDs from the other side of the planet are for sale alongside produce and kerosene.

The contemporary marketplace in America gives aesthetic form to the forces at work in the market principle and mass production. A great number of our market environments, whether they are titularly toy stores, supermarkets, department stores, drug stores, hardware stores, stationery stores or multipurpose stores such as KMart and WalMart, are very similar rather sterile places where mass-produced goods fill large overlit warehouse-like structures owned by conglomerates, located on strips in the suburbs which can be reached only by car. A sense of scale and place has been lost, the products are homogenized, and there is almost no personal transaction in their purchase. Perhaps even more impersonal is the catalogue as market, through which people, alone at home, pick up the phone 24 hours a day and shop for cookie-cutter items made by anonymous makers all over the world, sent to them from a placeless firm with nameless operators. Eros is reduced to the sale of look-of-sensual lifestyles through Williams-Sonoma or the squeaky-clean safe-sex world of J. Crew.

In contrast, traditional markets vividly express our genuine erotic interdependence. Aesthetically, that translates into direct physical experiences: centralized spaces on a human scale, a walking scale, that create intimacy, contact, interaction, responsiveness; spaces that express and create a sense of place, gathering, social exchange, excitement, possibility. Intensely sensual and physical places, with jostling and puddles and honking and smells of palm oil and kerosene and orange peels and perfume, strains of pop music or bamboo flutes, women carrying babies and baskets, men playing board games and smoking small cigarettes, cows walking by, slapping their tails. Many bodies, bales, baskets, beasts of all descriptions, all interconnected, all engaged in transactions, big and small. It means products that have a sense of place, that come from somewhere or someone, be it a craftsman selling his wares, a dealer who knows the maker, or the young farmer who grew her own squash. It implies a human sensibility at work, whether in the organic irregularities or the remarkably skillful exactitude of making, and a sense of accountability for the quality of the product, with the careful making that implies. It translates into a sense of abundance, a sense of nourishment created by genuine exchange, and the refreshment of fluidity, purposeful activity and surprise.

Aesthetics, of course, is not actually about the surface appearance of things, but about how the true nature of an undertaking, or an object, or an intention is embodied and expressed in form. It is the direct visceral way we understand what something means, what it is. As such, the traditional marketplace is a true expression in form of life's vitality and of the necessity of transaction, economic, social, sexual and spiritual, for its sustenance. What can we bring from the marketplace of the past into our late-20th-century cyberspace economy? Can we reclaim in some ways the fundamental aesthetic satisfactions inherent in the act of commerce? Must access by definition be at the expense of feeling?

5 EDWARD ROBBINS

'Thinking Space/Seeing Space: Thamesmead Revisited'

from *Urban Design International* (1996)

INTRODUCTION

For most of this century, space and spatial practices have been viewed as epiphenomenal to history and other social and political–economic practices and structures. It has become fashionable, though, in the 1990s to speak about the importance of spatial form, spatial encounters, spatial practices and spatialization in social and planning theory ... As Fredric Jameson has stated:

> I think that it is at least empirically arguable that our daily life, our psychic experience, our cultural languages, are today dominated by categories of space.

Yet with all the new emphasis on theorizing the spatial, little of such theorizing moves from the textual understanding of space to an analysis of space. Even though Jameson argued that we must read the past 'off the template of its spatial structures,' in most analyses that emphasize the spatial, looking at physical spaces and the spatial practices they induce plays only a secondary role, if it plays any role at all. Even where there are calls to use spatial theory in urban planning practice, space is usually conceived of as a condition which is reflected in social practices rather than a condition which might reveal things about social practices and social thinking otherwise hidden from view ... It may also be that we live within sites that delineate and reveal the many and complex relations and, at times, contradictory practices through which we construct our world.

Our understanding that social relations and social practices delineate site and that site is a reflection or embodiment of those relations and practices is commonly claimed by analysts looking at housing and the planning of housing. These analyses are not notable, however, for looking at the way housing and urban space reveal social relations and social thought. Mellor has an insightful analysis of the way social ideology and institutional and political–economic circumstances influence the nature of housing in urban planning but as insightful as it is the analysis moves from social relation to embodiment in spatial form. A number of architectural theorists and historians have taken a critical look at housing in the 20th century but although they discuss the social implications of spatial form or the history of a housing type, they are not centrally concerned with the dialectic of space and society ...

More specifically, when analysts try to understand the nature and roots of government housing or urban policy and its meanings and implications, they tend to look at the social and political construction of that policy rather than its spatial outcomes (even Lefebvre resorts to a more institutional and political discourse when writing of government interventions in urban planning). The representation of policy embodied in its actual spatial realization is rarely used as a means to analyse and understand the policy itself. Works on government policy look at the ideological underpinnings, the process of institutional decision making and policy formulation and the types of social and economic investment which define the outcomes the policy eventually puts in place. What they do not look at is the extent to which the production itself is a form of policy as well as a representation of that policy ...

THAMESMEAD REVISITED

What follows is an examination of Thamesmead, a major housing and urban development of the late 1960s and early 1970s in England, in order to explore what we might learn about policy from looking at a spatial realization of design. Projects such as Thamesmead are no longer undertaken, nor are they in intellectual vogue. It is probable that few architecture students even know about them. Revisiting Thamesmead is important, nonetheless, because it teaches us how even with the best of intentions our ideological premises may be hidden because we do not look critically at the spaces that such policies and programmes produce even when we are willing to engage exhaustively those policies and programmes.

The issue is not Thamesmead *per se* but what we can learn about the public sector's thinking by looking at a project's design and development. This issue was important in the past and it is still important today. Whether the public sector builds or not, it is still a crucial player in the realization of housing be it through subsidy, public–private partnerships, tax policies or other financial incentives. The public sector is still dominant in shaping urban design even if this **specific** role has changed.

Thamesmead was never meant to simply provide a stop-gap solution for the homeless or the underhoused. It was meant, rather, to offer a model for a technically rational way of producing housing while at the same time providing a better environment for those previously living in the inner city slums and run-down neighbourhoods. To begin, let us look by comparison at the areas that Thamesmead was to supersede and for which, it was assumed by the designers and policy makers, it would be a major improvement.

If we looked at those neighbourhoods what might we have found? Certainly, we would have found run-down and dilapidated houses, overcrowded flats, dense and overburdened residential streets with an old and often inadequate infrastructure, e.g. outdoor toilets, bad plumbing, little or no heat or decent ventilation. The streets would have been generally noisy, particularly those streets that abutted commercial areas or transport routes. Pollution and filth would have been more the rule than the exception as city services were pushed to the limit by the densities and inadequate infrastructural supports, e.g. narrow streets, closes and alleys, which are common in poorer neighbourhoods.

The helter-skelter of the streets with their ground floor stores, pubs and light industrial and craft shops intermixed with residences and exhibiting no obvious order or rationality might well have suggested to the outside observer a general problem of social chaos at worst, or inefficiency and insalubrity at best. Competition between pedestrian and motorist and between the mix of people – children, teens, adults – all with different intentions and reasons for using the street, would suggest a level of social contestation and even conflict which was both unavoidable and unpleasant. Idlers conflicting with working people, teens with storekeepers, children with adults, and drunks with the sober created a cacophony of problematic social confrontations. Social life could appear to the uninitiated to be endlessly enervating and debilitating; so much so that those living in such communities would find it difficult to improve their lot because of the surrounding environment. Poverty, idleness, and conflict would all appear to be in the nature of the place.

Yet, one could have a very different but equally evocative vision of these neighbourhoods. Life on the streets might have appeared to be active and alive. People could be seen to have developed quite a complex and supportive social interaction which overcame the apparent disorder of the streets. Harmony of purpose would, in this view, have transcended the differences of the moment.

The public spaces of these neighbourhoods for some observers could become living rooms where people socialized and passed the day because their homes are neither large nor nice enough to entertain indoors. Streets provided a place to live out both a public presence as well as private longings. Families fought on the street, youths strutted their stuff to exhibit their sense of self-worth. It is in public where boy meets girl, children learn about growing up and adults come to share their joys and woes. What appears to be the very chaos of the street is its attraction. Cacophonous though these streets may be, shared understandings of the rules of engagement make the street a most ordered and organized place.

Each of the two visions of the older working-class neighbourhoods of London are, granted, somewhat overstated. One vision is overly pessimistic while the other overly romantic. Both speak to the energetic public activity and social life in public places and spaces. Whatever else we might say about the lower-class neighbourhoods of London, they are alive, spontaneous, even chaotic if not necessarily disorganized or disorderly.

Thamesmead was built to respond to the failings of older urban neighbourhoods. Built by the Greater London Council [GLC] in the late 1960s, Thamesmead was a recipient of much praise – and a little criticism from a small but vociferous group of community planners – from the planning and architectural communities at the time. It was internationally honoured and won the Sir Patrick Abercrombie Award in 1969 for architecture and planning given by the *Union Internationale des Architectes*.

Thamesmead was the product of a particularly active period in British history when government social policy was developed to correct the inequities of society and provide the basis for economic and social mobility. It was a period in which the GLC spent £150 million on 26,000 dwellings in a four year period between 1965 and 1969 and was planning to invest upwards of £150 million more on the design and construction of Thamesmead . . . If anything, Thamesmead is important to revisit if only because it reminds us of a time when government was committed to large scale interventions on behalf of the poorer sectors of society and remedying the inequities which they suffered. It was also a time when architects, urban designers and planners believed that housing and its urban design and built form might play an important part in helping to overcome social and economic privation – an idea no longer holding much support.

It is also important to revisit Thamesmead because as a material realization of policy planning and architectural and urban design, it represents an attitude about the poor and working class on the part of policy-makers and designers which the stated policies and design programmes barely address, if they address them at all.

[. . .]

The overall site plan, dictated to a large degree by the location and the marshy subsoil of the site, called for a series of highrise buildings on the edge of the estate that would serve as noise and wind breakers for a series of lowrise and service areas set within a series of multiple levels. Residential occupancy was limited to floors above the ground level because regulations prohibited groundlevel residences in marshy areas. As a result, the designers, to their delight, were able to design all the houses with a car and storage area below and a living area above. This theme of functionally distinct levels was carried out throughout the site plan by keeping auto traffic below and pedestrian traffic above.

Emphasis on lowrise housing was a response to earlier GLC projects which had placed greater reliance on highrises. Thamesmead was a response to the expense of tower blocks and the emergence of some criticism of them on social grounds. What is clear from this comment is that the designers of Thamesmead at a conscious policy and programmatic level were concerned with the social uses and meanings that design produces. It also reveals that these same designers and planners were not at all naive about the importance of creating good social conditions if their design was to succeed.

Thamesmead, was not only a social experiment. It was a test for the GLC of industrialized systems building techniques and of cost efficiencies the techniques were supposed to provide. A factory to provide the building elements was to be built on site. Until it was built, what the British called at the time the 'Rad-Trad' technique was used – rational and traditional technologies intermixed. What the GLC wanted to accomplish was the building of a complex and largescale housing estate (or mini new town depending on who was defining Thamesmead) using a 'broad framework of standardization and rationalization of procedure' without resorting to a reliance on tower blocks.

Rationalization was the central concept guiding the plan. The rationalization of technological procedures, however, while crucial in determining the scale and initial building blocks of the plan did not entirely define all the aspects of the design nor exhaust all meanings ascribed by the GLC to rationalization. Rationalization also meant something more: it was the means to define the nature of Thamesmead as an urban place of social and cultural form.

Despite what were viewed as financial constraints at the time (today such a level of investment in social housing would be considered extremely generous) emphasis was placed on the design of a total environment. Residential areas would be interspersed with schools, play areas for children, adult recreation areas, pedestrian walkways, cycling paths, shopping and community centres, churches and even an area for the development of light industry . . . A total environment would not be provided at the expense of the community, the GLC argued. Community was to be designed into the project as well. As the GLC averred:

> A sense of community . . . can best be achieved in a compact development like Thamesmead by linking areas which might be other wise separate neighbourhoods by spines of flats.

Nothing would be left to chance. All aspects the GLC hoped would be adequately planned and provided for.

Thamesmead thus was designed, in plan, as a series of residential sections (what the GLC called 'neighbourhoods') intersected by pedestrian ways and accessed by roadways. These sections were joined by a long undulating spine of lowrise housing which linked the sectors and set them off as distinct formed elements of the plan. Each section was to be provided with its own play areas, churches, community centre and playing fields. All the sectors would be linked by a large shopping centre and a central community centre. This centre would border a yacht basin and marina – it was hoped that Thamesmead would attract middle-class as well as working-class residents on the Thames. Away from the centre, along the edges of the site but set in the middle of the estate area, two artificial lakes would be built for visual effect and recreation.

Particular attention was paid to the local contours of the site to accentuate visual interest and harmony. The contours of the site also allowed the designers to protect it from noise and pollution from the highway and railroad that border the estate. By separating pedestrian, auto and bicycle traffic the designers felt they provided even more protection while adding, through the use of layering, greater visual harmony and social order. Protection, visual harmony and the rationality of the plan would provide, the GLC argued, a sense of belonging 'through its [the estate's] buildings and landscape'.

For the GLC policy-makers and designers, Thamesmead offered the latest in housing form and social possibility. They were providing a clean, well-ordered, safe, functionally delineated and segregated, and well-defined space into which people would come and build meaningful and happy lives. Thamesmead would be diametrically opposed to the world of the slum, and the working-class neighbourhoods from whence most residents would emanate. It would present to its residents a vision of urbanity and social practice absolutely foreign to the traditional and dysfunctional neighbourhoods of working-class London.

Thamesmead's design formed a world of order and clarity over the helter-skelter and ambiguity of traditional working-class neighbourhoods. In Thamesmead everything has its place and was always functionally distinct. Work and play, auto and pedestrian, shopping and residing are segregated. Order and functionality though replaced the serendipity and spontaneity of traditional streets and public places. Movement and chance meeting were limited and the multiplicity of activities of the traditional street were replaced by unifunctional spaces. Turf was defined by its function and not by the people who used it. What practices were left unplanned for were left with no space, no turf and thus no place.

Conflict and competition over space was eliminated in Thamesmead. Happenstance was highly improbable and the creative disorder of older urban areas was planned away. There was no place to call your own, no space in which to intermix the public personae with the private self and no area where one could choose from an array of social and functional possibilities with which to define who one wants to be in public. Residents, in effect, were provided no space for the extension of self and no place for their own initiatives.

Looking at Thamesmead reminds us that spatial form is rarely without its social connotations and the design of space is rarely without its social intentions. Representation of social intention in space is rarely, however, a linear reflection or embodiment of a consciously stated idea or notion. Spatial forms, rather, represent a complex, at times even contradictory, set of both conscious and unconscious understandings and intentions.

The designers and planners of Thamesmead had, on the one hand, a strong faith in the efficiencies and potentials made possible by technically rational design and building techniques. This faith, though, cannot in itself explain the form and plan of Thamesmead. It would have been possible to create other forms and plans.

It has become popular in recent discourse to see in Thamesmead, on the other hand, the workings of an ideology of an hegemonic state or culture consciously trying to order its urban social worlds and control public life ... Certainly Thamesmead can be used as evidence for such an ideological bent but it also represents, in part, the work of policy-makers and architects who ardently wanted to build a better world for the underprivileged and the working class. Most, I would suggest, would have seen themselves as socialists and as allies of the underprivileged and working classes of the UK and anything but allies of those trying to destroy the social and political potential of those classes. So why did they produce a project which would become an example of cultural distopia

– it was the site used for the filming of *A Clockwork Orange*, Anthony Burgess' disputation on a debased society – rather than a working class utopia?

Underlying the design of Thamesmead is the assumption that physical environment is a strong cultural and social influence on behaviour, morality and well-being of people. The shift to a precise, ordered and rationalized spatial form comes out of a deep suspicion of traditional working- and lower-class neighbourhood form and possibly unconscious and unstated distrust of the life it is assumed to produce. The segregation of function we see at Thamesmead represents a deeply felt anti-urbanism and a distrust in the chaos of the streets because they were unsafe, unhealthy and threatening.

This attitude is deeply rooted in the English reform tradition. Social reformers, socialists and revolutionaries have been suspicious of what we might call 'the old city'. Identifying the filth, disease and dissipation of slum life with the spaces in which poverty was housed was commonplace in late 19th century socialism ... Even Engels, who argued that the root causes of degradation were caused by the capitalist mode of production, saw in the old city a planless and knotted chaos. From Dickens to Morris, Dore to Kingsley, the solution to the old city lay in the open spaces and single family homes of the middle class. A critique of the lifestyles found in a space unwittingly became a critique of the space itself. As a result, social reformers failed to look at the way their critique of space not only corrected the more unhealthy and problematic aspects of working- and lower-class neighbourhoods and uses of social space, but also unwittingly destroyed the social and cultural energy also associated with those neighbourhoods.

It was the same with the designers and planners of Thamesmead. What is embodied in its design but not in any stated policy is unwittingly a deep distrust of working- and lower-class life phrased as a critique of aspects of the physical condition of the neighbourhood. The working and lower classes cannot be allowed to reproduce their old spatial patterns. Without realizing it, planners and designers in condemning the space of working-class life were saying that the working class should not be allowed to reproduce any of the spontaneous, chaotic and serendipitous social interactions of the traditional street. The individual household is the route to success and economic well-being; a route that is encumbered when rooted in the social interactions and social body produced by traditional working-class spatial forms and relations. Images of the spaces of middle-class familialism and individualism replace images of the spaces of working-class solidarity and sociality.

The operative word is images. Evidence for this does not come from anything written or stated. It comes from the design itself. It is by looking at outcomes and forms, spatial properties and arrangements that one can hypothesize this distrust. Although Thamesmead represents a huge commitment to social equity through the design of space, what the space itself says about the design and by extension its designers and planners was and has never been faced by those very planners and designers.

Ironically, by not looking at what their spatial design says about them, the planners and designers of Thamesmead and other projects have no response to those who accuse them of paternalistic and hegemonic practices; practices which they were not consciously undertaking but whose implications were realized in their designs.

If the planners and designers of Thamesmead created effects subconsciously, and did not see how those effects embodied strongly-felt attitudes about others, the residents of Thamesmead clearly did. This was made clear in the study of resident complaints about Thamesmead done for the GLC. Two examples from the study are revealing.

Residents complained about heating, leakage, noise in the apartments and other individual problems. There were also complaints about the lack of decent jobs in the area, the distance from their old neighbourhoods and friends and family, about the placement of slurry pipes for an adjacent disposal plant in the centre of the estate, and about views of industrial Bermondsey from many of the homes. These complaints suggest that the residents perceived the design to be a constant reminder of class status and the fact that the designers were aware of that status. No middle-class estate would have slurry pipes put through it, nor would its residents be given views of belching industrial chimneys. In a way the residents were being told through the design of the space that their class needed to be moved out of its old environment but they should not forget their class through the design of the new environment.

Many residents told researchers that there was no place for their children to play, no areas for recreation, no place to go. For the designers this seemed to be just pure recalcitrance because they had so lovingly supplied play areas, fields, community centres and the like throughout the estate space. Areas for socializing and play were made available at Thamesmead but these areas were cut off from the residential areas on the estate. As such they were foreign to a culture for whom public life is a mix of uses. Public life meant spaces for socializing, functional intermixing, crowds, noise, seren-dipity and spontaneity. Segregated facilities prevent this. In this case the designers were not reminding the residents that they should forget their class and the culture with which it was associated. In so doing the designers were revealing their implicit if unconscious distrust of the residents' traditions which for them created a sense of spatial life by creating a lifeless space.

Ultimately, as the two examples reveal, Thamesmead for all its good intentions was a failure. It failed because well meaning planners and designers did not look at the spatial implications of their design. They did not understand that what they produced was ironically both a reminder to the residents of their class and an ideological condemnation of a lifeway associated with that class rather than the creation of the better place they desired.

CONCLUSION

It is commonplace today to speak of the mega-projects of modernist practice as failure. At the same time, it has become popular to speak of the necessity of understanding spatial practice. Most who criticize the great modernist projects do so by condemning the hegemonic, paternalistic, ideologically rooted practices of the modernists. What they see in their spatial designs is the reflection of these practices. I would rather see something very different. Those who designed Thamesmead were clearly unapologetic modernists who were motivated by a belief that architects and planners could help make the world a better place for all. They believed in the role of spatial form if they did not believe as postmodernists do in the necessity of spatial theorizing.

There is an irony here. The designers of Thamesmead had an abundant and strong faith in the power of spatiality to change the world. And, if anything, their belief that investment to create social equity is a practical and estimable goal is a salutary corrective to present postmodern cynicism about politics and professional practice. Yet neither those who criticize the modernist project nor the modernists themselves – even though both posit the critical importance of spatiality albeit in different ways – look to the spaces they create for answers to the questions about the world that they pose. Had they done so, each would have had a different critique of Thamesmead and other similar modernist projects. Those who designed and planned Thamesmead might not have been so eager to reframe the spatial world of the working and lower classes. Those who criticize the project from a postmodernist or post-structuralist perspective today would not be so quick to condemn the morality, ideology and measure of the modernist project. Rather each would look more carefully at the way a lack of conscious spatial theorizing and viewing spatial forms belied and undercut the modernist project.

What I hope this brief examination of Thamesmead teaches us is that spatial practices are important and that neither modernist nor postmodernist has any monopolistic claim to understanding their import. I also hope that this examination teaches us that we need to think not only about how we might understand space but how, by looking at our spatial practices and the spaces we produce, we can understand ourselves. Put another way, we need to design and plan, theorize and practice. We also need to look at the spaces which we plan, design, theorize and in which we live.

6 KEN-ICHI SASAKI

'For Whom is City Design: Tactility versus visuality'

from *City Life* (1998)

SUBJECT AND THESIS

My subject in this chapter is the aesthetics of
the city. By aesthetics I mean, here, the complex
of problems concerning the value of city and
urban life judged through our direct sensitive
experiences in the city. In other words, whether
we feel it agreeable and happy to lead our lives
in such and such a city.

So my subject is the aesthetics of city or
townscape. I have no intention of applying
notions of any ready-made aesthetics, but rather
should like to innovate aesthetics through the
question of the city, which has been kept outside
of modern Western aesthetics. It seems to me
that both our main words, aesthetics and townscape, have a strong tendency to be understood
in reference to visuality. The word 'townscape'
is especially curious. I do not know if English-
speaking people share my sense of this word,
namely the paradigmatic series of words:
'landscape', 'townscape', 'cloudscape', 'water-
scape' etc. which invite me to link the morpheme
common to these words, '-scape', with another
one, '-scope' – as in 'kaleidoscope', 'microscope',
'telescope', 'Cinema-Scope' etc. So in the
semantic constellation of words in my mind, the
word 'townscape', as well as 'landscape', belongs
to the group of visual words. In fact, however,
'landscape' shares its origin with the German
word *Landschaft*, so that the suffix '-scape' is a
variant form of another one: '-ship', which
corresponds to the German suffix '-*schaft*'; in
short, it has nothing to do with visuality.

This visual connotation of the word 'townscape' is strengthened by the basic orientation of
modern aesthetics. I am convinced that modern
Western aesthetics was constructed mainly with
reference to visual phenomena. I shall explain
this point a little later on. For the moment, it is
enough for me to put forward, briefly, the basic
line of my argument. My thesis, however, is this:
the most important factor in the aesthetics of the
city is not visuality but tactility.

I consider visuality as the viewpoint of the
visitor to a city, and tactility as that of its
inhabitants. And I also think that the deepest
and the most authentic understanding and
knowledge of a city is given, in general, by
inhabitants and not by visitors. I am indeed even
inclined to maintain that it is this tactile beauty,
if I may use this expression, which is reflected in
urban visuality as a profound beauty. Meanwhile, a special and important reference will be
made to the city of Tokyo and its history. Why
Tokyo? The opposition of my point of view to
that of modern Western aesthetics might have
been brought about and encouraged by my life in
Tokyo: Tokyo is the only city that I can say I
know in a strong sense. So, speaking of Tokyo, I
hope I will be able to explain my idea.

THE CARTESIAN CITY AND VISUALITY

First of all, I shall try to ascertain the privileged
status of visuality in modern Western aesthetics.
In the eighteenth century, when modern
aesthetics was inaugurated as a division of
philosophy, painting was the paradigm of every
art: poetry, theatre, dance, gardening and even
music. And this colouration of visuality
constituted one of the fundamental characteristics of modern aesthetics and remained so. But
before describing the new trends that appeared
in the urban space in the eighteenth century,

which I may call the modern visualization of the city, let me start with the so-called 'Cartesian city'. This term is used by historians of architecture and comes from a phrase of Descartes in his *Discourse on Method* (1637):

> there is very often less perfection in works composed of several portions, and carried out by the hands of various masters than in those on which one individual alone has worked. Thus we see that buildings planned and carried out by one architect alone are usually more beautiful and better proportioned than those which many have tried to put in order and improve, making use of old walls which were built with other ends in view. In the same way also, those ancient cities which, originally mere villages, have become in the process of time great towns, are usually badly constructed in comparison with those which are regularly laid out on a plain by a surveyor who is free to follow his own ideas.

The Cartesian city: the city which is 'regularly laid out on a plain by a surveyor who is free to follow his own ideas'. Our philosopher opposes it to the 'old towns' growing little by little through history, of which we may find a typical example in Maastricht. The essence of the Cartesian city consists in the geometrical principle which unifies and dominates the totality of the city, and which excludes all forms of contingency from its urban space. So the Cartesian city is visualized in the higher level of order of the disposition; in other words, not in the shapes of every building in the city but in their relationship. I think we may see Parmanova as an example of a realised Cartesian city.

Parmanova was constructed in 1593 under the direction of Giulio Savorgnan as a fortress for the Venetian Republic, for an advanced guard against possible attacks from Austrians and Turks. It was the sole case of a realised ideal city in Italy at the time of the Renaissance. I do not claim that Descartes speaks of this small city in the above-quoted paragraph, or even that he knew it. But I am sure that he knew about the plans for the ideal city, which were abundant then. What is important for us is the principle of the value judgement Descartes makes on cities. The excellence of a city consists, according to him, in the 'good form' (in the sense of Gestalt psychology) of the city realised on paper by ruler and compass.

What is curious to note is the fact that this Cartesian ideal city accords with a contemporary conception of townscape proposed by the American architect and urban designer Kevin Lynch. Lynch presents the idea that the beauty of a city consists in its 'legibility'. 'Legibility' means the facility of grasping it as an entity. Lynch names 'image-ability' as the objective character peculiar to the 'legible' parts or totality of a city, and proposes five standards of 'image-ability': 1) 'paths' or streets; 2) 'edges' or borders between two areas, or limits of an area, which restrict our movements; 3) 'districts', namely a stretch of surface consistent with a definite character, whether functional, socio-economical, or racial; 4) 'nodes', or important spots of traffic; 5) 'landmarks'.

Now, applying these standards to Parmanova, we find four of them clearly fulfilled by this 'ideal city'. To quote only the most important factors, as to 'paths', you have here three main streets leading from the city gates to the central circle; as to 'edges', it goes without saying that the most striking outer line of the city itself constitutes its 'edge', since the total area of the city was constructed at a higher level than its surroundings (according to my impression, the outer bank is about ten metres high). We cannot pass through this outer line except at the three gate points; these three gates, designed by V. Scamozzi, are without any doubt the most important architecture of the city and constitute its 'landmarks', as does the cathedral also attributed to Scamozzi; the gates are also 'nodes' of the city, as is the central circle.

All things considered, we can say that Parmanova is endowed with quite high 'imag-ability' and has a strong 'legibility'. According to Lynch, the beauty of a city is dependent on its legibility. Then should we say that Parmanova is very beautiful? I admit that it is beautiful to a certain extent, but in a quite superficial sense. What I mean is that it does not conform to the idea we have of a beautiful city. I find this small town charming, but do not feel like living there in any sense. It is beautiful just like a well-made earthwork. In order to enjoy its beauty, we must take a bird's eye view. In short, Parmanova is beautiful in an aerial photograph, but not in the eyes of its inhabitants. Significantly, among Lynch's five elements of 'image-ability', it is that of 'district' which is lacking in Parmanova, because it concerns not the design of the city but the life people lead there.

In short, the visuality of the Cartesian city was abstract, since it was addressed not to human eyes but to God's eye. The secularization of visuality in the modern history of Western civilization started, in my opinion, approximately during the first half of the eighteenth century. Let me quote just one fact as an index of the new trend in urban design at that time. It concerns the parvis of Notre-Dame in Paris.

Now we have a vast space in front of the cathedral which stretches over to the River Seine, so that we can enjoy the full aspect of the façade of this masterpiece of Gothic architecture, and even from the opposite bank across the river. I think this sight constitutes one of the best known townscapes. This point of view was, however, not the original one; it was procured little by little from the mid-eighteenth century on. According to J. Hillairet, 'this lake of asphalt, created by Haussmann in 1865, is six times vaster than the parvis of the Middle Ages, four times bigger than before the demolitions made during the Second Empire'. The demolition of the buildings preventing a view of the full aspect of the façade of Notre-Dame, started in 1747. Our interpretation that this demolition was made in order to get a full view of the cathedral is supported by the short article 'Aspect', written by the famous architect Jacques-François Blondel and published in the *Encyclopédie* in 1751. The fact that this monumental and synthesizing work of philosophy and sciences contains an article on such a visual concept as 'aspect', is already interesting in itself, because it expresses the increasing interest in visual phenomena. The clearest evidence of this interest is found in the appeal made by Blondel to demolish all barracks in the court of the Louvre in order to recover 'the sight of the peristyle and the interior façades of the Louvre', which is, in his opinion, the most beautiful example of the 'aspect' in architecture. In the same way, I maintain that the townscape, with its peculiar aesthetic requirements of vision, is an invention of modern times. Looking at a Gothic cathedral, lit up against the dark night sky, we are under the impression of catching a glimpse of a mediaeval scene. In fact, however, it is far from this. For, besides the electric lighting, it is our eye, our visual interest, that is modern. People before the mid-eighteenth century did not recognize the beauty of the façade of the Notre-Dame. Does this mean that they lacked in aesthetic needs or aesthetic sense? Not at all. Let me quote Hillairet once more. He says, with regard to the original design of this cathedral, that the Notre-Dame 'was constructed in order to be seen from the foot of its towers, and not from the end of the present empty space; this view minimises it.' In other words, the original manner of appreciating this architecture, according to him, consisted in standing not in front of the façade but beneath the towers, physically receiving the effect of the stone mass and feeling as though the towers reached as high as the sky. This notion is quoted without a source reference in an old edition of *Guide Vert*, and I even hear echoes of it in some Japanese guide books on Paris. I think it is not appropriate to speak of vision, view or sight, and use verbs such as 'see' or 'look' in relation to this style of aesthetic experience. We realise here that the visual aesthetics of the townscape are not universal.

TOKYO AND ITS MOUNTAIN EXPERIENCE

To begin with, let me quote my personal experience. In the summer of 1973, I had my first chance to go abroad. Having won a French government scholarship, I spent about three months in Bordeaux attending an intensive course in French, before starting studies at the University. Of course, the culture shock was strong, and a great many impressions of the foreign culture swooped down upon me. In several weeks, the first big wave had gone and I started to notice a certain uneasiness in this mixture of all kinds of impressions. One day I found it: I felt a strong desire to see mountains. I do not mean that I was a fan of mountainscapes or had frequented mountains on my holidays – in fact, I had never felt this desire in Japan. Needless to say, the campus of the University of Bordeaux is in a suburb; there had formerly been a vast, spreading pine forest, and the lands are very flat. Soon, having the chance to visit the Dordogne valley, one of the most beautiful spots in France, my desire was satisfied. Now I think I am ready to somehow explain this curious feeling I had many years ago. I have two solutions and I cannot tell which is correct; rather, I am inclined to think that both of them are correct.

In Tokyo we cannot see mountains every day, but, sometimes, we do. Tokyo is situated in the southern part of the plains of Kanto, and faces Tokyo Bay. The mountains are at least fifty kilometres from the centre of the city, but if there are no obstacles in between, we can catch sight of them from Tokyo. When I was a schoolboy, from time to time I could enjoy the beautiful sight of Mount Fuji from the window of the first floor of my primary school. Even now, we can see the mountains on the horizon when it is fine and windy, if we go to a high place. So, especially in winter, it is not so rare that one has the opportunity to catch sight of the mountains from Tokyo, without making any special effort.

This fact reflects an aspect of Tokyo's geography: Tokyo is very hilly, so there are several high spots from which we can see mountains. In this respect, there are typical place-names such as '*Fujimi-chô*' or '*Fujimi-dai*' or '*Fujimi-zaka*', which mean 'place or heights or slope from where one can see Mount Fuji.' These are stock place-names; I find, in the list of the officially registered names of quarters in Tokyo (1963), four '*Fujimi-chô*' and one '*Fujimi-dai*'. I am sure there are several other places named the same way: among the names of railway stations I know at least one '*Fujimi-dai*', one '*Fujimi-ga-oka*' ('*oka*' means hill) and one '*Fujimi-chô*', (in the metro). It is important to note that these higher view-points are in themselves kinds of mountains. If we move through Tokyo on foot or by bicycle our bodies feel these 'mountains' in the city intensely.

So, in Tokyo we have two forms of mountain experience: one is visual and concerns distant mountains, Mount Fuji in particular; the other is tactile and concerns mountains in the city. I think it was my own experiences of mountains in Tokyo which evoked in me that strong need to see mountains.

Needless to say, my theme is not the mountain but visuality and tactility in Tokyo. What I have just been discussing may already suggest how these two elements permeate the urban structure peculiar to Tokyo.

Now, let us consider some *ukiyoe* woodprints by Hiroshige, who painted a series of views of Edo (the ancient name for Tokyo). The first one is a view of the Suruga-chô quarter. Here, the proportions are not natural: Mount Fuji is drawn too big. Nonetheless, the scenery is real.

As I mentioned above, Mount Fuji, about a hundred kilometres south-west of Edo, was one of the city's most important landmarks, so we very often find it figuring in the townscape prints of Hiroshige.

In views of northern Edo, Hiroshige very often draws Mount Tsukuba in the background. This mountain, about fifty-five kilometres north of Edo, was much lower and less beautiful than Mount Fuji, but because of its particular shape as well as its isolated location, it was also a second natural landmark for the city. So these two mountains were basic landmarks for Edo, which served as principles of its city design. I should like to add that the seascape was also very much appreciated. It is said that there were eight slopes in this hilly city, called '*Shiomi-zaka*', which means 'slope from where one can see the sea'.

From these facts, it follows that the main interest of visuality was thrown far from the centre of the city, to the surrounding nature. This was a natural attitude for city design in the context of Japan's cultural tradition, for we had a very weak inclination for enclosing space. Even a city like Heian-kyô (ancient Kyoto) which was a Cartesian city, did not have a city wall. There is a basic strategy in gardening, '*shakkei*', meaning 'borrowing scenery from outside the space of the garden'. If we make a garden near mountains, we design it with these mountains as its background. We could say that Edo was designed with the views of Mount Fuji and Mount Tsukuba as its basic *shakkei*.

To continue on the subject of the experience of mountains in the city space, I would like briefly to mention the aspect of tactility. Of course, from a higher spot in the city, we can enjoy the sight of mountain or sea. We may say that vision dominates in a geographical as well as a political sense. In fact, most of uptown Edo was occupied by the mansions of provincial governors ('*daimyô*') and barons ('*hatamoto*'). But there are no mountains without valleys, and I think we also have an experience of the mountain in the valley. In the valley, we can feel we are being guarded by the mountains surrounding us. We have an idiomatic phrase in Japanese 'being embraced in the bosom of the mountain'. It speaks of the quasi-bodily feeling of leaning against the slope of a mountain. The more up-and-down a city is, the more we have the chance of feeling the security of 'being embraced in the bosom of the mountain'. And I wonder if it was this sense of security which I lacked in Bordeaux, prompting in me a strong desire to see mountains.

So my point here is as follows: in Tokyo, our vision is, or at least was, led to a distant object outside of the city, whereas city space was mainly experienced through bodily contact, almost subconsciously.

TOKYO AND THE MINIMAL ORDER

On the city map of Edo we notice that the city also had a landmark in its centre. This was the tower of the castle, which was the centre of political power, and its tower symbolized this power. But, as a landmark, the castle is very different from the city hall and the church, with their public squares (or circles), as we have in European cities. While the city hall and the church were constructed and maintained by the citizens, the castle was imposed upon them: in Japan the ruler of the city was not its citizens, and the castle, as the centre of government, was not a common institution shared by the citizens.

So the centripetal ordering power is not strong in Japanese cities. This does not mean there lacked any kind of order, on the contrary, the city of Edo had a comprehensive design. We have a traditional notion, '*machi-wari*', which means 'division of towns'. Roughly speaking, the total land of the city was divided into three categories. The best parts of the hill-tops were given to the most important governors, the rest of uptown to the middle classes, and the downtown to the common people. For the areas of these last two categories, especially, the government imposed quite a strict and geometrical design: streets and paths were straight and the division of quarters respected a large module. I think this geometry came from the needs of government: the ruler had to be able to grasp the total state of the city so its design had to respect the 'good form' of the plan.

We must admit that the city of Edo lacked an overall unity, but there was a degree of order in each quarter, and it was this partial order which dominated everyday urban life. What is very significant in this respect was the existence of the '*kido*', or town gate. Besides the city gates,

controlled, like the customs, by the government, each section of the quarter had such a small gate, which expressed the smallest unit of a community. The gate was closed after ten o'clock at night. So the city of Edo was like a state, and each of these small units of the community was like a city. Indeed, each inhabitant belonged, in the first place, to this small community, and people belonging to the same community were like a big family. They shared their interests as well as the space of their lives – keeping secrets from neighbours was particularly disliked – so they were very open to each other but quite closed to the outside. This is what I meant by the 'minimal order' mentioned in the subtitle of this chapter.

Minimal order is not only found at the level of social life. Let me point to two other facts. The first concerns the sense of space. As I noted in chapter 1, Japanese culture is anti-urban. With this notion, I am referring in particular to the fact that even a small city house has a little bit of ground reserved for the garden, which reflects nature and its seasonal cycle; Japanese cities keep nature within them. In other words, we wish to keep a small portion of nature in the form of a private garden, rather than sharing the larger natural space of the public garden. I find it very interesting, in this respect, that Japanese has two different words corresponding to the English word 'wall': '*Kabe*' refers to the interior wall of a building, while '*hei*' means the exterior wall which serves as fence. In the traditional style of Japanese house, which is based upon the baronial house of the Edo era the grounds are enclosed on all sides by the *hei*. In general, the *hei* is a solid construction of mud, wood or stone which protects the interior space from curious eyes, as well as from thieves or robbers.

The characteristic feature here is that the exterior fencing wall, or *hei*, is a much stronger barrier than the wall of the house. Unlike the exterior walls of the Western house (built with masonry), in the traditional Japanese house (post and beam construction) the equivalent walls (*kabe*, which are considered as interior walls) have no structural function. The total construction is supported not by the walls, but by posts and beams, making it possible for the spaces to be kept open rather than being walled in. In fact, in the traditional house most of the divisions between rooms, as well as between the interior and exterior spaces, are made with sliding doors. Consequently, while it is almost impossible for there to be any kind of privacy between family members, the privacy of the family as a whole is firmly guarded. Now, let me remind you that the traditional Japanese city had no city wall around it; instead, every house had a wall around its grounds. We can see, now, that the enclosed spaces of the houses of the warrior class corresponded to the small unit of the downtown community of the common people. So the Japanese city was and probably is always constituted with these minimal orders. In my opinion, it is the moral and ultimate reason why the development of the spirit and concept of the city and citizen did not develop in the same way as they did in Western civilization.

The second fact of minimal order concerns religion. In a classical European city, the church and the city hall were the central landmarks. In Edo, on the contrary, the government adopted the policy of keeping the temples away from the city centre, removing them to its periphery. From the point of view of the inhabitants, this policy was admissible because, since there was a family altar ('butsu-dan') in every house, they could perform daily prayer and other services in their own homes. For big ceremonies it was not unusual for the priest to be called to the house, rather than the believers visiting the temple. Here we find the same structure of order as we have just verified concerning the attitude to nature, namely, what constituted a social order at the level of the community in Western cities, was dispersed between every family in Edo. It also means that the social rules which the inhabitants of a Japanese city like Edo had to observe, remained merely those of this small society.

What consequence, then, did this minimalized social order have for visual design in Japanese cities? In fact, this lack of the notion of the total city order has produced a townscape which is peculiar to Japanese cities. I deeply regret that we do not have a sense for collaborating with our neighbours in order to form a harmonious townscape. In a photograph of a main street in Tokyo, we immediately perceive the contrast between this townscape and the big Western cities such as Paris, Amsterdam or London. In Tokyo, the buildings lack stylistic unity. What is most remarkable is the number and variety of signs on the walls of the buildings. They make an impression of noise and energy. Indeed there are many rich and pretentious buildings, but because of the lack of a sense of urban context, the townscapes in Tokyo are generally poor in terms of visuality. The next photograph, a view of Kyoto, is even more striking. The houses here are facing the Kamogawa river. The scene is frankly ugly, and the houses appear quite poor. In fact, however, most of them are very expensive restaurants, luxuriously decorating the street. The rooms facing the river are the best rooms in these houses. We Japanese appreciate waterscapes very much, and there are many clients who prefer such a spot facing the river. The owners of these houses are indifferent to the exterior appearance of the river side, merely because it concerns the place from which to see the outside, and not one to be looked at. They do not regard the townscape as their affair.

In conclusion, I must say that Japanese cities are generally very weak as regards the visual appearance of their townscapes. I think, however, this comes from the fact that city life in Japan has long been formed according to a principle other than visuality, namely: tactility. The basic space of the daily lives of the inhabitants of a Japanese city such as Tokyo, is a small, closed and private or rather familial space, which gives them a certain feeling of ease and peace. It is also a place from which we can watch the world, without being watched. This ease and peace is felt by the body rather than perceived by the eye. I am sure that Western people know this feeling as well. For example, someone enters a restaurant or a coffee shop. All the tables are free, so they choose a table next to the wall and watch, at their ease, the other clients coming in; they too choose tables at the side of the room. Finally, we find that the tables in the centre of the room are the last ones people want to take.

VISITORS AND INHABITANTS

I have concentrated on Tokyo a little too long. Having no intention of promoting Tokyo as a model city design, I merely wanted to present the tactile principle at a concrete level. I think, however, that Tokyo is not exceptional among megalopolises. After the demolition of town walls, carried out in many cities in the Western world throughout the nineteenth century, these big cities lost their former unity and their centres were dispersed in several areas. City life has become more and more complex, and the modern city is constituted by so many layers that there are no longer any citizens who might know them all. Through this tendency toward a greater diversity, the urban order inevitably becomes more and more minimal. I have the impression that the present megalopolises are structured in a way which is closer to Tokyo, than to a city as Parmanova.

I propose tactility as the principle of city design at the level of this minimal order. Let me come back to Parmanova. For whom was its form beneficial? I am sure, at least, that it was not for its inhabitants; it may have been for God, or for the political ruler of the city. (Above, I have mentioned the relation between visuality and political power with respect to the design of Edo). The oldest townscape print of Parmanova, published in 1589, is very indicative: all of the urban space is filled with houses, and at the centre of the circle is a watch tower, which was never actually constructed. Why did the artist find it necessary to put a watch tower at the very centre of the city? I have the impression that its purpose was to survey the inhabitants, rather than the enemy, and it was his notion of the city which oriented his pen. However, the real formation of the city was completely different. We have here three maps of the city at different times. Through them, we know that the houses were being constructed little by little. This development was made not according to the visual principle of good form, but the tactile principle of contact: a new house was always constructed next to an old one. Tactility is not an ideal principle but a real one.

I do not think that vision is the most important element for those who lead their lives in the city. If someone wishes to catch the totality of his city with a single glance, he must go outside it and choose a suitable spot – a resident of Maastricht might climb St. Peter's Hill, for example; a night view is particularly favoured for this purpose. But such views are sought by outsiders, and are not part of daily life: in general it is lovers or tourists who look for these townscapes. Most sightseeing spots in the city are constituted by landmarks, including monuments, temples or churches, a palace or a castle, or some other historical building. They are visual objects. Generally speaking, such places are not visited by the city's inhabitants; often tourists know these spots better than the local inhabitants. On the other hand, the inhabitants know the differences between the 'districts' (in Lynch's sense): one area is very poor, another is bourgeois and rich, and yet another is dangerous. They know the backstreets and short cuts, and have a sense of each particular atmosphere. They know people and manners, and how they should behave in each area. This kind of knowledge is not described in guide books. It includes things which are impossible to describe, because they are concerned with knowledge as the sum of accumulated experiences. The inhabitants of the city become acquainted with it very naturally, on foot and through the body, much in the way that we learn our mother tongue – yet they do not always know the famous sightseeing spots very well. This leads us to our final question concerning the nature of knowledge of the city.

TACTILE KNOWLEDGE OF THE CITY

Personally, I know Tokyo, Paris, Aix-en-Provence, New York, Maastricht, Venice, Gent, and so on, but each city to a different degree – a difference which is so great as to become a difference of nature. Let me describe my experience of Gent, because it was a recent one and my impressions are fresh. What I noticed very strongly was the gap between a knowledge acquired through maps, photographs and guide books and that gained through my eyes and body. Looking at a map of Gent, I would like to stress the fact that the lines on the map speak to those who know the city, and that this is completely different from the abstract information encountered by someone who does not know the town. Before my visit, I looked at this map, and I understood that the centre of the city is quite far from the station, and the streets are winding so that the city is similar not to a Parmanova but to a Maastricht. I also learned the disposition of some sightseeing spots. However, finding myself finally in the city, I realised how poor and empty this knowledge was, and I wondered if it had anything to do with the reality I was feeling.

The same thing happens with photographs. The perception of a photograph is a complex problem that needs to be dealt with independently. Here, I must simplify it and pick up on just one point, namely the great distance between the image formed through a photograph and the real impression given by the object. It happens quite often that it takes some time for me to recognize that a piece of architecture I am looking at in reality is the same thing that I saw a photograph of in my guide book. Once the recognition is established, I realise that this building is just the same as the photograph in terms of its visual form.

Through these examples of a map and a photograph, I want to talk about how knowledges differ. After reading the map, I knew something about Gent. But this knowledge remained abstract in comparison with the other knowledge I gained in the city itself. What is the difference between the abstract knowledge and the concrete one? I immediately recall the answer given by the eighteenth century philosophers who made investigations into the nature of vision. According to someone like Condillac, or Herder, vision does not work without the aid of the sense of touch: that is to say, we cannot differentiate between objects which occupy different places in space. A photograph catches only the optical mosaic in the visual field. What is materialized on the surface of the photographic print is not the totality of original vision, but merely its optical ingredient. The difference between the visual image and the optical one seems to correspond to that between three dimensions and two. But it is only a superficial impression, for real vision, by which I mean vision supported by the sense of touch, catches a much richer reality: the different feelings of building materials, stone, wood, metal, glass, mud etc.; the effect of imposing mass, or that of lightness; impressions of open or closed space; the feeling of security like the effect of 'bosom of the mountain', and so on. A photograph conveys to us only a faint shadow of these ways of knowing a townscape.

I find a decisive explanation of this kind of rich knowledge in a phrase of Bergson: 'Objects reflect, like a mirror, to my body its eventual influence'. What he means is that our bodies are always projecting their 'influence' upon surrounding objects, and the objects echo back in their turn this influence to our bodies. A metal renders a sharp echo, a cushion a soft one, etc.

This philosophical theory reveals a hidden but essential factor in our contact with the world. Our sensory perception is not a purely passive bodily process as most philosophers before Bergson had thought; in fact, perception presupposes that our bodies exercise their influence, that is, a latent action upon the object. So our bodies are actively as well as passively involved in perception. Now we can understand the nature of the knowledge we get directly from a townscape, and also the reason why a photograph can hardly provide this knowledge – but this is not all.

Until now, we have only been concerned with the material aspect of the townscape. Let us remember the case of Parmanova. In this city, each aspect of 'image-ability' was easily satisfied except one, that of 'districts'. 'Districts' concern the social organization of urban space, which is very often the result of history. Consequently, it is impossible to give it a visual expression. A guidebook may warn us not to enter a certain quarter but from this information we can not really know the 'district'. The knowledge provided by a photograph may be a little more concrete and alive, and if we visit the city, we can get some idea of the moral, social and historical reality of the place, for history and life are woven into the urban texture. But, needless to say, only the inhabitants can have a true knowledge of 'district'; this involves a refined knowledge obtained through daily, repeated experiences; history and society thus permeate the body.

These are different forms of knowledge in relation to the city or townscape, ranging from the more superficial to the more profound or substantial. The most profound knowledge of the city is a tactile one, but this excludes neither vision nor intelligence. Tactile knowledge requires merely that our bodies are involved. Here I am naturally reminded of the fact that the word 'tact' means adroitness in dealing with persons or circumstances, as well as the sense of touch. Tactile knowledge is 'tact-like' knowledge, an adroit and acute knowledge acquired by and programmed in the body.

CONCLUSION

To conclude, I should like to make some remarks on city design, as I proposed in the title of the chapter. My idea is simple and clear: city design should take the viewpoint not of the visitor but of the inhabitant, and should not pursue a 'good' form on the planning sheet, but a good feeling of tactility recognized by inhabitants, and even visitors. Of course, visual design is easier to make than tactile design, but this difficult project is not in vain. In order to show the practical possibilities, I would like to look at one just particular topic. This concerns the disposition of human space. As I pointed out above in relation to the mountains in the city of Tokyo, we feel at ease in enclosed and hollow spaces, and these are one of the most basic conditions of good cities.

Japanese architects have two technical terms designating types of corner: '*de-zumi*', literally 'projecting corner', and '*iri-zumi*', or 'drawn back corner'. *De-zumi* is aggressive, cool and rejecting; *iri-zumi*, on the other hand, is agreeable, warm and receptive. I have the impression that the English word 'nook' and the French '*coin*' designate this inner corner with its particular qualities.

The Japanese architect Yoshinobu Ashiwara insists on the use of this 'drawn back corner', and proposes construction of many spaces in the city according to this principle. For this to be fully realised, a vast field of research on the relation of urban space needs to be undertaken. I must be satisfied with this suggestion of the possibility of applying my idea of the tactile city to a real city design.

CULTURES AND URBAN POLITICS AND ECONOMICS

INTRODUCTION

Why should a book on urban culture be concerned with politics and economics? Traditionally in disciplines like geography and sociology, politics, economics and culture have been hived off into distinct sub-disciplines. However, such divisions have increasingly broken down in recent years. Studies of economics and politics have taken more cultural turns, and traditionally 'cultural' sub-disciplines have begun to scrutinize economic and political formations. This section draws together some of the best examples characteristic of this cross-fertilization of concern, which highlight the deeply entangled relationships between urban cultures, economics and politics.

In looking at the relationship between culture, economics and politics in an urban context we have identified six key issues of note:

- The effects of economic change on social groups within cities
- The forms of economic change and their effects on cities
- Variations in the economic activity of cities between the developed and the developing world
- The relationship between cities and the global economy
- Fundamental political identities raised by the city
- Different approaches to the study of the city

Each of these important questions is examined in the extracts collected in this section.

Sociologist Sharon Zukin recognizes that there are two predominant approaches to the study of the built environment: the political economy approach and the symbolic economy approach. The former is concerned with the material conditions of groups in urban society resulting from the process of local and global urban development. However, the latter is concerned with the relationship between dominant representations of the city, through architecture, urban design and advertising, and what Rosalyn Deutsche (Section VII) calls the 'rights to the city', the inclusion and exclusion of certain groups in urban society. As Zukin argues, the symbolic economy approach is concerned with the relationship between culture and power. The two are strongly interconnected in the reshaping of cities through urban development. Zukin argues in this extract from 'Space and Symbols in an Age of Decline' (1996) that 'a vibrant symbolic economy attracts investment capital from the global portfolios of real estate investors, banks, property developers and large property owners'. She takes a slightly different approach here looking at the position of the symbolic economy during times of urban decline, drawing on the example of New York during the recession of the early 1990s. It is interesting to compare the argument and observations outlined by Sharon Zukin with those discussed by the various authors of the extracts in Section VII Representations of the City.

David Harvey's (1989) contribution is more typical of the political economy approach, being concerned with the ways that city governance has changed since the late 1970s in response to changes in the international economy and the attitudes of central governments. Harvey argues that the local public sector authorities of cities in much of the urban 'first' world have typically taken on a number of activities formerly characteristic of the private sector, including profit generation, place promotion and the active encouragement of local economic development over their previous concerns for managing and distributing urban resources. This he characterizes as a greater risk-taking and entrepreneurialism within the local state. This shift in urban governance has been a response to the erosion of the local tax base through the widespread collapse of the manufacturing sector, heightened competition between cities and regions and the reduction in central government funding to local authorities. It has been a shift characteristic, although in different forms and to different extents, of much of urban Europe and North America. This has had a number of profound implications for the internal geographies of cities and the relations between them, some of which, most notably heightened levels of social and economic polarization, Harvey carefully outlines in an important essay.

Two further extracts situate cities and urban development within global urban systems. Saskia Sassen is concerned with the reciprocity between cities and the global economy, both with the roles of cities within processes of globalization and the impacts these have had on the form, economies and social geographies of cities. The extract from *Cities in a World Economy* (1994, Pine Forge Press) is concerned specifically with the last of these, the ways that the incorporation of cities into the global economy has generated new patterns of inequalities for different social groups within cities. Jane M. Jacobs introduces and demonstrates the relevance of a dimension widely neglected in debates on the relationships between culture and urban change: the role of imperialism. Rather than being a dead and historic issue, Jacobs cites examples from London and Perth demonstrating that this remains a very real and powerful force shaping contemporary urban development in first world cities. Her argument draws on the differing memories of empire among contrasting groups to point out how these and the political projects of these groups have influenced the shaping of space and identity near the hearts of first world cities. Jacobs's contribution is a telling corrective to those accounts of urban change that have failed to recognize the salience of empire. This omission is beginning to be addressed with a growing body of work focusing on the concerns Jacobs raises. This extract comes from Jacobs's first book *Edge of Empire* (1996, Routledge). The arguments raised in this book are likely to remain influential over the coming years.

Whilst Jacobs deals with emerging issues in Australia, Jeremy Seabrook, in an extract from *In the Cities of the South* (1996, Verso), offers a broad consideration of the cities of the non-affluent world (or southern hemisphere). The extract included here begins with an account of life in Bangkok. Seabrook, in common with many women writers on cities, uses personal narrative as a means to investigate urban questions. This lends a particular kind of reality to experiences of urban culture often described in more neutral, and generalized, terms. Seabrook's text moves also to Dhaka and Bombay. In Dhaka, a poet tells him why people move into cities from the land, and how important it remains to be rooted somewhere. In Bombay, the insecurities of living on the margin of development are brought home, again through a personal story. The text ends on a note more of endurance than hope.

These extracts are complimented by that from Jonathan Charley (1995) 'Industrialization and the City: Work, Speed-up, Urbanization'. Charley reminds us that the reciprocity between economic transformation and urbanization has a long history, by examining the modernization process and its associated processes of urbanization, and also that the effects on the city are profound. In ways that bear some parallel with David Harvey's analysis of the postmodern city, Charley looks at the forging of the modern city under the expansion of capitalism in the late nineteenth and early twentieth centuries. The rise of capitalism brought with it enormous changes – in the labour process, in the nature of everyday life and in the spaces of the city. Charley reminds us that the modernization associated with the expansion of capitalism forged a city, in every sense of the word, just as it is being reforged and transformed by the remaking of capitalism examined by Harvey.

It is also worth remembering that the city raises fundamental political and economic questions. Some of these Murray Bookchin raises in the extract from *The Limits of the City* (1986, Black Rose Books). Bookchin's basic thesis is that the city creates its own world and own sets of meaning that transcend the sociological abstractions of class and nation-state. He cites, for example, the case of the Paris Communards who regarded themselves first and foremost as *Parisians* rather than French or part of any class alignment. Recognizing such a position, Bookchin argues, raises fundamental and wide-ranging possibilities and problems for the practice and examination of politics and economics, some of which are explored in this extract.

7 MURRAY BOOKCHIN

'Introduction'

from *The Limits of the City* (1986)

I am concerned not only with the uniqueness and limits of various cities but also with the city as a distinctly human and cultural terrain. The city is more, in my view, than an epiphenomenon of a broad division of labor between agriculture and crafts or barter and commerce; it is a world *in its own right* that goes beyond familial, tribal, economic, and social ties to establish a uniquely political universe of its own. For all its collectivism and strong bonds of solidarity, tribal society was surprisingly parochial. Based on kinship, however fictitious its reality, the tribe rooted its affiliations and loyalties in lineage ties or what I call the "blood oath" in *The Ecology of Freedom*. The city, by contrast, over a long process of development, created a more universal terrain – the realm of the citizen. Civic rights depended upon residence rather than a shared ethnic background, even upon wealth rather than common descent. In any case, it formed the arena for the emergence of a common "humanity" rather than a parochial "folk." Here, the "stranger" could first find a home and the protection of laws, and, later, citizenship as one among equals, not the arbitrary treatment that characterizes the status of visitors to tribal communities. From a distance of millenia, it is hard for people today to realize what a social and cultural revolution this step out of the lineage system proved to be. Aside from the sense of universality it produced, the variety and openness to different cultural stimuli it created made the city the most powerful *civilizing* factor in human history. The origin of the word "civilization" from *civitas* is not accidental: it authentically reflects the emergence of a distinctly human culture – universal in its scope – from city life as such.

What is equally important, the city has always been the most immediate human environment that people experience: the locus of our most intimate social and personal concerns beyond the family circle or the workplace. And, indeed, for much of humanity, particularly women, children, and the aged, not to speak of those who still live outside the orbit of western capitalist relationships, it is still the most immediate of the human environments they encounter. Existentially, the city forms the most direct sphere of our lives as public beings: it is the place where we live out our daily lives, rear our young, enjoy the amenities of life as well as its disasters, the place where we work, the locus of "home," and the terrain which gives authentic meaning to the word "environment." Aside from acid rain, our greatest environmental concerns are urban ones, not those which are related to rural areas and wilderness. What impresses us most as environmentally concerned individuals is the cultivation of gardens *in* the city or the use of solar collectors on urban dwellings (as I can attest from my innumerable lectures on the subject) rather than food cultivation in the countryside, where we obviously expect these innovations to occur. What fascinates people most is when we try to bring the country into the city as gardens or when we use alternate energy sources on apartment houses. The failure of environmentalists to see this distinctly urban bias has done much to marginalize many of their ideas and efforts. This bias is not to be condemned: it speaks to the centrality of the town and city in our lives and the functions they play as centers of human activities.

In dealing with these issues, I have not found an overwhelming emphasis on "class analyses" useful. To the contrary: they often prove to be an impediment. The more I look into the great revolutionary movements that opened the modern era, the more they appear as urban movements, not simply class movements – notably, the Great Revolution in France and the Paris Commune of 1871. All his Marxism aside, Henri Lefebvre was to veer more sharply from his dogmatic proclivities than he realized, I suspect, when he saw the Commune of 1871 as an urban movement *par excellence*. Manuel Castells' recent interpretation of that movement as a transclass phenomenon in which "the use of the term 'worker' is misleading when we really want to determine if the *Commune* was in fact a major episode in the class struggle between the bourgeoisie and the proletariat over the control of industrialization" comes as a welcome relief from tendentious interpretations of the Commune as the "model" of a "proletarian dictatorship." What the Communards agreed upon, first and foremost, was the fact that they were *Parisians*, not simply "citizens" of a nation-state called "France." And they were "citizens" of a municipality, not of classes called "proletarians," "petty bourgeois," or "capitalists." Hence, as Castells concludes: "the *Commune* was primarily a municipal revolution, with the qualification that such an orientation does not imply any parochial view; on the contrary, the transformation of the state as a whole was at stake, with the municipal institutions as the keystone of a new political constitution."

Castells' observations on the Commune apply, I believe, to all the so-called "bourgeois" revolutions of the past with their centers in London during the 1640s, Boston during the 1770s, and, again, Paris during the 1790s. The importance of the municipality as a revolutionary center should not be taken as a deprecation of the revolutionary role of the countryside. Indeed, the agrarian world is a reservoir of massive social discontent, particularly in highly transitional periods: the impact of industrial, even urban, life on peasants and craftspeople who drift into cities explains in great part the revolutionary discontent of the so-called "proletariat" in Russia and Spain. Caught in the forcefield created by industrial rationalization on the one hand and a leisurely seasonal world shaped by nature on the other, the worker-peasant is more authentically the voice of "proletarian" upsurges than his or her proletarian heirs for whom the factory has already become a way of life and a school for hierarchical obedience. Marx's famous "antithesis" between "town and country," in fact, raises the problem of "proletarian hegemony" in ways he could never have anticipated. We must turn to more deep-seated sources of discontent besides the "self-interest" of specific classes: cultural factors which make for municipal solidarity over and beyond class factors, a unique sense of municipal identity, the powerful role of the neighborhood in fostering collectivist ties, citizenship itself conceived as an ethical compact, and, over the long run, the importance of municipal confederations as an alternative to the nation-state. These issues carry us far beyond Marx's largely economistic approach to the "antithesis" between "town and country." Indeed, they raise questions not only of our interpretation of the past, but our attempt to formulate programmatic guides for the future. What they ultimately involve is a recovery of politics as an activity that must be distinguished from statecraft: politics in the Hellenic sense of wide public participation in the management of the municipality.

8 DAVID HARVEY

'From Managerialism to Entrepreneurialism: The Transformation in Urban Governance in Late Capitalism'

from *Geografiska Annale* (1989)

[. . .]

THE SHIFT TO ENTREPRENEURALISM IN URBAN GOVERNANCE

A colloquium held at Orleans in 1985 brought together academics, businessmen, and policy-makers from eight large cities in seven advanced capitalist countries. The charge was to explore the lines of action open to urban governments in the face of the widespread erosion of the economic and fiscal base of many large cities in the advanced capitalist world. The colloquium indicated a strong consensus: that urban governments had to be much more innovative and entrepreneurial, willing to explore all kinds of avenues through which to alleviate their distressed condition and thereby secure a better future for their populations. The only realm of disagreement concerned how this best could be done. Should urban governments play some kind of supportive or even direct role in the creation of new enterprises and if so of what sort? Should they struggle to preserve or even take over threatened employment sources and if so which ones? Or should they simply confine themselves to the provision of those infrastructures, sites, tax baits, and cultural and social attractions that would shore up the old and lure in new forms of economic activity?

I quote this case because it is symptomatic of a reorientation in attitudes to urban governance that has taken place these last two decades in the advanced capitalist countries. Put simply, the "managerial" approach so typical of the 1960s has steadily given way to initiatory and "entrepreneurial" forms of action in the 1970s and 1980s. In recent years in particular, there seems to be a general consensus emerging throughout the advanced capitalist world that positive benefits are to be had by cities taking an entrepreneurial stance to economic development. What is remarkable, is that this consensus seems to hold across national boundaries and even across political parties and ideologies.

Both Boddy (1984) and Cochrane (1987) agree, for example, that since the early 1970s local authorities in Britain "have become increasingly involved in economic development activity directly related to production and investment," while Rees and Lambert show how "the growth of local government initiatives in the economic field was positively encouraged by successive central administrations during the 1970s" in order to complement central government attempts to improve the efficiency, competitive powers and profitability of British Industry. David Blunkett, leader of the Labour Council in Sheffield for several years, has recently put the seal of approval on a certain kind of urban entrepreneurialism:

"From the early 1970s, as full employment moved from the top of government priorities, local councils began to take up the challenge. There was support for small firms; closer links between the public and private sectors; promotion of local areas to attract new business. They were adapting the traditional economic role of British local government which offered inducements in the forms of grants, free loans, and publicly subsidised infrastructure, and no request for reciprocal involvement with the community, in order to attract industrial and commercial concerns which were looking for suitable sites for investment and trading … Local government today, as in the past, can offer its own brand of entrepreneurship and enterprise in facing the enormous economic and social change which technology and industrial restructuring bring" (Blunkett and Jackson, 1987).

In the United States, where civic boosterism and entrepreneurialism had long been a major feature of urban systems the reduction in the flow of federal redistributions and local tax revenues after 1972 (the year in which President Nixon declared the urban crisis to be over, signalling that the Federal government no longer had the fiscal resources to contribute to their solution) led to a revival of boosterism to the point where Robert Goodman (1979) was prepared to characterise both state and local governments as "the last entrepreneurs." An extensive literature now exists dealing with how the new urban entrepreneurialism has moved center-stage in urban policy formulation and urban growth strategies in the United States.

The shift towards entrepreneurialism has by no means been complete. Many local governments in Britain did not respond to the new pressures and possibilities, at least until relatively recently, while cities like New Orleans in the United States continue to remain wards of the federal government and rely fundamentally on redistributions for survival. And the history of its outcomes, though yet to be properly recorded, is obviously checkered, pockmarked with as many failures as successes and not a little controversy as to what constitutes "success" anyway (a question to which I shall later return). Yet beneath all this diversity, the shift from urban managerialism to some kind of entrepreneurialism remains a persistent and recurrent theme in the period since the early 1970s. Both the reasons for and the implications of such a shift are deserving of some scrutiny.

There is general agreement, of course, that the shift has something to do with the difficulties that have beset capitalist economies since the recession of 1973. Deindustrialisation, widespread and seemingly "structural" unemployment, fiscal austerity at both the national and local levels, all coupled with a rising tide of neoconservatism and much stronger appeal (though often more in theory than in practice) to

market rationality and privatisation, provide a backdrop to understanding why so many urban governments, often of quite different political persuasions and armed with very different legal and political powers, have all taken a broadly similar direction. The greater emphasis on local action to combat these ills also seems to have something to do with the declining powers of the nation state to control multinational money flows, so that investment increasingly takes the form of a negotiation between international finance capital and local powers doing the best they can to maximise the attractiveness of the local site as a lure for capitalist development. By the same token, the rise of urban entrepreneurialism may have had an important role to play in a general transition in the dynamics of capitalism from a Fordist-Keynesian regime of capital accumulation to a regime of "flexible accumulation". The transformation of urban governance these last two decades has had, I shall argue, substantial macro-economic roots and implications. And if Jane Jacobs (1984) is only half right that the city is the relevant unit for understanding how the wealth of nations is created, then the shift from urban managerialism to urban entrepreneurialism could have far reaching implications for future growth prospects.

If, for example, urban entrepreneurialism (in the broadest sense) is embedded in a framework of zero-sum inter-urban competition for resources, jobs, and capital, then even the most resolute and avantgarde municipal socialists will find themselves, in the end, playing the capitalist game and performing as agents of discipline for the very processes they are trying to resist. It is exactly this problem that has dogged the Labour councils in Britain. They had on the one hand to develop projects which could "produce outputs which are directly related to working people's needs, in ways which build on the skills of labour rather than de-skilling them" (Murray, 1983), while on the other hand recognizing that much of that effort would go for nought if the urban region did not secure relative competitive advantages. Given the right circumstances, however, urban entrepreneurialism and even inter-urban competition may open the way to a non zero-sum pattern of development. This kind of activity has certainly played a key role in capitalist development in the past. And it is an open question as to whether or not it could lead towards progressive and socialist transitions in the future.

[. . .]

ALTERNATIVE STRATEGIES FOR URBAN GOVERNANCE

There are, I have argued elsewhere, four basic options for urban entrepreneurialism. Each warrants some separate consideration, even though it is the combination of them that provides the clue to the recent rapid shifts in the uneven development of urban systems in the advanced capitalist world.

1. Competition within the international division of labour means the creation of exploitation of particular advantages for the production of goods and services. Some advantages derive from the resource base (the oil that allowed Texas to bloom in the 1970s) or location (e.g. favoured access to the vigour of Pacific Rim trading in the case of Californian cities). But others are created through public and private investments in the kinds of physical and social infrastructures that strengthen the economic base of the metropolitan region as an exporter of goods and services. Direct interventions to stimulate the application of new technologies, the creation of new products, or the provision of venture capital to new enterprises (which may even be cooperatively owned and managed) may also be significant, while local costs may be reduced by subsidies (tax breaks, cheap credit, procurement of sites). Hardly any large scale development now occurs without local government (or the broader coalition of forces constituting local governance) offering a substantial package of aids and assistance as inducements. International competitiveness also depends upon the qualities, quantities, and costs of local labour supply. Local costs can most easily be controlled when local replaces national collective bargaining and when local governments and other large institutions, like hospitals and universities, lead the way with reductions in real wages and benefits (a series of struggles over wage rates and benefits in the public and institutional sector in Baltimore in the 1970s was typical). Labour power of the right quality, even though expensive, can be a powerful magnet for new economic development so that investment in highly trained and skilled work forces suited to new labour processes and their managerial requirements can be well rewarded. There is, finally, the problem of agglomeration economies in metropolitan regions. The production of goods and services is often dependent not on single decisions of economic units (such as the large multinationals to bring a branch plant to town, often with very limited local spillover effects), but upon the way in which economies can be generated by bringing together diverse activities within a restricted space of interaction so as to facilitate highly efficient and interactive production systems. From this standpoint, large metropolitan regions like New York, Los Angeles, London, and Chicago possess some distinctive advantages that congestion costs have by no means yet offset. But, as the case of Bologna and the surge of new industrial development in Emilia Romagna illustrates, careful attention to the industrial and marketing mix backed by strong local state action (communist-led in this instance), can promote powerful growth of new industrial districts and configurations, based on agglomeration economies and efficient organisation.

2. The urban region can also seek to improve its competitive position with respect to the spatial division of consumption. There is more to this than trying to bring money into an urban region through tourism and retirement attractions. The consumerist style of urbanisation after 1950 promoted an ever-broader basis for participation in mass consumption. While recession, unemployment, and the high cost of credit have rolled back that possibility for important layers in the population, there is still a lot of consumer power around (much of it credit-fuelled). Competition for that becomes more frenetic while consumers who do have the money have the opportunity to be much more discriminating. Investments to attract the consumer dollar have paradoxically grown

a-pace as a response to generalised recession. They increasingly focus on the quality of life. Gentrification, cultural innovation, and physical up-grading of the urban environment (including the turn to postmodernist styles of architecture and urban design), consumer attractions (sports stadia, convention and shopping centres, marinas, exotic eating places) and entertainment (the organisation of urban spectacles on a temporary or permanent basis), have all become much more prominent facets of strategies for urban regeneration. Above all, the city has to appear as an innovative, exciting, creative, and safe place to live or to visit, to play and consume in. Baltimore, with its dismal reputation as "the armpit of the east coast" in the early 1970s has, for example, expanded its employment in the tourist trade from under one to over fifteen thousand in less than two decades of massive urban redevelopment. More recently thirteen ailing industrial cities in Britain (including Leeds, Bradford, Manchester, Liverpool, Newcastle and Stoke-on-Trent) put together a joint promotional effort to capture more of Britain's tourist trade. Here is how *The Guardian* (May 9th, 1987) reports this quite successful venture:

"Apart from generating income and creating jobs in areas of seemingly terminal unemployment, tourism also has a significant spin-off effect in its broader enhancement of the environment. Facelifts and facilities designed to attract more tourists also improve the quality of life for those who live there, even enticing new industries. Although the specific assets of the individual cities are obviously varied, each is able to offer a host of structural reminders of just what made them great in the first place. They share, in other words, a marketable ingredient called industrial and/or maritime heritage." Festivals and cultural events likewise become the focus of investment activities. "The arts create a climate of optimism – the 'can do' culture essential to developing the enterprise culture," says the introduction to a recent Arts Council of Great Britain report, adding that cultural activities and the arts can help break the downward spiral of economic stagnation in inner cities and help people

"believe in themselves and their community". Spectacle and display become symbols of the dynamic community, as much in communist controlled Rome and Bologna as in Baltimore, Glasgow and Liverpool. This way, an urban region can hope to cohere and survive as a locus of community solidarity while exploring the option of exploiting conspicuous consumption in a sea of spreading recession.

3. Urban entrepreneurialism has also been strongly coloured by a fierce struggle over the acquisition of key control and command functions in high finance, government, or information gathering and processing (including the media). Functions of this sort need particular and often expensive infrastructural provision. Efficiency and centrality within a worldwide communications net is vital in sectors where personal interactions of key decision makers is required. This means heavy investments in transport and communications (airports and teleports, for example) and the provision of adequate office space equipped with the necessary internal and external linkages to minimise transaction times and costs. Assembling the wide range of supportive services, particularly those that can gather and process information rapidly or allow quick consultation with "experts", calls for other kinds of investments, while the specific skills required by such activities put a premium on metropolitan regions with certain kinds of education provision (business and law-schools, hightech production sectors, media skills, and the like). Inter-urban competition in this realm is very expensive and peculiarly tough because this is an area where agglomeration economies remain supreme and the monopoly power of established centres, like New York, Chicago, London, and Los Angeles, is particularly hard to break. But since command functions have been a strong growth sector these last two decades (employment in finance and insurance has doubled in Britain in less than a decade), so pursuit of them has more and more appealed as the golden path to urban survival. The effect, of course, is to make it appear as if the city of the future is going to be a city of pure command and control

functions, an informational city, a post-industrial city in which the export of services (financial, informational, knowledge-producing) becomes the economic basis for urban survival.

4. Competitive edge with respect to redistributions of surpluses through central (or in the United States, state) governments is still of tremendous importance since it is somewhat of a myth that central governments do not redistribute to the degree they used to do. The channels have shifted so that in both Britain (take the case of Bristol) and in the United States (take the case of Long Beach–San Diego) it is military and defense contracts that provide the sustenance for urban prosperity, in part because of the sheer amount of money involved but also because of the type of employment and the spin-offs it may have into so-called "high-tech" industries. And even though every effort may have been made to cut the flow of central government support to many urban regions, there are many sectors of the economy (health and education, for example) and even whole metropolitan economies where such a cut off was simply impossible. Urban ruling class alliances have had plenty of opportunity, therefore, to exploit re-distributive mechanisms as a means to urban survival.

These four strategies are not mutually exclusive and the uneven fortunes of metropolitan regions have depended upon the nature of the coalitions that have formed, the mix and timing of entrepreneurial strategies, the particular resources (natural, human, locational) with which the metropolitan region can work, and the strength of the competition. But uneven growth has also resulted from the synergism that leads one kind of strategy to be facilitative for another. For example, the growth of the Los Angeles–San Diego–Long Beach–Orange County megalopolis appears to have been fuelled by interaction effects between strong governmental redistributions to the defense industries and rapid accrual of command and control functions, that have further stimulated consumption-oriented activities to the point where there has been a considerable revival of certain types of manufacturing. On the other hand, there is

little evidence that the strong growth of consumption-oriented activity in Baltimore has done very much at all for the growth of other functions save, perhaps, the relatively mild proliferation of banking and financial services. But there is also evidence that the network of cities and urban regions in, say, the Sunbelt or Southern England has generated a stronger collective synergism than would be the case for their respective northern counterparts. Noyelle and Stanback (1984) also suggest that position and function within the urban hierarchy have had an important role to play in the patterning of urban fortunes and misfortunes. Transmission effects between cities and within the urban hierarchy must also be factored in to account for the pattern of urban fortunes and misfortunes during the transition from managerialism to entrepreneurialism in urban governance.

Urban entrepreneurialism implies, however, some level of inter-urban competition. We here approach a force that puts clear limitations upon the power of specific projects to transform the lot of particular cities. Indeed, to the degree that inter-urban competition becomes more potent, it will almost certainly operate as an "external coercive power" over individual cities to bring them closer into line with the discipline and logic of capitalist development. It may even force repetitive and serial reproduction of certain patterns of development (such as the serial reproduction of "world trade centers" or of new cultural and entertainment centers, of waterfront development, of postmodern shopping malls, and the like). The evidence for serial reproduction of similar forms of urban redevelopment is quite strong and the reasons behind it are worthy of note.

With the diminution in transport costs and the consequent reduction in spatial barriers to movement of goods, people, money and information, the significance of the qualities of place has been enhanced and the vigour of inter-urban competition for capitalist development (investment, jobs, tourism, etc.) has strengthened considerably. Consider the matter, first of all, from the standpoint of highly mobile multinational capital. With the reduction of spatial barriers, distance from the market or from raw materials has become less relevant to locational decisions. The monopolistic elements in spatial competition, so essential to the workings of Löschian theory, disappear. Heavy, low value items (like beer and mineral water), which used to be locally produced are now traded over such long distances that concepts such as the "range of a good" make little sense. On the other hand, the ability of capital to exercise greater choice over location, highlights the importance of the particular production conditions prevailing at a particular place. Small differences in labour supply (quantities and qualities), in infrastructures and resources, in government regulation and taxation, assume much greater significance than was the case when high transport costs created "natural" monopolies for local production in local markets. By the same token, multinational capital now has the power to organise its responses to highly localised variations in market taste through small batch and specialised production designed to satisfy local market niches. In a world of heightened competition – such as that which has prevailed since the post-war boom came crashing to a halt in 1973 – coercive pressures force multinational capital to be much more discriminating and sensitive to small variations between places with respect to both production and consumption possibilities.

Consider matters, in the second instance, from the standpoint of the places that stand to improve or lose their economic vitality if they do not offer enterprises the requisite conditions to come to or remain in town. The reduction of spatial barriers has, in fact, made competition between localities, states, and urban regions for development capital even more acute. Urban governance has thus become much more oriented to the provision of a "good business climate" and to the construction of all sorts of lures to bring capital into town. Increased entrepreneurialism has been a partial result of this process, of course. But we here see that increasing entrepreneurialism in a different light precisely because the search to procure investment capital confines innovation to a very narrow path built around a favourable package for capitalist development and all that entails. The task of urban governance is, in short, to lure highly mobile and flexible production, financial, and consumption flows into its space. The speculative qualities of urban investments simply derive from the inability to predict exactly which package will succeed and which will not, in a world of considerable economic instability and volatility.

It is easy to envisage, therefore, all manner of, upward and downward spirals of urban growth and decline under conditions where urban entrepreneurialism and inter-urban competition are strong. The innovative and competitive responses of many urban ruling class alliances have engendered more rather than less uncertainty and in the end made the urban system more rather than less vulnerable to the uncertainties of rapid change.

THE MACRO-ECONOMIC IMPLICATIONS OF INTER-URBAN COMPETITION

The macro-economic as well as local implications of urban entrepreneurialism and stronger inter-urban competition deserve some scrutiny. It is particularly useful to put these phenomena into relation with some of the more general shifts and trends that have been observed in the way capitalist economies have been working since the first major post-war recession of 1973 sparked a variety of seemingly profound adjustments in the paths of capitalist development.

To begin with, the fact of inter-urban competition and urban entrepreneurialism has opened up the urban spaces of the advanced capitalist countries to all kinds of new patterns of development, even when the net effect has been the serial reproduction of science parks, gentrification, world trading centers, cultural and entertainment centers, large scale interior shopping malls with postmodern accoutrements, and the like. The emphasis on the production of a good local business climate has emphasised the importance of the locality as a site of regulation of infrastructural provision, labour relations, environmental controls, and even tax policy vis-a-vis international capital. The absorption of risk by the public sector and in particular the stress on public sector involvement in infrastructural provision, has meant that the cost of locational change has diminished from the standpoint of multinational capital, making the latter more rather than less geographically mobile. If anything, the new urban entrepreneurialism adds to rather than detracts from the geographical flexibility with which multi-national firms can approach their locational strategies. To the degree that the locality becomes the site of regulation of labour relations, so it also contributes to increased flexibility in managerial strategies in geographically segmented labour markets. Local, rather than national collective bargaining has long been a feature of labour relations in the United States, but the trend towards local agreements is marked in many advanced capitalist countries over the past two decades.

There is, in short, nothing about urban entrepreneurialism which is antithetical to the thesis of some macro-economic shift in the form and style of capitalist development since the early 1970s. Indeed, a strong case can be made that the shift in urban politics and the turn to entrepreneurialism has had an important facilitative role in a transition from locationally rather rigid Fordist production systems backed by Keynesian state welfarism to a much more geographically open and market based form of flexible accumulation. A further case can be made that the trend away from urban based modernism in design, cultural forms and life style towards postmodernism is also connected to the rise of urban entrepreneurialism. In what follows I shall illustrate how and why such connections might arise.

Consider, first, the general distributive consequences of urban entrepreneurialism. Much of the vaunted "public-private partnership" in the United States, for example, amounts to a subsidy for affluent consumers, corporations, and powerful command functions to stay in town at the expense of local collective consumption for the working class and poor. The general increase in problems of impoverishment and disempowerment, including the production of a distinctive "underclass" has been documented beyond dispute for many of the large cities in the United States, Levine, for example, provides abundant details for Baltimore in a setting where major claims are made for the benefits to be had from public-private partnership. Boddy (1984) likewise reports that what he calls "mainstream" (as opposed to socialist) approaches to local development in Britain have been "property-led, business and market oriented and competitive, with economic development rather than employment the primary focus, and with an emphasis

on small firms". Since the main aim has been "to stimulate or attract in private enterprise by creating the preconditions for profitable investment", local government "has in effect ended up underpinning private enterprise, and taking on part of the burden of production costs". Since capital tends to be more rather than less mobile these days, it follows that local subsidies to capital will likely increase while local provision for the underprivileged will diminish, producing greater polarisation in the social distribution of real income.

The kinds of jobs created in many instances likewise militate against any progressive shift in income distributions since the emphasis upon small businesses and sub-contracting can even spill over into direct encouragement of the "informal sector" as a basis for urban survival. The rise of informal production activities in many cities, particularly in the United States, has been a marked feature in the last two decades and is increasingly seen as either a necessary evil or as a dynamic growth sector capable of reimporting some level of manufacturing activity back into otherwise declining urban centers. By the same token, the kinds of service activities and managerial functions which get consolidated in urban regions tend to be either low-paying jobs (often held exclusively by women) or very high paying positions at the top end of the managerial spectrum. Urban entrepreneurialism consequently contributes to increasing disparities in wealth and income as well as to that increase in urban impoverishment which has been noted even in those cities (like New York) that have exhibited strong growth. It has, of course, been exactly this result that Labour councils in Britain (as well as some of the more progressive urban administrations in the United States) have been struggling to resist. But it is by no means clear that even the most progressive urban government can resist such an outcome when embedded in the logic of capitalist spatial development in which competition seems to operate not as a beneficial hidden hand, but as an external coercive law forcing the lowest common denominator of social responsibility and welfare provision within a competitively organised urban system.

Many of the innovations and investments designed to make particular cities more attractive as cultural and consumer centres have quickly been imitated elsewhere, thus rendering any competitive advantage within a system of cities ephemeral. How many successful convention centres, sports stadia, disney-worlds, harbour places and spectacular shopping malls can there be? Success is often short-lived or rendered moot by parallel or alternative innovations arising elsewhere. Local coalitions have no option, given the coercive laws of competition, except to keep ahead of the game thus engendering leap-frogging innovations in life styles, cultural forms, products and service mixes, even institutional and political forms if they are to survive. The result is a stimulating if often destructive maelstrom of urban-based cultural, political, production and consumption innovations. It is at this point that we can identify an albeit subterranean but nonetheless vital connection between the rise of urban entrepreneurialism and the postmodern penchant for design of urban fragments rather than comprehensive urban planning, for ephemerality and eclecticism of fashion and style rather than the search for enduring values, for quotation and fiction rather than invention and function, and, finally, for medium over message and image over substance.

In the United States, where urban entrepreneurialism has been particularly vigorous, the result has been instability within the urban system. Houston, Dallas and Denver, boom towns in the 1970s, suddenly dissolved after 1980 into morasses of excess capital investment bringing a host of financial institutions to the brink of, if not in actual bankruptcy. Silicon Valley, once the high-tech wonder of new products and new employment, suddenly lost its luster but New York, on the edge of bankruptcy in 1975, rebounded in the 1980s with the immense vitality of its financial services and command functions, only to find its future threatened once more with the wave of lay-offs and mergers which rationalised the financial services sector in the wake of the stock market crash of October, 1987. San Francisco, the darling of Pacific Rim trading, suddenly finds itself with excess office space in the early 1980s only to recover almost immediately. New Orleans, already struggling as a ward of federal government redistributions, sponsors a disastrous World Fair that drives it deeper into the mire, while Vancouver, already booming,

hosts a remarkably successful World Exposition. The shifts in urban fortunes and misfortunes since the early 1970s have been truly remarkable and the strengthening of urban entrepreneurialism and inter-urban competition has had a lot to do with it.

But there has been another rather more subtle effect that deserves consideration. Urban entrepreneurialism encourages the development of those kinds of activities and endeavours that have the strongest *localised* capacity to enhance property values, the tax base, the local circulation of revenues, and (most often as a hoped-for consequence of the preceding list) employment growth. Since increasing geographical mobility and rapidly changing technologies have rendered many forms of production of goods highly suspect, so the production of those kinds of services that are (a) highly localised and (b) characterised by rapid if not instantaneous turnover time appear as the most stable basis for urban entrepreneurial endeavour. The emphasis upon tourism, the production and consumption of spectacles, the promotion of ephemeral events within a given locale, bear all the signs of being favoured remedies for ailing urban economies. Urban investments of this sort may yield quick though ephemeral fixes to urban problems. But they are often highly speculative. Gearing up to bid for the Olympic Games is an expensive exercise, for example, which may or may not pay off. Many cities in the United States (Buffalo, for example) have invested in vast stadium facilities in the hope of landing a major league baseball team and Baltimore is similarly planning a new stadium to try and recapture a football team that went to a superior stadium in Indianapolis some years ago (this is the contemporary United States version of that ancient cargo cult practice in Papua, New Guinea, of building an airstrip in the hope of luring a jet liner to earth). Speculative projects of this sort are part and parcel of a more general macro-economic problem. Put simply, credit-financed shopping malls, sports stadia, and other facets of conspicuous high consumption are high risk projects that can easily fall on bad times and thus exacerbate, as the "overmalling of America" only too dramatically illustrates (Green, 1988), the problems of over-accumulation and overinvestment to which capitalism as a whole is so easily prone. The instability that pervades the U.S. financial system (forcing something of the order of $100 billion in public moneys to stabilise the savings and loan industry) is partly due to bad loans in energy, agriculture, and urban real estate development. Many of the "festival market places" that looked like an "Alladin's lamp for cities fallen on hard times", just a decade ago, ran a recent report in the Baltimore Sun (August 20, 1987), have now themselves fallen on hard times. "Projects in Richmond, Va., Flint, Mich. and Toledo, Ohio, managed by Rouse's Enterprise Development Co., are losing millions of dollars", and even the "South Street Seaport in New York and Riverwalk in New Orleans" have encountered severe financial difficulties. Ruinous inter-urban competition on all such dimensions bids fair to become a quagmire of indebtedness.

Even in the face of poor economic performance, however, investments in these last kinds of projects appear to have both a social and political attraction. To begin with, the selling of the city as a location for activity depends heavily upon the creation of an attractive urban imagery. City leaders can look upon the spectacular development as a "loss leader" to pull in other forms of development. Part of what we have seen these last two decades is the attempt to build a physical and social imagery of cities suited for that competitive purpose. The production of an urban image of this sort also has internal political and social consequences. It helps counteract the sense of alienation and anomie that Simmel long ago identified as such a problematic feature of modern city life. It particularly does so when an urban terrain is opened for display, fashion and the "presentation of self" in a surrounding of spectacle and play. If everyone, from punks and rap artists to the "yuppies" and the haute bourgeoisie can participate in the production of an urban image through their production of social space, then all can at least feel some sense of belonging to that place. The orchestrated production of an urban image can, if successful, also help create a sense of social solidarity, civic pride and loyalty to place and even allow the urban image to provide a mental refuge in a world that capital treats as more and more place-less. Urban entrepreneurialism (as opposed to the much more faceless bureaucratic managerialism) here meshes with a search for local identity and, as such, opens up a range of mechanisms for social control. Bread and circuses was the famous Roman formula that now stands to be reinvented and revived, while the ideology of locality, place and community becomes central to the political rhetoric of urban governance which concentrates on the idea of togetherness in defense against a hostile and threatening world of international trade and heightened competition.

The radical reconstruction of the image of Baltimore through the new waterfront and inner-harbour development is a good case in point. The redevelopment put Baltimore on the map in a new way, earned the city the title of "renaissance city" and put it on the front cover of Time Magazine, shedding its image of dreariness and impoverishment. It appeared as a dynamic go-getting city, ready to accommodate outside capital and to encourage the movement in of capital and of the "right" people. No matter that the reality is one of increased impoverishment and overall urban deterioration, that a thorough local enquiry based on interviews with community, civic and business leaders identified plenty of "rot beneath the glitter", that a Congressional Report of 1984 described the city as one of the "neediest" in the United States, and that a thorough study of the renaissance by Levine (1987) showed again and again how partial and limited the benefits were and how the city as a whole was accelerating rather than reversing its decline. The image of prosperity conceals all that, masks the underlying difficulties and projects an imagery of success that spreads internationally so that the London Sunday Times (November 29th, 1987) can report, without a hint of criticism, that "Baltimore, despite soaring unemployment, boldly turned its derelict harbor into a playground. Tourists meant shopping, catering and transport, this in turn meant construction, distribution, manufacturing – leading to more jobs more residents, more activity. The decay of old Baltimore slowed, halted, then turned back. The harbor area is now among America's top tourist draws and urban unemployment is falling fast". Yet it is also apparent that putting Baltimore on the map in this way, giving it a stronger sense of place and of local identity, has been successful politically in consolidating the power of influence of the local public-private partnership that brought the project into being. It has brought development money into Baltimore (though it is hard to tell if it has brought more in than it has taken out given the absorption of risk by the public sector). It also has given the population at large some sense of place-bound identity. The circus succeeds even if the bread is lacking. The triumph of image over substance is complete.

[. . .]

REFERENCES

Blunkett, D. and Jackson, K. (1987): *Democracy in crisis: the town halls respond.* London.

Boddy, M. (1984): "Local economic and employment strategies," in Boddy, M. and Fudge, C. (eds): *Local socialism.* London.

Cochrane, A. (ed) (1987): *Developing local economic strategies.* Open University, Milton Keynes.

Goodman, R. (1979): *The last entrepreneurs.* Boston.

Green, H.L. (1988): "Retailing in the new economic era", in Sternlieb, G. and Hughes, J. (eds): *America's new market geography,* New Brunswick, NJ.

Jacobs, J. (1984): *Cities and the wealth of nations.* New York.

Levine, M. (1987): "Downtown redevelopment as an urban growth strategy: a critical appraisal of the Baltimore renaissance", *Journal of Urban Affairs,* 9(2), pp. 103–23.

Murray, F. (1983): "Pension funds and local authority investments", *Capital and Class,* 20, pp. 89–103.

Noyelle, T. and Stanback, T. (1984): *The economic transformation of American cities.* Totawa, NJ.

9 SASKIA SASSEN

'The New Inequalities within Cities'

from *Cities in a World Economy* (1994)

What is the impact of the ascendance of finance and producer services on the broader social and economic structure of major cities? And what are the consequences of the new urban economy on the earnings distribution of a city's work-force? We know that when manufacturing was the leading sector of the economy, it created the conditions for the expansion of a vast middle class because (1) it facilitated unionization; (2) it was based in good part on household consumption, and hence wage levels mattered in that they created an effective demand; and (3) the wage levels and social benefits typical of the leading sectors became a model for broader sectors of the economy.

We want to know about the place of workers lacking the high levels of education required by the advanced sectors of the economy in these major cities. Have these workers become superfluous? We also want to know about the place in an advanced urban economy of firms and sectors that appear to be backward or to lack the advanced technological and human capital base of the new leading sectors. Have they also become superfluous? Or are such workers, firms, and sectors actually articulated to the economic core, but under conditions of severe segmentation in the social, economic, racial, and organizational traits of firms and workers? We want to know, finally, to what extent this segmentation is produced or strengthened by the existence of ethnic/racial segmentation in combination with racism and discrimination.

Remarkably enough, we can see general tendencies at work on the social level just as we can on the economic level. Recent research shows sharp increases in socioeconomic and spatial inequalities within major cities of the developed world. This finding can be interpreted as merely a quantitative increase in the degree of inequality, one that is not associated with the emergence of new social forms or class realignments. But it can also be interpreted as social and economic restructuring and the emergence of new social forms: the growth of an informal economy in large cities in highly developed countries; high-income commercial and residential gentrification; and the sharp rise of homelessness in rich countries.

[. . .]

TRANSFORMATIONS IN THE ORGANIZATION OF THE LABOR PROCESS

The consolidation of a new economic core of professional and servicing activities needs to be viewed alongside the general move to a service economy and the decline of manufacturing. New economic sectors are reshaping the job supply. So, however, are new ways of organizing work in both new and old sectors of the economy. The computer can now be used to do secretarial as well as manufacturing work. Components of the work process that even ten years ago took place on the shop floor and were classified as production jobs today have been replaced by a combination of machine/service worker or worker/engineer. The machine in this case is typically computerized; for instance, certain operations that require a highly skilled crafts-person can now be done through computer-aided design and calibration. Activities that were once consolidated in a single-service retail establishment have now been divided between a service delivery outlet and central headquarters. Finally, much work that was once standardized mass production is today increasingly characterized by customization, flexible specialization, networks of subcontractors, and informalization, even at times including sweatshops and industrial homework. In brief, the changes in the job supply evident in major cities are a function of new sectors as well as of the reorganization of work in both the new and the old sectors.

[. . .]

THE EARNINGS DISTRIBUTION IN A SERVICE-DOMINATED ECONOMY

What we want to know next is the impact that the shifts have had on the earnings distribution and income structure in a service-dominated economy. A growing body of studies on the occupational and earnings distribution in service industries finds that services produce a larger share of low-wage jobs than manufacturing does, although the latter may increasingly be approaching parity with services; moreover, several major service industries also produce a larger share of jobs in the highest-paid occupations.

Much scholarly attention has been focused on the importance of manufacturing in reducing income inequality in the 1950s and 1960s. Central reasons typically identified for this effect are the greater productivity and higher levels of unionization found in manufacturing. Clearly, however, these studies tend to cover a period largely characterized by such conditions, and since that time the organization of jobs in manufacturing has undergone pronounced transformation . . . [T]he most detailed analysis of occupational and industry data found that earnings in manufacturing have declined in many industries and occupations . . . [others] found that a majority of manufacturing jobs in the sunbelt are low wage, and . . . [a] growth of sweatshops and homework in several industry branches in New York and Los Angeles.

There is now a considerable number of studies with a strong theoretical bent which argue that the declining centrality of mass production in national growth and the shift to services as the leading economic sector

contributed to the demise of a broader set of arrangements. In the postwar period, the economy functioned according to a dynamic that transmitted the benefits accruing to the core manufacturing industries onto more peripheral sectors of the economy. The benefits of price and market stability and increases in productivity could be transferred to a secondary set of firms, including suppliers and subcontractors, but also to less directly related industries. Although there was still a vast array of firms and workers that did not benefit from the shadow effect, their number was probably at a minimum in the postwar period. By the early 1980s the wage-setting power of leading manufacturing industries and this shadow effect had eroded significantly.

Scholarship on the impact of services on the income structure of cities is only now beginning to emerge in most countries. There are now several detailed analyses of the social impact of service growth in major metropolitan areas in the United States . . . [one study] found that from 1970 to 1980 several service industries had a significant effect on the growth of under-employment, a label . . . define[d] as employment paying below poverty-level wages in the 199 largest metropolitan areas. The strongest effect was associated with the growth of producer services and retail trade. The highest relative contribution resulted from what the authors call "corporate services" (finance, insurance and real estate, business services, legal services, membership organizations, and professional services), such that a 1 percent increase in employment in these services was found to result in a 0.37 percent increase in full-time, year-round, low-wage jobs. Furthermore, a 1 percent increase in distributive services resulted in a 0.32 percent increase in full-time, year-round, low-wage jobs. In contrast, a 1 percent increase in personal services was found to result in a 0.13 percent increase in such full-time jobs and a higher share of part-time, low-wage jobs. The retail industry had the highest effect on the creation of part-time, year-round, low-wage jobs, such that a 1 percent increase in retail employment was found to result in a 0.88 percent increase in such jobs.

But what about the impact of services on the expansion of high-income jobs? . . . To establish why male earnings are more unequal in metropolises with high levels of service sector employment, . . . [one study] measured the ratio of median earnings over the 5th percentile to identify the difference in earnings between the least affluent and the median metropolitan male earners; and . . . measured the ratio at the 95th percentile to establish the gap between median and affluent earners. Overall, . . . [it was] found that inequality in the 125 areas appeared to be the result of greater earnings disparity between the highest and the median earners than between the median and lowest earners. Furthermore . . . [it] found that the strongest effect came from the producer services and that the next strongest was far weaker (social services in 1970 and personal services in 1980).

The conditions for ongoing inequality can also be seen in projections for educational requirements. In the United States, the evidence for 1988 shows that 17 percent of jobs required less than a high school diploma and over 40 percent would require only a high school diploma. Only about 22 percent of jobs required at least a college degree. By the year 2000 the expectation is that there will be very little change in these levels, with 16.5 percent of jobs requiring less than high school and only 22.9 percent requiring at least a college degree. That is to say, by the year 2000 over half of all jobs will require only a high school diploma or less. The change is somewhat sharper when we consider only net new jobs, with only 13 percent of jobs requiring less than high school and almost 30 percent requiring at least a college degree. The expansion of low-wage service jobs in large cities and the downgrading of many manufacturing jobs suggest that a good share of jobs in cities will be among those requiring only a high school education or less.

In their own distinct form, these trends are evident in many highly developed countries. Of interest here is Japan, since little seems to be known or reported in general commentary about growing casualization of employment. We now turn to a brief description of the growth of service jobs in Japan.

The growth of low-wage jobs in Japan

Japan also has seen considerable growth of low-wage service jobs, the replacement of many full-time male workers with part-time female workers, and the growth of forms of sub-contracting that weaken the claims of workers on their firms. Over half of the new jobs created in Tokyo in the 1980s were part-time or temporary jobs.

There are other indications of structural change in Japan in the 1980s. Since the mid-1980s, average real earnings in Japan have been decreasing, and the manufacturing sector has been losing its wage-setting influence. Furthermore, with few exceptions, most of the service industries that *are* growing have significantly lower average earnings than do manufacturing, transport, and communications. Hotel and catering had among the lowest average earnings, along with health services and retail. Many of the industries that are growing either pay above-average wages – as in finance, insurance, and real estate – or pay below-average wages. The same trends found in many western cities are becoming evident in Tokyo . . .

Data from the Labor Force Survey in Japan show that the share of part-time workers increased from under 7 percent of all workers in 1970 to 12 percent in 1987, or 5 million workers. Among female workers, this share almost doubled, from about 12 percent in 1970 to 22 percent in 1985 and over 23 percent in 1987, or a total of about 3.65 million women. Almost 24 percent of part-time female workers were in manufacturing, an indication of the growth of a casualized employment relation in that sector.

By the late 1980s, 58 percent of all firms surveyed employed part-time workers. Part-time work in Japan is defined by the Ministry of Labor as a job with scheduled hours per week "substantially shorter than those of regular workers." Part-time work is measured as regular employment of under thirty-five hours a week. This definition excludes seasonal and temporary employment and is thus an undercount of all jobs that are not year-round, full-time jobs. What distinguishes them is the lack of various benefits and entitlements, or, in the terms used here, a casualized employment relation.

Of interest also is the situation of home-workers, a growing category in most developed countries. Official counts of legal homeworkers in Japan show gradual decline over the last decade. In 1987 there were over 1 million such workers, almost all women. The largest share of homework, 34 percent, is in clothing and related items, followed by 18.6 percent in electrical/ electronic equipment (including assembly of electronic parts) and almost 16 percent in textiles. The remaining share includes a very broad range of activities, from making toys and lacquerware to printing and related work. It is quite possible that the existing regulations protecting homeworkers and providing them with fringe benefits are eroding. Official figures describe a decline in the fully entitled share of homeworkers but possibly do not register an absolute increase among homeworkers with no protection. There are some indications that the latter category may be increasing.

The growth of low-wage and part-time jobs is likely to facilitate the employment of illegal immigrants. In Japan, where immigration, both legal and illegal, is not part of the cultural heritage as it is in the United States, there is now a growing illegal immigration from several Asian countries. The evidence on detected illegal immigration from the Ministry of Justice ... shows that over 80 percent of men apprehended from 1987 to 1990 held construction and factory jobs. Clearly, factories and construction sites lend themselves to apprehension activity, unlike small service operations in the center of Tokyo or Osaka. Thus we cannot assume that this level is an adequate representation of the occupational distribution of illegal immigrants; but it does indicate that factories are employing illegal immigrants.

According to a study of illegal immigrant employment in the major urban areas in Japan carried out by the Immigration Office of the Ministry of Justice, factories employing illegal immigrants are in a broad range of branches: metal processing, plastic processing, printing and binding, plating, press operating, materials coating. Most recently a growing number of women have been apprehended in factories in metals and plastic processing and in auto-parts manufacturing. Most illegal immigrants were found in medium-size and small factories. The figures for 1991 point to a continuation of these patterns; almost half of illegals detected by the government were in construction, followed by 14 percent in manufacturing and certain jobs in the retail industry, in particular back-room jobs in restaurants.

Estimates about the evolution of illegal immigration for unskilled jobs vary considerably, but all point to growing demand. The Ministry of Labor estimates the labor shortage will reach half a million by the end of the decade. Japan's most powerful business organization, Keidaren, puts the shortages at 5 million. Specialists estimate the shortage will range between 1 and 2 million by the end of the decade. Currently the largest shortages are in manufacturing, particularly small- and medium-size firms. But there is considerable agreement that the service sector will be a major source of new shortages. As the current generation of Japanese employees in low-skill service jobs retires and young highly educated Japanese reject these jobs, there may well be a gradual acceptance of immigrant workers. Although later than most advanced economies, Japan now has a growing labor demand for low-wage, unskilled jobs in a context where Japanese youth are rejecting such jobs.

The restructuring of urban consumption

The rapid growth of industries with a strong concentration of high- and low-income jobs has assumed distinct forms in the consumption structure, which in turn has a feedback effect on the organization of work and the types of jobs being created. In the United States, the expansion of the high-income workforce in conjunction with the emergence of new cultural forms has led to a process of high-income gentrification that rests, in the last analysis, on the availability of a vast supply of low-wage workers. As I have argued at great length elsewhere, high-income gentrification is labor intensive, in contrast to the typical middle-class suburb that represents a capital-intensive process: tract housing, road and highway construction, dependence on private automobiles or commuter trains, marked reliance on appliances and household equipment of all sorts, large shopping malls with self-service operations. Directly and indirectly, high-income gentrification replaces much of this capital intensity with workers. Similarly, high-income residents in the city depend to a much larger extent on hired maintenance staff than does the middle-class suburban home, with its concentrated input of family labor and machinery.

Although far less dramatic than in large cities in the United States, the elements of these patterns are also evident in many major Western European cities and, to some extent, in Tokyo. For instance, there has been considerable change in the occupational composition of residents in central Tokyo. Not unlike what we have seen in other major cities, there is a tendency for growing numbers of upper-level professional workers and of low-level workers to live in central cities: . . . [one study] found that the share of the former grew from 20 percent of all workers in 1975 to over 23 percent in 1985 and (although difficult to measure) we know that the numbers of low-wage legal and undocumented immigrants also grew and certainly have grown sharply since then. The share of middle-level workers, on the other hand, fell: for instance, the share of skilled workers fell from 16 percent in 1975 to 12 percent in 1985. Similar patterns hold for other areas of the city. The total size of the resident workforce stayed the same, at about 4.3 million in 1975 and in 1985. We know that there was a sharp growth of high-income professional and managerial jobs in the second half of the 1980s, which could only have reinforced this trend.

The growth of the high-income population in the resident and commuting workforce has contributed to changes in the organization of the production and delivery of consumer goods and services. Behind the delicatessens and speciality boutiques that have replaced many self-service supermarkets and department stores lies a very different organization of work from that prevalent in large, standardized establishments. This difference in the organization of work is evident both in the retail and in the production phase. High-income gentrification generates a demand for goods and services that are frequently not mass-produced or sold through mass outlets. Customized production, small runs, speciality items, and fine food dishes are generally produced through labor-intensive methods and sold through small, full-service outlets. Subcontracting part of this production to low-cost operations and to sweatshops or households is common. The overall outcome for the job supply and the range of firms involved in this production and delivery is rather different from that characterizing the large department stores and supermarkets. There mass production is prevalent, and hence large standardized factories located outside of the region are the norm. Proximity to stores is of far greater importance with customized producers. Mass production and mass distribution outlets facilitate unionizing.

The magnitude of the expansion of high-income workers and the high levels of spending contribute to this outcome. All major cities have long had a core of wealthy residents or commuters. By itself, however, this core of wealthy people could not have created the large-scale residential and commercial gentrification in the city. As a stratum, the new high-income workers are to be distinguished from this core of wealthy or upper class. The former's disposable income is generally not enough to make them into major investors. It is, however, sufficient for a significant expansion in the demand for highly priced goods and services – that is, to create a sufficiently large demand so as to ensure economic viability for the producers and providers of such goods and services. Furthermore, the level of disposable income is also a function of life-style and demographic patterns, such as postponing having children and larger numbers of two-earner households.

The expansion in the low-income population has also contributed to the proliferation of small operations and the move away from large-scale standardized factories and large chain stores for low-price goods. In good part, the consumption needs of the low-income population are met by manufacturing and retail establishments that are small, rely on family labor, and often fall below minimum safety and health standards. Cheap, locally produced sweatshop garments, for example, can compete with low-cost Asian imports. A growing number of products and services ranging from low-cost furniture made in basements to "gypsy cabs" and family daycare is available to meet the demand for the growing low-income population.

There are numerous instances of how the increased inequality in earnings reshapes the consumption structure and how this reshaping in turn has feedback effects on the organization of work. Some examples are the creation of a special taxi line for Wall Street that services only the financial district and an increase of "gypsy" cabs in low-income neighborhoods not serviced by regular cabs; the increase in highly customized woodwork in gentrified areas and low-cost rehabilitation in poor neighborhoods;

the increase of homeworkers and sweatshops making either very expensive designer items for boutiques or very cheap products.

CONCLUSION: A WIDENING GAP

Developments in cities cannot be understood in isolation from fundamental changes in the larger organization of advanced economies. The combination of economic, political, and technical forces that has contributed to the decline of mass production as the central driving element in the economy brought about a decline in a broader institutional framework that shaped the employment relation. The group of service industries that constitute the driving economic force in the 1980s and into the 1990s is characterized by greater earnings and occupational dispersion, weak unions, and a growing share of unsheltered jobs in the lower-paying echelons, along with a growing share of high-income jobs. The associated institutional framework shaping the employment relation diverges from the earlier one. This new framework contributes to a reshaping of the sphere of social reproduction and consumption, which in turn has a feedback effect on economic organization and earnings. Whereas in the earlier period this feedback effect contributed to reproduction of the middle class, currently it reproduces growing earnings disparity, labor market casualization, and consumption restructuring.

All these trends are operating in major cities, in many cases with greater intensity than in medium-size towns. This greater intensity can be rooted in at least three conditions. First is the locational concentration of major growth sectors with either sharp earnings dispersion or disproportionate concentration of either low- or high-paying jobs. Second is a proliferation of small, low-cost service operations made possible by the massive concentration of people in such cities, in addition to a large daily inflow of nonresident workers and tourists. The ratio between the number of these service operations and the resident population is probably significantly higher in a very large city than in an average city. Furthermore, the large concentration of people in major cities tends to create intense inducements to open up such operations, as well as intense competition and very marginal returns. Under such conditions, the cost of labor is crucial, and hence the likelihood of a high concentration of low-wage jobs increases. Third, for these same reasons together with other components of demand, the relative size of the downgraded manufacturing sector and the informal economy would tend to be larger in big cities like New York or Los Angeles than in average-size cities.

The overall result is a tendency toward increased economic polarization. When we speak of polarization in the use of land, in the organization of labor markets, in the housing market, and in the consumption structure, we do not necessarily mean that the middle class is disappearing. We are rather referring to a dynamic whereby growth contributes to inequality rather than to expansion of the middle class, as was the case in the two decades after World War II in the United States and the United Kingdom, and into the 1970s in Japan. In many of these cities the middle class represents a significant share of the population and hence represents an important channel through which income and life-style coalesce into a social form.

The middle class in the United States is a very broad category. It contains prosperous segments of various ethnic populations in large cities as well as longtime natives. What we can detect in the 1980s is that certain segments of the middle class gain income and earnings, becoming wealthier while others become poorer. In brief, we see a segmenting of the middle class that has a sharper upward and downward slant than has been the case in other periods. The argument put forth here is that while the middle strata still constitute the majority, the conditions that contributed to their expansion and politico-economic power – the centrality of mass production and mass consumption in economic growth and profit realization – have been displaced by new sources of growth. This is not simply a quantitative transformation; we see here the elements for a new economic regime.

The growth of service employment in cities and the evidence of the associated growth of inequality raises questions about how fundamental a change this shift entails. Several of these questions concern the nature of service-based urban economies. The observed changes in the occupational and earnings distribution are outcomes not only of industrial shifts but also of changes in the organization of firms and of labor markets. A detailed analysis of service-based urban economies shows that there is considerable articulation of firms, sectors, and workers who may appear to have little connection to an urban economy dominated by finance and specialized services but in fact fulfill a series of functions that are an integral part of that economy. They do so, however, under conditions of sharp social, earnings, and often racial/ethnic segmentation.

10 JONATHAN CHARLEY

'Industrialization and the City: Work, Speed-up, Urbanization'

from Iain Borden and David Dunster (eds), *Architecture and the Sites of History: Interpretations of Buildings and Cities* (1995)

LABOUR PROCESS IN SPACE

Industrialization, as a material entity, cannot exist independently of space or time. As matter is transformed, so is space. This would suggest that the transformation of the labour process is mutually interdependent with urbanization and the rapid physical transformation of cities. The space of work (the factory) and the space of home life (such as the mass housing schemes of the twentieth century) represent revolutions in the spatial organization of social life that are inseparable from the revolution within the labour process itself.

Similarly, at a more general level, industrialization of building not only involves the rationalization of the labour process, but is accompanied by the rationalization of space through urban design – both processes demonstrate the desire for order and control with all the discipline a ruling class can muster. This truly gets underway with the nineteenth century replanning of Paris, Budapest, Glasgow, and indeed of virtually every other major European city. Even at this early stage there is an inescapable correspondence between the rationalization of space through the boulevard, public square and urban grid, and the increasing division of labour. Both urban space and labour become fragmented into easily controlled and packaged parts, a process which gathers pace in the transition to the zoned city of the late twentieth century, with its mass-produced middle class suburbs and working class estates, and reaches its apogee in the globalization of capitalism. As the truly international character

of communications, travel and capitalism increases, the world appears to shrink, and we experience an increasing convergence of scientific knowledge and of economic and political practice, and we witness the simultaneous construction of almost identical built environments, be they in New York, London or Tokyo, whether as office developments, out-of-town residential sectors or the enclosed world of the shopping mall.

If this is one of the most important features of the modern industrialized landscape, the spatial segregation of society along class, gender and ethnic lines is another, and is similarly replicated wherever capitalist urbanization gathers pace. Historically this has been reinforced by the class disparities in the quality of building, and in this respect the *barriadas* and *favelas* of twentieth century Latin American cities are merely the latest version of the speculative housing developments such as those of London's nineteenth century East End.

But without doubt the most celebrated examples of the industrialization of building are the seas of housing blocks that punctuate the peripheries of virtually every metropolitan centre, whether it be Manchester, Berlin or Paris, and which continue to be built in Latin America, the former USSR and Africa. This aspect of urbanization sharply reveals the contradictory character of building production. On the one hand, we can see a quantitative expansion in building production brought about by technological innovation. Faced with periodic destruction through war, and the associated pressures of migration and population growth, governments everywhere had a moral, social and political imperative to build as much and as fast as possible.

But there were clearly other forces at work. Post-war Fordist building production offered hitherto unimagined opportunities, not just for solving the housing crisis, but for the rapid accumulation of capital and surpluses. Within the west, peoples' needs were met with a state-regulated building industry where the land and technologies of construction lay in private hands. In the Soviet bloc, needs were met through a centrally-planned and completely state-owned industry.

Despite their differing patterns of ownership, the speedy realization of the housing programme was vital for the ruling classes and state bureaucracies in both kinds of society to maintain their ruling class hegemony. The housing and welfare needs of subject peoples were ultimately subordinate to financing the arms race, propping up the military industrial complex, and lining the coffers of capitalists, bureaucrats and Communist Party hacks.

It is precisely during this classic post-war industrialization that the Soviet building industry became almost completely dependent on the prefabrication of concrete components, employing thirteen million workers and operating over two thousand concrete factories. Similarly in Britain, construction output tripled between 1948 and 1964 and doubled between 1955 and 1970, allowing the giant construction firms of Wimpey, Laing and Taylor Woodrow to consolidate their position at the forefront of the British building industry, a position that they still enjoy, not least because of the speculative mass production of new housing suburbs in the late 1980s and early 1990s.

When we think of industrialized building, it is often the concrete panel housing project that comes to mind. Although many countries have now ceased this kind of system-built production, the process of industrialization is accelerating rather than slowing down. Indeed, almost all contemporary buildings use factory pre-fabricated building components. It is simply that the process has changed to using more sophisticated machines and computer-based technologies, and to the production of building types that differ in appearance from those normally associated with industrialized building. The timber-framed house, the populist Wimpey and Barratt home in the United Kingdom, the Levittown suburb of America, the speculative office block, the McDonald's fast food outlet, the supermarket store, and the light industrial factory unit are all heavily dependent on factory prefabrication of components. All are equally reliant on a site labour force that is mobile, and has been largely retrained with 'fitter' type skills – building workers who do not so much make buildings, as assemble them. Buildings and space can now be produced and transformed with ever increasing speed in a situation where capital turns over with equal rapidity.

11 JANE M. JACOBS

'Conclusion'

from *Edge of Empire* (1996)

In 1988, the year in which Australia 'celebrated' the bicentenary of its founding as a settler colony, an Aboriginal activist visited Britain. On a windy day on Brighton beach, surrounded by invited journalists, Burnam Burnam raised the Aboriginal flag and declared the British Isles to be Aboriginal territory. This colonial return mimicked Governor Phillip's hoisting of the Union Jack at Sydney Cove on the east coast of Australia 200 years before. Both were symbolic events. The first marked the 'beginning' of the British territorialisation of the continent that came to be known as Australia. The second successfully parodied the audacious banality of that event. In 1788 those few Aborigines who witnessed the raising of the Union Jack could not imagine what would happen – to them, to all Aborigines – as a result. In 1988 the millions of Britons to see the media reports of the Aboriginal flag being hoisted on Brighton beach could assume it did not mark the 'beginning' of anything much at all. One was an inaugural event which, through the force of desire and sheer might, opened out into a history of colonisation. The other was a memorial event which, despite the force of desire, could neither claim Britain nor undo Australia's history of colonisation. The embedded unevenness of power, which is the legacy of imperialism, meant that these events did their symbolic work in quite different ways. Like these events this book has brought Australia and Britain back into 'contact'. But this is not a cause-and-effect encounter. It is, like many 'first contacts', one in which the two sides see each other, impinge upon each other, but do not recognise the full weight of the histories and geographies implied by their meeting.

GEOGRAPHICAL ENCOUNTERS

This book has sought to create a productive encounter between new theorisations of imperialism and postcolonialism and the specific space of the contemporary city. I have undertaken this task in order to give geographical expression to a theoretical field that is rich in its allusions to space but often poor in its elaboration of the real worlds in which this spatiality operates. The location of much of the current theory within the fields of literary criticism and history has meant that its relevance to the conditions of everyday life in the present is often oblique. And while much of this theory is about difference, about deconstructing master narratives, about space, these concerns are often expressed through grand theory and not through the fundamentally deconstructive space of the local. I am not simply suggesting that the local provides exceptions to the rules, although this is often the case. I am proposing that through attending to the local, by taking the local seriously, it is possible to see how the grand ideas of empire become unstable technologies of power which reach across time and space.

The generic 'local' that has formed the focus of this book is the contemporary First World city. This focus has not been in pursuit of a new model of global urbanisation or to embellish the nature of the 'postmodern' urban. Indeed, these urban studies self-consciously work away from both possibilities. Old models of urban development which placed the colonial city as a mid-point in an evolution from pre-modern to modern have outlived their usefulness. It is not that the distinction between core and periphery, haves and have nots, has gone away – it is devastatingly present. But the 'where' of this geography is increasingly confused: First World cities have their Third World neighbourhoods, global cities have their parochial underbellies, colonial cities have their postcolonial fantasies. Urban transformations such as gentrification, consumption spectacles and heritage developments, are regularly understood as postmodern. But these spectacles of postmodernity are entwined in a politics of race and nation which cannot be thought of constructively without recourse to the imperial inheritances and postcolonial imperatives that inhabit the present.

The 'real worlds' of this book are of course not simply material worlds. Imaginary and material geographies are not incommensurate, nor is one simply the product, a disempowered surplus, of the other. They are complexly intertwined and mutually constitutive. Together they gave energy and drive to the territorialisations that constitute imperialism. Together they have created the most painfully uneven geographies of advantage and disadvantage. The social construction of space is part of the very machinery of imperialism. In the name of the imperial project, space is evaluated and overlain with desire: creating homely landscapes out of 'alien' territories, drawing distant lands into the maps of empire, establishing ordered grids of occupation. These spatial events did not simply supplement the economic drive of imperialism, they made it make sense; they took it from the visioned to the embodied, from the global reach of desire to the local technologies of occupation. They established the beginnings of that most permanent legacy of imperialism: the contest between that which, through space itself, has been 'naturalised' and that which has been made 'illegitimate'.

UNRULY IMPERIALISM

Imperialism has neither a uniform realisation nor a static persistence. Imperialist constructs have taken an unruly passage across time and space. The hold of imperialist regimes of power is tied to the very uncertainty they face in their manifestation on the ground: in their encounter with the unpredictability of the Other and the inconsistency of the Self. In the face of this uncertainty imperialism must always reinscribe

its frames of power and difference and this is what helps to give it its tenacity. Space is a crucial component of this anxious articulation of imperial authority.

The ordered spatialities of here and there, Self and Other, which were imagined in the inaugural moments of the imperial project, were regularly unsettled by the disorderly encounters forged during colonisation. In Australia the imperial gaze imagined a land unoccupied but encountered a land most surely peopled. Colonial constructs of space, such as the ordered plans for the city, struggled to contain the subversion created merely by Aborigines being present. That some 200 years later the Waugal Dreaming can 'appear' in Perth and unsettle contemporary urban redevelopment plans points to the failure of such technologies of containment to realise comprehensively the imaginary construct of *terra nullius*. In British cities, too, imperialism meant change and in particular new levels of industrialisation and urbanisation. While there was pride in this imperial growth, it also spawned anti-urban organisations which are still present over a century later, and which through their conservation efforts regularly activate the memory of empire.

In the contemporary moment the always contingent order of imperial geographies has been further undone. Formal decolonisation and postwar migration and settlement have brought an embodied edge of empire into the heart, while the demise of empire has meant that Britain now has different global and regional affiliations. These changes have not marked the end of empire – a pure postcolonialism – but established the conditions for revised imperial articulations. Imperialism lingers in the present as the idea of empire itself, as a trace which is memorialised, celebrated, mourned and despised. This is a potent memory which can shape trajectories of progress, drive nostalgic returns and establish the structures of difference through which racialised struggles over territory operate. In the City of London, for example, imperial nostalgias are not simply present as a residual past. They are sanctioned and activated by conservation practices which influence, in most material ways, the very course the City takes into the future. Here as in the various other case studies, tradition is not simply about an escape from modernity but about negotiating modernity, about being modern. Indeed, it is precisely the desire to memorialise empire that has helped to drive the City beyond its traditional boundaries and into areas like Spitalfields where Bengali Britons are now struggling to make a home-space in a new nation.

Imperialism also comes into the present through 'new' forms of exploitation and domination, what might be thought of as neo-colonialism. Nineteenth-century imperialism has given way to new regimes of desire. Otherness is no longer a repressed negativity in the constitution of the Self, but a required positivity which brings the Self closer to, say, a multicultural present or an ecological future. Contemporary spaces of consumption seek out Otherness. But it is not just any Otherness that is required. In tourism developments in Australia it is an essentialised Aboriginality, honed for the spectacle, which is taken up into these systems of commodification. In these new tourist spectacles Aboriginality is, once again, returned to Nature. Early imperial constructions of Aboriginality and Nature confirmed the division between an 'uncivilised Native' and a 'civilised Culture'. In the current unison, Aboriginality serves as a template for the 'proper' return of Civilisation to Nature, the source of an ecological Self. But these eco-desires also serve as a means for settler Australians to develop their own sense of being in the land and, as such, of a more final and complete colonisation. Of course, such returns are still likely to be regulated by the requirements of capital. In the city of Perth, for example, the proposition to return development land to Nature was seen as 'unproductive' and deeply subversive to city development aspirations.

In Spitalfields too the desire for Otherness became part of the politics of place. Gentrifiers and developers regularly celebrated the distinctive 'multicultural' history of the area. But this celebrated cohabitation could not be too promiscuous or unpredictable. Here a multi-culturalism of convenience emerged based on a properly placed (spatially segregated) Bengali community. Ordered and domesticated, the Bengali residents of Spitalfields could become a safe, present-day supplement to the narrative construction of Spitalfields as the emblematic place of an embracing, tolerant Englishness.

In the mapping of current imperialisms, appropriation is often posited as a key dynamic in the exercise of power. A prime example of this is the way in which tourism and heritage industries reproduce essentialised constructions of identity and take these up into systems of commodification. It is as if the stark territorial appropriations of nineteenth-century imperialism have given way to more fractured patterns of cultural appropriation. As I have shown, such appropriations do occur and they can have imperialist effects. Essentialist constructions of Otherness do work to categorise colonised groups within desired and confining templates. Nevertheless, it is a form of imperial nostalgia, a desire for the 'untouched Native', which presumes that such encounters only ever mark yet another phase of imperialism. Colonised groups do not enter passively into such systems of commodification, nor are they ever neatly stitched into place within them. Essentialised constructs of identity and place are open to a range of reinventions, adaptations, invigorations and reappropriations at the hands of both colonisers and colonised. The encounters that occur through new processes of commodification might well help colonised and diasporic groups to articulate a sense of self in productive new ways: giving local political struggles a global reach; providing new arenas for the elaboration of tradition; opening up new economic opportunities; and establishing influential systems of pedagogy. Commodification is not simply a process by which the colonised, the 'native', 'tradition' is corrupted.

The idea for Banglatown in London shows precisely how problematic it is to locate imperialism in the processes of appropriation/ commodification and the construction of essentialised subjectivities. Banglatown was not imposed on the Bengali businessmen, nor did they produce this idea unaware of their strategic toyings with the category 'Bengali'. Here the Bengali businessmen actively engaged in elaborating essentialist constructions of identity through commodified systems in an attempt to wrest control of power and space. Similarly, the Aboriginal artists who entered into the 'reconciliation' place-making project in Brisbane inventively redeployed 'tradition' in order to establish new protocols of influence over the modern city. In both cases, different though they may be, the notion of sure-footed imperialist appropriation is fundamentally unsettled. Also in these cases, essentialised constructions, although reaching back into the past, are produced in the present in order to negotiate the inequities of power produced in the modern.

POSTCOLONIAL POSSIBILITIES

By problematising the dynamics of appropriation and essentialism I do not simply strive towards an emergent postcolonialism by way of the Native. The tenacity of an adaptive imperialism is a reminder of the fantastic optimism of the term 'postcolonialism'. While indigenous land rights claims and postwar settlements in Britain might be thought of as postcolonial formations, their truly *post*colonial effect is still faintly traced. It is indeed hard to imagine a moment that is beyond imperialism. In this sense the postcolonial is not so much about being beyond colonialism as about attending to the social and political processes that struggle against and work to unsettle the architecture of domination established through imperialism. This includes the 'space clearing' gestures of intentional anti-imperial politics. But it also includes the various uncontained excesses and unsettlingly hybrid outcomes of the cohabitation produced by imperialism itself. For example, the anticolonial intention of the Aboriginal claims over Perth were amplified by the anxious reactions of non-Aboriginal Australians to the 'appearance' of the Waugal Dreaming in the secularised city. Similarly, the Banglatown plans for Spitalfields actively sought to create alliances with 'big' capital in pursuit of creating a Bengali place in Britain. This created a crisis of affiliation between the Bengali community and the very groups who saw themselves as their guardians and their anti-racist allies.

Within the current discussion of the postcolonial much emphasis is given to the disruptive power of hybridity. In placing hybridity as a key signifier of postcoloniality the inevitable vulnerability of colonial structures of power and categorisation is properly exposed. Imperialism undoes itself, against itself. Nevertheless, in thinking through the postcoloniality of the hybrid form it is often too easy for agency and intentionality to be displaced: that is, for the postcolonial to be posited merely as the 'subject-effect' of the colonial. In the example of the Brisbane 'reconciliation' place-making project a most decidedly and self-consciously hybrid place was created out of a creative redeployment of 'tradition' and a destabilising appropriation of the colonial construct of the map. Understanding the postcolonial effects of this project required more than relishing its visual hybridity. Here hybridity was not about the dissolution of difference, or even only about destabilising the constructs of Self and Other, but about renegotiating the structures of power built on difference. The political weight of this hybrid space resides not simply in its surface form and in its effects, but also in the intentional politics of its production, its desire to reassert Aboriginal authority over the space of the city. Attending to such intention, even while it might fracture into a disorderly effect, is a necessary component of a postcolonial critique.

Throughout this account of the politics of identity and place there have been various versions of what might be thought of as a return to origins – the claim that identity is 'given' through some uncontested inheritance or static place-based genesis. While the fractured and contingent nature of identity is undeniable, so too is the necessity of temporary fixings of identity around such essentialised notions. Such claims are deployed by the powerful to legitimate their rights over territory, to categorise Otherness as 'outside' and to domesticate difference. That is, claims of origin can be hegemonic. But claims of origin, such as strategies of fixing identity in place, are also important for marginalised groups who want to distinguish their claims from the hegemonic. Proposing that essentialist notions of identity and place are social constructs, and strategic ones at that, destabilises a whole range of claims for rights over space which are argued through the idea of origin. It is one thing for such a generalisation to unsettle nostalgias for empire or the violence of some nationalisms; it is another thing when it also compromises the claims for land made by colonised groups who are still intensely marginalised. That is, theoretical generalisations about the socially constructed nature of essentialised identity have an uneven political consequence which is far from incidental to ongoing political struggles.

In a contemporary world, constituted out of

complex processes of deterritorialisation and reterritorialisation, movement and cohabitation, it may well be that what Kristeva calls the 'cult of origins' needs to give way to a sense of place which is built around fractured vectors of connection and histories of disconnection. There is little doubt that the present has seen a radical 'unbuilding' of the geographies of imperialism, but it is also true that these new geographies do not surpass their past. They are made out of the continued negotiation of lingering and newly formed imperialisms as much as they are made out of the hope of postcolonial futures. It is because of the undeniable persistence of domination that claims of origin, the strategic formation of fixed identity, continue to be part of the politics of the present. That geography which Doreen Massey refers to as a 'progressive sense of place' could surely only flourish within a more even, more democratic, terrain of power than that which characterises the present.

If postcolonialism is always haunted by colonialism, can there be a postcolonial geography? Can the spatial discipline of geography move from its positioning of colonial complicity towards producing postcolonial spatial narratives? The postcolonial geographies traced in these pages have not sought out a 'pure' postcoloniality. Instead, they have concerned themselves with the unruly fortunes of colonial constructs as they fold in on themselves, are recharged by new contexts or fundamentally challenged by their encounter with the colonised. They have not simply sought out Otherness in order to elaborate the nature of the imperial core or the inappropriateness of its colonial projects. Rather, they have engaged with anticolonial political imperatives. These postcolonial geographies have replaced the security of the maps of the past with the uncertainty of touring the unsettled spatialities of power and identity in the present.

12 JEREMY SEABROOK

'People of the City I'

from *In the Cities of the South* (1996)

THE EXILE, BANGKOK

I met Henry on the way to Bangkok, where he goes at least twice a year to see what he calls his 'Thai wife'. What he didn't say in the beginning was that he also has an English wife, who stays at home in their little house in south London. For the past five years Henry has regularly spent six or eight weeks a year in Bangkok in pursuit of his mildly polygamous life.

Henry in his mid-sixties, corpulent, with grey-white hair and pale-blue eyes. He wears smart cream-coloured trousers, a blue shirt and casual white shoes. This is, as it were, the uniform of infidelity, for he dresses far more conservatively at home. He worked for over twenty years in southeast London as an engineer with a multinational company that made machinery for cigarette manufacture. He was made redundant ten years ago when the company shifted its operations, significantly, to the Third World; Henry has made the same journey in its wake, even though in pursuit of different satisfactions. In many ways Henry exemplifies the relationship between the Western working class and its equivalent in the South.

In 1983 he received a severance gratuity, which pays for his twice-yearly trips to Thailand. 'They were good employers,' he says. 'They used to provide a turkey at Christmas for every employee, and Havana cigars.' Whether this act of charity is now extended to the employees in its present location may be doubted.

Henry's wife – his English wife – is not keen on vacations and does not care to accompany him to Thailand. She doesn't like leaving home, and she knows nothing of the existence of the Thai wife. 'I'm not taking anything away from my wife,' he says. 'When I married her I gave up a lot of things for her sake. I had to sacrifice my love of opera and ballet, because she couldn't stand the music. For thirty years I behaved myself. Even now I wouldn't fool around on my own doorstep. I suppose I'm still puritanical at heart. But I've made sure that if I die my wife will be all right; the bank balance is sound. While my daughter was young and I was working, I never looked at another woman. I kept on the straight and narrow until I was sixty, and surely that's long enough for anybody.

'The first time I went to Thailand, it was just for a holiday. My daughter had been, and she suggested it. I met Pia in a bar. She was wiping the glasses at the back of the bar. She was thirty-six then, and considered to be an old lady: too old to satisfy the punters who come to Thailand for the young girls. Their working life is finished by the age of thirty. So in a way I rescued her. And now she looks after me. She can't do enough for me. That's the way Thai women are. They feel good if they have someone to look after. Put it like this: when you go into the bathroom in the morning, you find the toothpaste already squeezed onto the brush waiting for you. I call that caring. The women in Thailand have something that Western women don't have.

'It's taken years off me, meeting Pia. I've some money in my account which my wife doesn't know about. I feel I've earned it, and now I deserve a bit of life. Of course, when I go home I feel a pang of conscience – I might decorate a room or dig the garden.

'My wife doesn't like me going off on my own. But she doesn't know how to enjoy herself. She has her own friends. They go to the shops, visit the supermarket, hunt for bargains. If they come back with some cardigan or hat they got a bit cheap, you'd think it was a trophy of war.

That's their life, and they're welcome to it. But for me there has to be something more, or what's the point? To have lived and died, done nothing, gone nowhere: that isn't a life, it's an existence.

'I need to get away. There was an old couple near where we live. Darby and Joan everybody called them. One day the women met my wife in the street. She told her she hated the sight of him. Why don't you leave him? she said. "Oh I couldn't leave him. I could murder him, but I couldn't leave him."

'It's wrong to live in one another's pockets. Put it like this. I'd never leave my wife, but if anything happened to her I'd come and live in Thailand tomorrow. In Thailand you meet a lot of ex-pats, people who've packed up and left. There's a cop I know from Brooklyn; he's been shot three times on duty. He came to Thailand because it's peaceful. It's safe. You can walk the streets without fear of being mugged. There was one fellow from Liverpool I knew, his wife had left him. He came to Thailand to spend his last few years. As a matter of fact he died there. His Thai wife looked after him till the end.

'Britain has lost a lot. I wouldn't want to be young in our country now. We've seen our best years. Where we live, there's too many blacks. It's like living in a foreign country. You're always looking over your shoulder to see who's following you, who's going to threaten you, pull a knife, rob you of what you've got. The wife would like to move. You feel you're under siege. They deal in drugs just a couple of doors away. It's not Britain any more. But it's the same everywhere. We gave all these countries their freedom, and what have they done with it? They've all gone to rack and ruin since we left.

'I share my Thai lady with a German. He takes his son with him to visit her, he's quite open about it. I could never do that. He's separated from his wife. He sent for Pia to go to Germany. She went for six months. She said, "If I like it, I'll marry you." But it didn't work out, so she came home. I've met him, I've shaken hands with him.

'Pia was married, an arranged marriage. She was nineteen, he was twenty-eight. He was very brutal to her sexually, and that was when she left him to go to Pattaya to make a career in the bars. Thai women like older men. Youngsters are too selfish. Older men know how to treat them decently. I put two thousand pounds in her

account. She has a smallholding in her village, pigs and chickens, and spends most of her time there. I write when I'm coming to Bangkok, and she'll be in the bar, waiting.

'Once, when I arrived, she wasn't there. I went to the bar, and her friend said she would send a telegram. In the meantime I had to have someone to take care of me. I went to the bar and said, "Would anyone like to take care of me for a couple of days until Pia gets here?" A 22-year-old offered herself. She was very nice, but when she asked me if I wanted to make love to her I couldn't. It would be like my own granddaughter. When Pia arrived at the hotel they rang from reception, because they knew I was upstairs with Dim. They said, "Shall we keep her down here?" I said, "No, let her come up. It's no secret." She didn't like it. But by God, next time she was there, ready and waiting. Two big tears rolled down her cheeks. She said, "If you do that again, you will not see my love any more."

'They know how to show affection. They're not afraid to. Taking care of you. I realize I'm lucky. I've had a good life. There's no justice in the world, I know that. Going to Thailand, I don't call it a holiday. I call it investing in memories for the future.'

THE POET, DHAKA

Syed Shamsal Haq for some years worked for the BBC World Service. His two children are grown up and remain in Britain; he himself is too deeply rooted in Bengal to stay away for long.

'All people tell stories about themselves, both to outsiders and to each other. For most Bengalis it is axiomatic that we are all village people. Even the inhabitants of Dhaka and Calcutta insist they are essentially rural, and many of their responses to the city are in reality the reactions of simple country people. We look upon the car in the same way we look upon our bullock cart. This is why we drive so dangerously. After all, we had no part in the invention of the car. If it stalls, we kick it in the same way we might kick a bullock. A Westerner might also kick his car if it breaks down, but the Westerner will do it in rage at the mechanical breakdown; we will do it as though it were a living creature.

'The decorations on the cycle rickshaws and auto rickshaws are the same as those that once adorned the horse-drawn landau; the pennants and floral paintings, the country scenes, are replicas of paintings that would have ornamented the silks and fabrics of the landau.

'This is how cultural continuity exists alongside technological change. You cannot see modernization as simply usurping tradition. There is a more subtle interplay between them, whereby cultural traits survive, adapt themselves even to what appear radical displacements.

'Even our emotional responses are nonurban. If a friend tells me that he has placed his mother in an old people's home, I will be shocked, even though in the middle class now the nuclear family is more often the norm. But our response is carried over from the extended village family. When we are introduced to people we call them *Bhai* (Brother), or Auntie or Uncle; we want to integrate everyone into a kinship system that actually, for many people, has perished.

'People come to Dhaka not only because they cannot find a livelihood or have no land, although of course those are also reasons for coming. There is another important factor, and that is insecurity in the village, in the sense that robbers and bullies can get away with injustice against the poor, steal their land, make life hell. To be a nobody is to be a somebody in the city. The city gives a kind of respect to people. In the centre of Dhaka you cannot behave badly to a rickshaw puller; whereas only thirty or forty kilometers outside the city, you can slap him, beat him, not pay the fare, and you can get away with it.

'The first question people always ask you in Dhaka is, Where do you come from? To establish where your roots are is the most basic thing. In the West the most usual question is, What do you do? Your function is more important than your origins. These apparently insignificant differences are deeply indicative of cultural characteristics, and they illuminate ways in which Dhaka is, and is not, a city. People here will not even ask your name; after where you come from, they will ask, Do you have a family? Is your father living or dead? Do you have children? Only much later will you be asked what you do. Then they will ask how much you earn, a question considered indiscreet in the West, where people's money and their

relationship with it have become very secret. When I take my leave of someone, only then he or she may ask, And what is your name?

'We are living in a time of great disillusionment. Many intellectuals do not believe the country has a future. Just as a former generation looked to Britain, many young people want to go the USA or Australia.

'Why has it become like this? In 1971, after the liberation, we dreamed of "Golden Bengal", that is an expression of Tagore, from the national anthem, "I love you, O my golden Bengal." That was very much current in 1970–71. The poet Das also wrote songs on the beauty of the natural landscape, the flora and fauna of Bengal. The air was full of myths of the soil, *ruposhi* Bangla, beautiful Bengal. The educated middle class believed in a golden Bengal and imagined that by liberating the country from Pakistan we would be walking straight into this Bengal, rather than creating it. When people fought the war, they were fighting for what they imagined was there, only the door was barred by the Pakistanis. We were naïve.

'People knew they were living under a colonial power, and had to throw them out. They wanted inspiration to help get rid of the oppressors. The common people suffered; their houses were burned, women were raped, they had to hang on to something. After such horrors, we thought tomorrow must be easy. After the war and the liberation, after the euphoria, we found to our shock that we were inheriting a devastated, ugly Bengal, of inequality and poverty. The people who had thrown off the colonial masters assumed their place – in business, administration, government. Industrialists were as bad as or worse then the Pakistanis. They took on the mantle of a controlling power. What the middle class were actually doing was making way for the rule of *goondas* and plunderers. We were duped. I accept responsibility for my part in this. We failed in our duty. We should have been able to foresee, to tell the people, but we didn't realize. We were carried away on a wave of emotion when we should have looked critically at what was happening. I'm shocked at myself now. Yet at the time we needed a myth, we needed the belief in ourselves to get rid of the occupying power. History is not changed by emotional outbursts.

'So we were left with shattered dreams: intellectual, political and economic ruin. Now nobody believes in Golden Bengal. Politicians, government take stopgap measures to deal with each emergency or problem. There is no vision, no programme, no long-term plan, no idea of how we create a decent, humane society, let alone a golden one.

'Politics are governed by negative responses. During the Raj, India was divided between Hindus and Muslims. The Muslim League had a majority only in Bengal. The Muslims were very poor, they did the tilling, the weaving, they were the poorest class. The educated and affluent were the Hindus; therefore people came to the conclusion that we had been exploited by the Hindus. We should of course have realized that the poor are exploited by the rich, the lower classes by the upper, the illiterate by the literate. But for political reasons it became Muslims exploited by Hindus. We didn't say we don't like exploitation, we said we don't like Hindus. Hence the Muslim League majority in Bengal. We voted for a party that led directly to the division of India; and that was how we became part of Pakistan. We, the articulate middle class, soon realized that the British colonists had been replaced by colonists of our own colour and religion.

'The majority do not really believe in the country. Even the rickshaw *wala* on the street would rather go to the Gulf. Many students want to go out. They see the half-educated, or the downright ignorant, do better than they ever can.

'Of course nationalist sentiment is there. It is a defensiveness against foreigners. Since we don't believe in our nation, we talk of nationalism in flowery terms, but that should not be confused with commitment. It is a kind of poetic fiction, a debased version of Golden Bengal for a generation that has experienced a Bengal that is anything but golden.

'I worked for the BBC World Service from 1972 until 1979. I came back to Bangladesh because this, after all, is where I belong. It is what nourishes me. In the UK, I studied two hundred years of British colonial domination of Bengal; what I came back to was a shambles of a country.

'When I go to my village now, people say, "He is not one of us." They feel I should go there, stay there. Even within this country I have a sense of exile. If I live in London, people say, "He left us." If I live here in Dhaka, people from my village in the north who also live here will say, "Will you do this for me, that for me, will you expose this or that injustice?" If I go to live in the north, people from my village will say, "He was born here, our landscape and our people are in his fiction, yet he doesn't come here to live."

'I don't feel I have a duty to return; it is a question of my existence, my identity. You don't ask to be born. You grow up and you realize you're a member of a given society, a certain culture. You try to fit yourself into that. The acquisition of a cultural identity is more important for me than anything else. You can only express that identity in that culture.'

I met Syed in the Dhaka Club, an oasis of calm in a frantic city, a colonial relic with several acres of lawns, tennis courts, reading rooms. We ate breakfast in a cavernous room, a vast stone chamber with a ceremonial stone laid by Sir Lancelot Hare in 1911. The area had been known as the Shahbhag, but had become neglected and overgrown. The British rehabilitated it in the late nineteenth century, and created a racecourse on what is now the park, where they made a lake which they called the Serpentine.

As we sat in splendid solitude on comfortable leather chairs in the great dining room, an official approached Syed. Placing his hand before his ear, he conducted a theatrical whispered conversation. Later Syed told me he had said that if guests come into the club, shirts with full-length sleeves are essential. I had been wearing a T-shirt, which is considered incorrect dress. I felt I too had become, in a very minor way, a victim of an archaic and defunct colonialism.

But only for an instant. Immediately afterwards my status was restored and I became once more a member of the reconstituted Raj. I was suffering from a subcutaneous skin infection, and a patch of menacing violet was spreading from my foot to my ankle. Syed took me to the nearby hospital, whose medical director was a friend of his wife. The hospital specializes in diabetes, but uses this as a mechanism to trace many other diseases. The medical director has a daughter who lives in Gants Hill, east of

London. The doctor who saw me had also studied in London. To reach his consulting room we passed through a throng of people, some on crutches, some with bandaged and wounded limbs. It did not feel good to pass through a crowd in far worse shape than I was, and to gain admittance to the doctor's consulting room. But neither did I offer to wait my turn. Inside were a poor man whose tongue was covered with sores, and another whose neck was distended and misshapen.

The doctor diagnosed cellulitis and gave me a prescription. I asked him how much, and he said it was complimentary. I felt I had stolen something from this poor, sad, beautiful country; but that, too, is in keeping with its long history of domination and plunder.

THE CHILD, BOMBAY

Maneka was born on a strip of waste ground that separates the *koli* (fishing) community from the soaring blocks of apartments in south Bombay; the place where her family settled was one of the last remaining enclaves of poor people in that part of the city. The huts were mostly of wood and bleached palm leaves, roofs of polythene, held down by stones and old tyres so they should not blow away in the firece monsoon winds.

When Maneka's parents came to Bombay from their native Solapur, they envied the fishing people their livelihood; but Maneka saw the waters of the Arabian Sea turn to poisonous colours of cobalt and sulphur with the effluents that poured into them. She watched the fishing people become poorer, coming back sometimes with strangely diseased sea creatures in their nets.

The only work for Maneka's mother was as a domestic. Her father pulled a cart, drawing loads of building materials on the long rectangular vehicle, running between the shafts he held in his hands.

Maneka's mother worked for a Parsee family in Colaba. She washed and cleaned the house each morning. After that, she worked in a second house. From each, she earned Rs250 a month. She had discovered on the night of her wedding that her husband was a drinker. He neglected the family, and even himself, and soon became too

weak for manual labour. Maneka had one brother, Munnu, three years younger than herself. There was a school in the slum run by Catholic nuns whom the children called 'Auntie'. Maneka enjoyed sitting in the hut the nuns rented for classes. She loved to chalk the letters of the Marathi alphabet, and the metal globe that the teacher sent spinning with one movement of her finger.

When she was twelve Maneka left school. For some months her mother had been losing weight, and she could not do her work properly without her daughter's help. Just going up the four flights of stairs to where her elderly Parsee employer lived left her breathless. She would place one trembling hand on the peeling blue-washed wall to steady herself before ringing the bell. Maneka began to go with her; while her mother cleaned and dusted indoors, the child went to the market for fish, milk, vegetables. She was fascinated by the Parsee woman, the melancholy eyes in a face the colour of wax.

Life became more oppressive at home. In the evenings Maneka took over the housework from her mother, who lay on the bedroll in the corner, her eyes closed. Her daughter made a fire, cooked rice and dal, and occasionally a few vegetables she had withheld from the employer. When her father came in he was always unsteady and smelled of daru. His wife turned her face to the wall and pretended to sleep.

In the Parsee colony Maneka's mother met a woman who worked as a live-in servant. She had a son of twenty-two, and her husband was employed as a maintenance worker on the building. Maneka's mother looked enviously at the one room and kitchen in which the woman lived with her husband and son. She knew she had not long to live, and she thought if she could leave her daughter with a roof over her head this would be her most valuable gift. The place where they were squatting offered no security. All around, more big buildings were coming up, more glass and stone office blocks guarded by uniformed men with rifles.

By this time the monthly income was earned solely by the labour of Maneka, now fourteen. Some days her mother didn't get up at all, but remained in the hut, coughing in the smoke from the cooking fire, sweating under the thin cotton blanket. To see her daughter married before she died became her obsession. A room in a building.

A dowry for which her daughter would thank her for the rest of her life.

One morning, while the people were away at their work, the demolition workers from the municipality came. The men were on construction sites or the docks; the women were domestic workers or in the wholesale fish market. Many older children also worked: in hotels, on stalls, selling cigarette lighters or garlands.

The demolition workers came with police in big blue vans with wire windows. The police stood leaning on their lathis, bulging stomachs hanging over the khaki belts. The workers simply advanced upon the front line of huts, shouted a warning to those inside to come out, and set about the destruction with bars and picks. The belongings of the people were thrown outside the compound. The houses came apart quite easily. The building materials were piled onto the back of waiting trucks. Some of the police poked around in the pathetic piles of possessions and helped themselves to some cooking vessels, a clock, a radio, a little money that fell out of a bedroll.

The only people at home at that time were those too old to work, a girl who was deaf and dumb, the boy who had lost his legs in a railway accident, some children left behind to look after the babies. They could do nothing. By the time they had run to fetch their parents, the work of destruction was finished. Maneka's mother struggled out of the hut. She begged the workers not to destroy their home. One of them looked at her with compassion and said, What can we do. We too are poor people. Into her hand he pressed a 50-rupee note.

Maneka's mother sat huddled under her quilt at the roadside, shivering in spite of the sun and the warm sea wind. She watched as the house she had constructed with her own hands was torn down. All around, the air was thick with the dust that had gathered in the folds of the polythene, and the beaten earth of the slum. Within less than an hour the piece of land that had been a dense maze of tiny narrow streets, where almost a thousand people had lived, was empty. Nothing was left but a few old shoes, some splinters of wood, fragments of broken glass. Some of the men who had been called away from their place of work tried physically to prevent the destruction, but they were arrested and thrown into the back of police vans.

That night, dejected but resolute, the people set up temporary shelters made of bamboo sticks, palm leaves and jute sacks on the edge of the road outside the compound which, by this time, had been cordoned off with barbed wire. The wind of cars and trucks passing by fanned the scarlet embers of cooking fires, made the canvas flap and the dry palm branches rustle.

For Maneka's mother it became more urgent than ever that her daughter should marry. The woman with whom she had worked was not averse to the idea of Maneka as a daughter-in-law. Maneka was submissive and hard-working. Someone in the slum had warned Maneka's mother against Shantibhai. She was said to be cruel and calculating. Maneka's mother could not listen, and heard only jealousy in the warning.

Maneka did not want to marry. She dreamed of returning to school, although for the moment she had to continue working, while her mother grew weaker every day. Shantibhai said her son had eyes for no one but Maneka, with her small hands, slight figure and corn-coloured skin.

One hot June morning, just before the monsoon, while the coppery clouds gathered and the city simmered in a metallic heat, Maneka's mother died on the pavement among the fumes and noise of the traffic. She was carried by her husband and some neighbours to the pyres at Charni Road. Maneka and Munnu walked behind the sad procession, the shrunken body wrapped in cerements and covered with petals of roses and marigolds.

Maneka insisted she was too young to marry. Shantibhai called the *panchayat* (the local community committee) together and the *panchayat*, all men, deemed in favour of the boy's family. Shantibhai made a virtue of accepting a girl with no dowry.

The people who had been cleared were considered an eyesore as they squatted on the edge of the road in front of big buildings where important people, businessmen and tourists, came and went each day, to whom their ragged presence might cause offence. After some months, and a campaign supported by some human rights lawyers and even film stars like Shabana Azmi, they were relocated on a piece of ground 30 kilometres away, a rough rocky piece of land without work, without amenity. Maneka's father and Munnu moved to the remote settlement. 'See how lucky you are', Shantibhai said to her daughter-in-law. 'These people will have to travel in the train to their work, while you can reach it in two or three minutes. You can clean at another house now that you do not have responsibility for your father and brother.'

The room at the top of the Parsee colony was hot and crowded. The stone seemed to absorb all the heat of the sun by day and return it at night. Maneka could not sleep. But when the monsoon came she watched the long steel rods of rain in the road below, and she thanked her mother for providing her with a secure place to live.

Maneka now worked in three houses. She gave all the money she earned to her mother-in-law. But even with the extra work the income seemed insufficient. She was made to feel guilty that she had brought no dowry, and shame that her father was a drunkard and her mother a tuberculosis patient. What kind of marriage gift is that to bring to honest, hard-working people? Shantibhai asked bitterly. Not only that. Maneka was aware that she inspired hostility, if not revulsion, in Suresh, her husband. She could not understand it; she made herself as compliant and pleasant as she could. At the same time she was obscurely aware that her father-in-law had been looking at her with a strange compelling light in his eye.

After three months she was pregnant. Suresh immediately denied that the child was his. Maneka understood that he was accusing her of having allowed his father to usurp his place in the bed.

She could not bear to sleep in the enclosed space of the room. She slept on the landing outside. She continued to work, but now the money was paid directly to Shantibhai by her employers. She had nothing of her own. Her husband would not listen to her. Only the father-in-law remained silent; and no word of blame was directed at him.

One evening when Maneka was four months pregnant, she went from her place of work and took a train from Churchgate to Santa Cruz. She had no money for the fare, and dreaded that she would be arrested by the railway police. She walked from the station, crossed the main highway, and followed the dusty winding road with the new buildings on either side until she came to the site where the people from Colaba had been relocated. Her father had another wife, who nevertheless welcomed Maneka. Munnu was working for a timber merchant near Film City.

Now Maneka travelled with the rest of the people, back to south Bombay. The journey took almost two hours each way. She rose in the steely dawn to get water from the crowded public tap, then took a bus ride to the station and travelled the 40 minutes to Churchgate; then she had the long walk to work. One day, as she was coming out of her place of work she was met by her father-in-law and two *goondas*, who seized her and tried to take her forcibly back with them. Her cries aroused other servants in the houses and they rescued her, beating off the kidnappers with sticks.

Maneka decided she would have the child aborted. The women's compartments of the railway carriages were full of advertisements offering termination of pregnancy for 90 rupees. She told Suresh of her intention. He flew into a rage and hit her. She returned to her father's house and decided that she would keep the child after all, bring it up by herself.

The baby was born in the spring of 1993, a girl. Maneka was transformed by the child. She says she will make sure her daughter never has to work as she has done. Her daughter will go to school, get an education, escape the cycle of work and want.

Maneka herself is tormented by a dry persistent cough. She is very thin. Some days, at work, she is very tired. The baby remains for the moment with her father's second wife, a kindly, compassionate woman.

Maneka gets up at five in the morning and can be seen joining the shadowy figures waiting at the bus stop in the thin dry dust of the colourless mornings, on their way to Bombay Central and Colaba to labour as peons, servants, menials, vendors; the queue is a frieze of servitude, all of them are bearers of silent stories of survival. Maneka says her cough comes from the foul air of the Bombay traffic. It is nothing. When she has provided her baby with everything she needs, then she will save some money for medicine. She is sixteen.

13 SHARON ZUKIN

'Space and Symbols in an Age of Decline'

from Anthony King (ed.) *Re-presenting the City* (1996)

There are two schools of critical thought about the city's built environment. One, identified with political economy, emphasizes investment shifts among different circuits of capital that transfer the ownership and uses of land from one social class to another. Its basic terms are land, labor and capital. The other school of thought, identified with a symbolic economy, focuses on representations of social groups and visual means of excluding or including them in public and private spaces. From this view, the endless negotiation of cultural meanings in built forms – in buildings, streets, parks, interiors – contributes to the construction of social identities. Few urban scholars at this point would defend using only one of these ways of looking at the city. The most productive analyses of cities in recent years are based on interpretations and interpenetrations of culture and power.

But one person's 'text' is another person's shopping center or office building, both a lived reality and a representational space of financial speculation. The ambiguity of urban forms is a source of the city's tension as well as of a struggle for interpretation. To ask 'Whose city?' suggests more than a politics of occupation; it also asks who has a right to inhabit the dominant image of the city. This often relates to real geographical strategies as different social groups battle over access to the center of the city and over symbolic representations in the center. At stake are not only real estate fortunes, but also 'readings' of hostility or flexibility towards those groups that have historically been absent from the city center or whose presence causes problems: women, racial minorities, immigrants, certain types of workers, and homeless people. Occupation, segregation and exclusion on every level are conceptualized in streets and neighborhoods, types of buildings, individual buildings and even parts of buildings. They are institutionalized in zoning laws, architecture and conventions of use. Visual artifacts of material culture and political economy thus reinforce – or comment on – social structure. By making social rules legible, they re-present the city.

The material reproduction of urban society depends on the continual reproduction of space in a fairly concentrated geographical area. Certainly the prime factors have to do with land, labor and capital. Yet the production of space depends in turn on decisions about what should be visible and what should not; concepts of order and disorder; and a strategic interplay between aesthetics and function. It is notable that as cities have developed service economies, they have both propagated and been taken hostage by an aesthetic urge. On the one hand, there is a tendency to take a connoisseur's view of the past, 'reading' the legible practices of cultural discrimination through a reshaping of the city's collective memory. Historic preservation connects an ecology of urban buildings and streets with an ecology of images of the city's past. Historicist post-modern architecture instantly makes the present part of a classical age. On the other hand there is a desire to humanize the future, by viewing artists and art work as symbols of a postindustrial economy. Office buildings are not just monumentalized by height and facades; they are given another embodiment by video artists' screen installations and public concerts. Every well-designed downtown has a mixed-use shopping center and a nearby artists' quarter. The derelict factory district or waterfront has been converted into a marketplace for seasonal produce, cooking equipment, restaurants, art galleries and an aquarium. Economic redevelopment plans have focused on museums, from Lowell, Massachusetts to downtown Los Angeles. Less successful attempts to use museums to stop economic decline in North Adams, Massachusetts and Flint, Michigan only emphasize the influence of cultural strategies in reshaping urban forms. Thus the symbolic economy of cultural meanings and representations implies real economic power.

For at least 100 years, since Engels visited Manchester and Haussmann rebuilt Paris, modern projects of urban renewal have tended to aestheticize the social problems they displace. Since the 1960s, as artists have developed more self-conscious political organization, and allied themselves with elected political officials, they have been co-opted into these projects as actors and beneficiaries, both developers of an aesthetic mode of producing space and investors in a symbolic economy. There are, moreover, special connections between artists and an urban service economy. A concentration of unemployed and underemployed artists represents a labor pool for service employers who want part-time, non-union, 'flexible' and creative employees for jobs ranging from the restaurant and hotel industry to advertising and television production. Artists also represent the claims of a service economy to cultural hegemony. Their visibility in forms of the built environment, in public art, art galleries, museums and studios, emphasizes the moral distance from old, dirty uses of space in a manufacturing economy. In specific cities, the presence of artists documents a claim to these cities' cultural hegemony. The display of art, for public improvement or private gain, represents an abstraction of economic and social power. Among local business elites, those from finance, insurance and real estate are generally big patrons of art museums and firmly committed to showing public art in their commercial spaces, as if to emphasize the prominence of these sectors in the city's symbolic economy.

It was no accident that the boom in these sectors of business services that lasted for most of the 1980s, influenced sharp price rises in the real estate and art markets in which their leading members were so active. Investment in art, for prestige or speculation, represented a collective means of social mobility. At the same time, a collective belief in the growth of the symbolic economy of art represented belief in the growth of the city's economy. Visual representation became a means of financially re-presenting the city.

The symbolic economy thus features two parallel production systems that are crucial to a city's economic growth: the production of space, with its synergy of capital investment and cultural meanings, and the production of symbols, which construct both a currency of commercial exchange and a language of social identity. By the 1990s, it is understood that making a place for art in the city goes along with establishing a place identity for the city as a whole. No matter how restricted the definition of art that is implied, or how few artists are included, or how little the benefits extend to other social groups outside certain segments of the middle class, the visibility and viability of a city's symbolic economy plays an important role in the creation of place.

This is especially important for global cities, those large metropolitan centers where the major share of world financial trade is concentrated. Global cities share with regional and national urban centers a common cultural strategy that imposes a new way of seeing landscape: internationalizing it, abstracting a legible image from the service economy, connecting it to consumption rather than production. But in global cities, the processes of producing space for cultural hegemony are more intense and have greater effect. A vibrant symbolic economy attracts investment capital from the global portfolios of real estate investors, banks, property developers and large property owners. Partly this continues the patterns of the past, with global cities like London and New York transcending the limited life span of colonial and commercial empires. Partly, too, it results from a desire for comparative advantage over a city's rivals. The symbolic economy of a global city shapes the lingua franca of global elites and aids the circulation of images that influence 'climates' of opinion and investment and 'mentalities'. Artists from all over the world (as well as aspiring bohemians who want to live near well-publicized, cutting-edge artists) are attracted to the symbolic economy of global cities. Their presence helps the symbolic economy to continue growing, although whether this economy thrives on the making or the selling of art is a serious question.

The commercial culture of the built environment has specific ways of incorporating cultural capital from the symbolic economy, converting it into visual images, and circulating them to a wider public. Throughout the twentieth century, configurations of built form, glass and light have offered intimations of sacred realms in secular quarters from Times Square to Docklands. The stores, office buildings, and theaters of commercial culture have defined public space for an increasingly mobile public. In commercial spaces the public are simultaneously customers and viewers, spectators and workers, at leisure and on display. The commercial spaces of the largest metropolises have always been gaudiest and most permeated by contradictions. They bear a double burden of representing both a global city and differentiations of power within that city, both a landscape of power and a vernacular.

The economic recession and worldwide retrenchment of the 1990s halted construction of commercial spaces in global cities. Economic conditions also caused the degradation of existing spaces by making owners decrease maintenance services and increase security controls while vacancy rates rose. These changes call for rethinking the relation between commercial culture and public space in the leading urban centers: How does the symbolic economy change public perceptions of commercial spaces? What happens when such spaces no longer represent growth but economic decline? When leading business sectors suffer, what means of representation differentiate landscape from vernacular? Here I can offer no theory of representation that will for ever alter the way we see cities. However, I can present a few preliminary notes on public space and the symbolic economy from the vantage point of New York City in the recession of the 1990s.

SIGNS OF DECLINE

Crises like the present economic depression are a good time to take stock of cities. Because the overbuilding of the mid 1980s has abated, we can look at the forms that are left behind, judging their aesthetics and uses with more critical detachment. At the same time uses of space are changing, as firms economize or go bankrupt, vacancies rise, and the real estate industry pursues alternatives to new commercial construction. We can also look at the basic scales of the city's social life – buildings, streets and neighborhoods – and try to determine which forms, in which places, are likely to survive, and for which groups of people. Not only building and reuse, but also decline and disuse can tell us something about the resonance between symbols, space and social power.

Financial crisis is also a good time for secrets to come out. It is no longer a secret that a small percentage of people in New York City are exceedingly rich and a large number are very poor. Because all sides make use of this claim, it is also clear that the working people of New York cannot support the non-working population, the bureaucracy, the public services, and the tax benefits to powerful business interests that have deadened the city's responsiveness to economic change (for a liberal view, see Epstein, 1992). Over a million residents of New York City are on a welfare program of some kind, more than at any time since the Great Depression. Neither is it mysterious that the city's pursuit of the middle class – or really, 'people with money' – has improved the quality of private goods and services while it has done nothing to halt the reduction of public goods. In these senses New York is just a microcosm of the United States, with a bloated budget for 'defense', an inferior education system, and an inability to invest capital in physical infrastructure.

Conditions for changing infrastructure are more complex today than in the past. After the Second World War, when federal urban development strategies acknowledged long-term middle-class flight to the suburbs, money was poured into demolishing and rebuilding downtown commercial centers. These projects aimed to protect department stores, which had been losing business to the suburbs since the 1920s, and office-building developers, who from the 1950s faced increasing corporate flight to new offices in the suburbs. During the 1970s and 1980s, public-private partnerships leveraged a dramatically diminished base of federal funding to finance a new round of downtown commercial construction, often with mixed commercial uses. Mainly through the commercial reuse of historic buildings, this spatial redevelopment harnessed an old urban vernacular – wholesale food markets, ports, railroad terminals – to a new sensual appreciation of aesthetic diversity and a related appropriation of nature. Many of these projects resulted in similar festival marketplaces. Most, though not all, relied on a predominantly corporate-owned mix of businesses to minimize financial risk. In general this paradigmatic change in the program of downtown commercial space indicated that the middle class was more diverse than ever before, the investment climate was more competitive, especially within metropolitan regions, and planning was no longer controlled by public agencies.

Abandoned to market competition for an image-conscious public, urban infrastructure focuses on speciality stores, art and food. By this time image is so grounded in the symbolic economy that it serves as an agent of transition, representing potential uses of otherwise vacant space, saving it for another 'landscape'. Nothing could illustrate transition of this sort more visibly than 'Harry, If I Told You, Would You Know?', a display in a vacant bank branch at 58th Street and Madison Avenue, at the core of a high-rent commercial district in midtown.

Harry is the name of a set of contemporary abstract paintings by Al Held that have been exhibited since September 1991 in vacant ground-floor retail space. This empty storefront is across the street from the General Motors Building, a block away from the IBM and Sony (formerly AT&T) buildings, and around the corner from a cluster of art galleries on East and West 57th Street that represent well-known artists. The André Emmerich Gallery, which represents Al Held, is on 57th Street. Emmerich got the idea of renting the storefront to show the paintings, which don't fit into the elevator to his gallery. In the old days – the 1970s and 1980s art market – the gallery might have shown these paintings at their SoHo branch, which was opened in 1971 in a loft building configured for wide loads. But that gallery was closed some years ago, and the storefront on Madison Avenue was vacated by the 1991 consolidation of Chemical and Manufacturers' Hanover banks. The sight of the storefront, denuded of tellers' safety glass windows, with makeshift lighting on bare cement floors and thick columns, either evokes the success of an uptown SoHo, with a

potential for upgrading property through cultural uses, or suggests a scenario of doom, with unrented commercial spaces used only for symbolic display.

Since about 10 per cent of the ground floor spaces in this neighborhood of Madison Avenue are now unrented, doom it is. *Harry* signals a cross between the derelict urban spaces of *Blade Runner* and gentrification by cultural consumption. But art work could hardly fill all the vacant space in New York's commercial real estate markets. These markets include not only the vacant flagship department stores that have succumbed to falling sales and leveraged buyouts (Macy's, Bloomingdale's and Saks remain, but Altman's and Alexander's are empty), and the one-storey stores of various sizes that have been subdivided into 'indoor malls', but also unrented shops and offices in large mixed-use complexes – like the World Financial Center and Worldwide Plaza – that express a world city's great pretensions. Could 'Harry, If I Told You, Would You Know?' represent the overexpanded, debt-ridden city of the developers of the symbolic economy?

Look what has happened to Olympia & York. This Canadian firm of real estate developers and managers was responsible for some of the most significant new construction in the symbolic economy in the 1980s, including the World Financial Center in New York and Canary Wharf in London's Docklands. Traditionally cautious in diversifying and hedging investments, developing a blue-chip portfolio, engaging signature architects, owning desirable buildings outright, and negotiating favorable deals with government agencies, Olympia & York nonetheless ran out of cash to pay its lenders. Neither could it pay its agreed share of the costs of extending a rail line between the City of London and Canary Wharf, a necessary condition of getting the workforce to their jobs. It is not irrelevant to the overall circularity of the symbolic economy – '. . . If I Told You, Would You Know?' – that while the company's property leases stopped producing enough income to cover debt repayment, Olympia & York's cash-producing investments in oil, gas and newsprint also lost economic value. Moreover the company couldn't sell off two buildings at Canary Wharf as it had planned to do, when first a Japanese pension fund refused to loan money to Morgan Stanley & Co. to buy an office tower, and second the British government rescinded tax breaks for setting up investment trusts so that individual investors could buy another building. With the still incomplete Canary Wharf passing into the reluctant hands of lending banks, the symbolic value of Olympia & York's architectural projects is no greater than their value in the stock and real estate markets.

The purchase by the German media company Bertelsmann of an office building in Times Square is another story of big bank lenders, developers who couldn't hold on to an empty building, delays, and uncertainty surrounding the remaking of a landscape of power. Even when part of the symbolic economy – in this case publishing and music distribution – continues to flourish under foreign ownership, real estate markets operate at a loss.

But these are current events. Considering space, symbols, and power demands both a short and a long-term view. In the long run vacant and undervalued space is bound to recede into the vernacular landscape of the powerless and be replaced by a new landscape of power. Not so long ago the image and economic value of urban factories were diminished by the rise of the service economy and rose again – in the spatial form of living lofts. The construction starts and cultural critiques that focused attention on cities' polarized growth during the 1980s only reflect four conditions that have shaped New York City's decline from the 1950s. These are the flight of the middle class, the weakening of place identities of major businesses, the standardization of consumption experiences, and, as the countervailing strategy of patrician elites, the reassertion of centrality as a landscape of power. These conditions define the legibility of urban space.

LEGIBLE SPACES

Legibility and identity are interdependent. Spaces are formed by capital investment and sensual attachment; both who pays for building and rebuilding and the gut feeling of being in and of a specific city. Legibility also speaks to the greed and exclusion that underlie perennial plans to rid a downtown of 'dirty' manufacturing, low-rent tenants, and all infrastructure connected to the poor, workers, and ethnic and racial minorities outside of tourist zones. Nearly all cities use spatial strategies to separate, segregate and isolate the Other, inscribing the legible practices of modernism in urban form. But the legibility of New York City is an exercise in contradictions. On the one hand tendencies to segregate are thwarted by a diversity greater than in other cities: the block-to-block variations in uses and social classes across Manhattan, the uneasy coexistence of Latinos, Chasidic Jews, Italians and East Europeans in Williamsburg, of Caribbean immigrants, American blacks and Chasidic Jews in Crown Heights. On the other hand, while projects to segregate high value uses are often defeated by 'community' resistance, they result in establishing markets that undermine the integrity of place: the legalization of artists' lofts in SoHo, the potential for another high-income neighborhood in the mixed-use development on the old West Side railroad yards sponsored by real estate developer Donald Trump and a coalition of community groups. The basic processes of incorporating the unique identity of place into real estate markets is always taken farthest in New York because of the visibility and viability of its symbolic economy.

In Henri Lefebvre's framework, New York is an example of abstract space: simultaneously homogenous and fragmented, subordinated to the flows and networks of world markets, and divided into units of exchange by real estate developers. Viewed from the city, however, New York resembles a dismembered imperial space, an imperium whose tributaries owe less and less to the center, whose facilities are improved upon in the provinces, and whose utopian joining of freedom and power looks like a dystopia of dirt, violence and anarchy. Simultaneously utopia and dystopia, New York claims a place in America's and the world's moral economy. Does it exert this claim because it is changing, or has it always been this way?

Much of the change in the legibility of New York reflects the greater choice that people have about where to live, and the influence of corporate planning on both cities and suburban or exurban alternatives. Squeezed by the decentralization of mass individual choice and corporate planning, design in New York, as in other cities, has become a more important mark of distinction. Design emphasizes legibility. It enables cities to compete for 'people with money': a representation of a tax base, a class of employers, and consumers for private markets. The new importance of people with money is intimately related to the disappearance of urban planning in New York since the 1960s. Public officials were caught without answers to a series of questions. Having chased industry out of the city, how can the city government increase employment? Can the city survive with only a symbolic economy? How do you plan a post-industrial city with a non-European population? How do you plan for large businesses when telecommunications make it sensible for them to create their own 'footprint' in the suburbs and exurbs? How do you plan public goods for a low-income or even homeless and unemployed population when the rich won't pay for them and no one else can? Under these conditions it is no surprise that the landscape of cities has been reorganized for visual consumption, abstracting an image of freedom and power that commands – in its very abstraction – some degree of consensus.

Since the 1970s only a handful of public spaces in US cities have been conceived and built as truly public – for neither profit nor market-based consumption, for association rather than individualism, for spending time rather than spending money. Certainly notable public buildings have been, and are still being, built: Michael Graves's Portlandia Building and Phoenix Civic Center, the Harold Washington Library in Chicago. But the State of Illinois Center in Chicago, designed by Helmut Jahn in the early 1980s, signaled a decisive change in the definition of public space. In that mixed-use complex, ground floor spaces are occupied by stores while government agencies occupy the higher floors, accessible by a glass-enclosed elevator. The identity of government is subordinated to the shopping center. If a city's legibility is derived from the design of commercial spaces, then the identity of the urban public is negotiated in those commercial public spaces where ownership, work and consumption interact.

To see how space mediates these issues we should look at commercial spaces on three scales typical of New York City, that span landscapes of power and vernacular. These are the office-building lobby, the multi-block commercial complex, and the neighborhood shopping street. To begin with, New Yorkers spend a great deal of time in all three spaces. Moreover these urban forms are still uniquely New York spaces because you don't need a car to get to them, see them and use them.

a. The office-building lobby

Office-building lobbies are interiors for a public in transit. Neither so public as a street, nor so private as an office, office-building lobbies are passageways, waiting rooms and theaters of corporate image. In some ways they recall public spaces built for a newly mobile urban public in the late nineteenth century. Like the comfort stations installed in British underground railway stations in the 1890s and railroad terminals like Grand Central after 1900, office-building lobbies between 1910 and 1970 had shoeshine stands, barbershops, telephones and news-stands. Perhaps the apotheosis of this urban form is in great office building lobbies of the 1920s and 1930s like Rockefeller Center. But the lobby of the Equitable Life Assurance Company head-quarters, built a few blocks north of Rockefeller Center in the 1980s, is quite different. The comfort station – that in this case includes a shoeshine stand, xerox store, news-stand, hair salon, fitness center, McDonald's and a computer training school – has been relegated to a subterranean passageway or concourse leading to the subways. The concourse connects the Equitable headquarters to other corporate buildings in the area and to Rockefeller Center. In the lobby of the Equitable building, you see an art gallery run by the Whitney Museum, open exhibition areas for art work, sculptures commissioned by the building's owner, very large paintings (like *Harry*) that fit the lobby's scale, all surrounded by glass and marble. Outside, separate street entrances lead to two elegant restaurants, Le Bernardin and Palio, which lease space on the lobby floor but are not visible from inside the lobby. Although the Whitney also runs two gift stores in the lobby, the Scriptorium and the Treasury, Equitable did not renew the museum's lease when it ran out in 1992. The company planned to organize its own art exhibits.

Walking through this lobby, you feel you are

in a public space that is at once open but not anarchic, a non-profit space of cultural production that is dependent in some indefinable way on corporate finance. This perception continues as you walk through to the lobby of the adjoining office building, named for the investment firm PaineWebber, its prime tenant. This lobby contains a PaineWebber art gallery, which mounts exhibits in cooperation with local museums, and a separate gallery space occupied by the archives of the Smithsonian Institution. Together the two lobbies contain a variety of small spaces that humanize their high ceilings and glass walls. They seem to break down hierarchies of outside and inside, corporate and cultural, public and private. They create the sense of a street in their design and of a public good in their function. Indeed zoning incentives have been a major influence on the design of office-building lobbies as public spaces.

In the center of the Equitable lobby is a very tall grey marble bench. It is so tall that passing behind it, you can only see the top of the head of anyone sitting on it. The bench curves inward behind tall plants, providing a somewhat secluded sitting area. At 12.45 p.m. one recent day, two homeless men sat on this bench. They were surrounded by secretaries, a computer repairman, and several Indians, Chinese and Mexicans who were not building employees. Several businessmen in suits sat down for a few minutes. Aside from five uniformed security guards who checked people going into the elevators, these were the only stationary users of the lobby. Two non-rotating surveillance cameras are mounted above one of the entrances, one directed at the lobby and one at the door.

The Philip Morris headquarters building on Park Avenue also has a new type of lobby. This lobby also has a branch of the Whitney Museum. It is much smaller than the Equitable lobby and is set up with 12 round tables. At 11.45 a.m. one day, nine of the tables were occupied. Four men of various ages read newspapers, two older middle-class men ate lunch bought in the lobby cafe, two white-collar workers ate bag lunches together, an Asian mailcarrier ate a Chinese take-out lunch, two women who could have been students had spread books and papers all over the table and were studying. Most of these people were white and alone. On the benches sat two young people who could have been students, and several elderly men read newspapers. A sign posted in the middle of the front of the lobby prohibits sleeping, alcohol consumption, inappropriate attire, excessive packages, touching the art objects, distribution of leaflets and 'conduct which is inconsistent with a sculpture gallery'. A Whitney Museum employee sits outside the door of the gallery. Nearby are an espresso bar, a news-stand, an arts gift-shop and a chocolate shop. A piano for free public performances stands in front of one of the stores. Four pieces of sculpture are installed on the floor and another piece is mounted on a wall. Three uniformed security guards stand in the lobby, and rotating surveillance cameras overlook the lobby from opposite walls.

b. The commercial complex

During the last two decades the locus of the symbolic economy moved downtown. The crowded core of the financial district in Wall Street was expanded onto landfill in the Hudson River, capping a 25-year plan to revitalise the downtown commercial real estate market. Yet Battery Park City, the new addition managed by Olympia & York that parallels and rivals London's Docklands, is a curious product of historical compromises. Morphologically it extends the grid pattern of Manhattan's streets to the west, but its wide avenues and openness to the sea contrast with the narrow streets that are the only vestige of New York's colonial trading center. Architecturally it adds to the office space materialized by late Governor Nelson Rockefeller in the World Financial Center, but it contradicts the severe verticality of those modernistic towers by a variety of post-modern designs, with mansard roofs, terracotta colored facades and a glass-enclosed atrium with palm trees, a Winter Garden on the river designed by the architect Cesar Pelli, which the *New York Times*'s architecture critic has described as New York's living room. The mixed uses that were envisioned as enlivening community and supporting middle-class residents when Battery Park City was first planned, in the 1960s, were refashioned to suit an upmarket tenancy oriented towards financial firms and business visitors. Similarly the mixture of social classes that was initially planned – as part of the same new urban community – was shifted off the site itself. A linkage agreement between the city and the developers proposed to transfer a percentage of funds derived from bond sales to the construction of new low-income housing in the outer boroughs, where most poor people already live. By the 1990s it was discovered that none of this money had actually been delivered. The city government was also discussing how to divest itself of its interest in the property to generate a one-time infusion of funds.

As the managers of Battery Park City, Olympia & York organize a constant array of cultural events that are open to the public. From the beginning of these programs in 1987 to the end of 1991, attendance was estimated at over 700 000 people, with another 232 000 at marketing or community events and 29 000 at private events, when the building is rented for dinners or other occasions. The overwhelming attendance is at the free cultural events, which are subsidized by the World Financial Center itself. These events are chosen to appeal to a wide audience. In a typical three-month period (January–March 1992), they included separate ballroom and modern-dance performances, an environmental film festival and tropical rainforest sound installation (which appears to be the *sine qua non* of any 1990s cultural complex), and pop vocal concerts. On sunny weekend afternoons the public space along the private marina on the river attracts a large, well-mannered crowd of men and women and families of various ethnicities. A new playground, open to the public, is very well designed for children of different ages and features a Swedish-made merry-go-round powered by children on bicycles. The Battery Park City security force patrols the entrances to the playground and riverside esplanade. There is an indoor shopping center with an art bookstore, clothing stores and restaurants catering to a knowledgeable group of diners.

This is much the same story as for the Equitable and Philip Morris lobbies. Through a mixture of zoning incentives and a desire to build friendly faces of corporate power, commercial developments cloak corporate identity in consumer marketplaces, large gathering spaces and signature architecture. Prices in the retail stores are often high, and there are always conflicts over people –

especially homeless people – using the gathering spaces. If the building provides places to sit, they are not too comfortable, and the architects' work is frequently interchangeable with what they have built in other cities. Private security forces, consumption spaces, false gathering places, signature architectural styles: this is what creates the new urban legibility. But what urban identity does it shape? New Yorkers who admire the public spaces in the World Financial Center and Battery Park City say, 'It doesn't look at all like New York'.

New York has few centers that are not commercial attractions. At the beginning of the twentieth century, Times Square marked a shift in symbolic public spaces from civic arenas or forums, and marketplaces or agoras, to commercial projects. As Times Square declined – a slow decline from the Depression, when many legitimate theaters were shuttered and others replaced by vaudeville theaters, burlesque houses and other tawdry amusements, to the massage parlors and porno-theaters of the 1970s – it was displaced by other symbolic centers. Rockefeller Center, a planned and highly regulated small city without residences, anchored the office market in midtown from the 1930s. As the secular counterpart of St. Patrick's Cathedral across the street, it also demarcated the symbolic processional space of Fifth Avenue, much used for annual ethnic, religious, and patriotic parades.

During the 1970s upper Fifth Avenue became another symbolic center, represented by a new street name, 'Museum Mile'. Under the directorship of Thomas Hoving, the Metropolitan Museum presented blockbuster art exhibits that broke attendance records. The museum also expanded into the public space of Central Park by building new wings, including that which sheltered the Temple of Dendur, transported from Egypt. From the Guggenheim Museum to the Metropolitan, the public space of the Fifth Avenue side of Central Park became an adjunct of the private museums.

Central Park as a whole has a similar history. From the 1860s to the 1880s, as the definition of the public expanded to include working-class and immigrant groups, the public space of Central Park became more inclusive, less culturally hegemonic and more commercial. Probably the park changed more in those first 20 years than in the following century. Nevertheless public uses of Central Park expanded under the populist first administration of Major John Lindsay, in the mid to late 1960s, when Thomas Hoving and August Heckscher served as park commissioners. Central Park became a symbolic staging ground for various 'unifying' events, both paying and free, from rock concerts to protest demonstrations. Central Park is also the site of the start of the New York City Marathon. Similarly the large Prospect Park in central Brooklyn, also designed by Frederick Olmstead and Calvert Vaux, has become a symbolic center too. An annual Caribbean Day parade on Labor Day weekend marches to Prospect Park.

The common point of these symbolic spaces in that they blend public and private uses, and commercial and non-commercial functions. While they are structured by public or governmental incentives to include public uses in the sense of free events, they are also shaped by the fact that people are more active as consumers than as citizens. People are also sometimes afraid of gathering in their neighborhoods. While public space in centrally-located, commercial complexes has increased, and is policed by security guards, people feel that usable public space in ordinary neighborhoods – safe streets and parks – has decreased. Neighborhood public space is dirty or unsafe, or outside effective public controls.

The satisfaction of private needs increasingly drives the construction of significant spaces of public life. This displaces the locus of emotional attachment in the city from the home and the local community to the central commercial complex. It reduces support for commercial spaces that cater to local consumption needs, and attracts support to larger, more diversified but ultimately more standardized spaces of consumption. It drives consumption spaces to incorporate symbols of public fantasy, especially fantasies of public life.

c. Shopping streets

While an increasing number of central commercial spaces are controlled by private security staffs, neighborhood shopping streets are considered unmanageable. The city government no longer guarantees sufficient street-cleaning or sanitation pickups. Private carters are said to be connected to the Mafia. Especially in ghetto areas, merchants fear daily theft and occasional community opposition. Ethnicity is both promoted and reviled in neighborhood shopping streets, which can equally become symbolic centers of solidarity or resistance. Yet neighborhood shopping streets are the site of vernacular landscape. Sometimes local merchants represent the vernacular of the powerless against the corporate interests of chain stores and national franchises. The local real estate sector typically includes small landlords and developers. The periodic struggle in New York over commercial rent control – which the city lacks – is a struggle over maintaining the vernacular in a populist sense: small-scale, local ownership, amenities at prices local people can afford, variety. The transformation of shopping streets from vernacular diversity to corporate monoculture is also a reflection of the global and national economies. But as a coherent image of public space, the basis of both a cognitive map and social reproduction, neighborhood shopping streets suggest alternatives to the symbolic public spaces of the office-building lobby and commercial complex.

That shopping streets have also been incorporated into a landscape of power is shown by the rapid growth of business improvement districts (BIDs), which enjoy a special legal status in New York State. These are associations of local business owners who have the legal right to tax themselves over and above city taxes in order to provide their own 'improvements', including street lighting, sanitation, holiday decorations, and street festivals. They may hire their own security force. Local merchant associations but with more clout, the BIDs represent the privatized public space of the current period of public-private partnerships. They compensate for the budget restraints on city government. However by taking responsibility for providing public goods, they also take the opportunity to define them. Their control makes shopping streets into a liminal zone neither public nor private. Moreover the biggest and richest BIDs are the ones with the most ambitious plans for re-making their neighborhood.

Because the 34th Street BID and the Grand Central Partnership (on 42nd Street) represent some valuable commercial properties, including the Empire State Building and Grand Central

Terminal, their projects to improve the neighborhood dovetail with those of major property owners. The Grand Central Partnership has assumed responsibilities for the redevelopment of such public spaces as Grand Central Terminal (including lighting the exterior) and the reconstruction of Bryant Park behind the Public Library on 42nd Street. Both have been contentious projects, dragging out over years of city government approvals and requests for funding, and the pursuit of private entrepreneurs to operate public amenities. They share a problem of providing access, especially to the homeless, while sustaining an image of the city that attracts the local workforce and tourists. The homeless used to be swept up by periodic police patrols through Grand Central Terminal; now the doors are closed between late night and early morning. In Bryant Park, a neatly printed sign informs us, only homeless people affiliated with a church in the neighborhood have the right to pick among the trash cans. (Collecting the deposits on discarded beer and soft drink cans and bottles is a major source of income for some homeless people.) The Grand Central Partnership established a subsidiary to manage the redevelopment of the park, which oversees landscaping for a privately-owned restaurant, leases to concession stand operators, programs of public concerts and performances, and the general public order. The new park is less likely to harbor the drug sales for which the previous park became notorious.

It is expensive to manage construction projects, field sanitation crews, and to provide trash cans, benches and other 'street furniture' and new signage. Moreover the group has a mission and a vision. They want to make the midtown area look and work like Disney World. For this aim, the Grand Central Partnership's annual budget of $6 million is not large enough. With the mayor's approval the private group received authorization to issue bonds for up to $35 million (*New York Times*, 2 April, 1992). And the Partnership's credit rating is higher than the city's.

Outside of midtown, shopping streets also represent, by ownership and employment patterns, the social and economic integration of neighborhood residents. But here, in contrast to midtown and downtown, a tension between commercial legibility and community identity creates a highly contested representation that

sparks periodic violence. The demonstrations and boycotts on 125th Street in Harlem in the 1920s to protest lack of jobs for neighborhood residents in local stores were early confirmations of the power of this representation. In the last three years, demonstrations and boycotts on Church Avenue in Brooklyn, this time against Korean rather than Jewish storeowners, have shown that the representation still has symbolic force. Even without demonstrations, neighborhood resentment against storeowners is usually high when, as on Pitkin Avenue in Brooklyn, the owners and employees are Indian or Caribbean and most residents are native-born African-Americans. Where storeowners do represent the residential community, shopping streets do not become symbols of decline. By comparison, in the Los Angeles insurrection of 1992, people attacked stores rather than housing, and neither the supermarkets financed by community development corporations nor the 31 McDonald's restaurants in the South Central area were harmed.

In the real estate recession of the early 1990s, major shopping streets in New York's immigrant neighborhoods have not suffered from more vacant storefronts than in the high-rent districts of Manhattan. On Flatbush Avenue between Church Avenue and Hawthorne Avenue, in Brooklyn, 11 per cent of 191 stores are now closed, empty or for rent. Storeowners are Haitian, Chinese and Greek; customers are mainly Caribbean blacks, with some Latinos, African-Americans and Chinese. On a recent Saturday there were no white shoppers or walkers. On Main Street in Flushing, Queens, from Highway to Stanford Streets, 12 per cent of 149 stores were closed, empty or for rent. Most owners here are Chinese, but there are also many corporate franchises. On a recent Saturday mainly Asians were on the street, with some Latinos, Indians, African-Americans and whites. Vacancy rates are similar on Madison Avenue in Manhattan, where storeowners and customers are predominantly white and often European. In 174 stores between 80th and 96th Streets, the vacancy rate is almost 10 per cent. Vacancies decline to 6 per cent in the 90s, which have more neighborhood services, but in the 70s, the vacancy rate is 15 per cent. Between 57th and 69th Streets, where there is a high concentration of European designer clothing and luxury-goods stores, as well as 'Harry, If I Told You, Would You Know?', the vacancy rate of just under 10 per cent is about the same as on Flatbush Avenue.

While these vacancy rates do not indicate a good situation, they compare favourably with vacancy rates in office buildings, which were as high as 16 per cent in Class A buildings in midtown in 1992. This suggests a contradiction between the public space constructed by two symbolic economies, the global and the local. The legibility of central commercial complexes and office-building lobbies differs from the identity negotiated in neighborhood shopping streets. Despite their problems, these streets produce the quality of life that New Yorkers prize, the public space that makes neighborhoods livable, and attaches people to place. A period of economic decline provides an opportunity to think creatively about these representations. When recession makes large projects prohibitively expensive, it is time to bring culture into public space in a different way.

SECTION III

THE CULTURE INDUSTRIES

INTRODUCTION

The culture industries include all forms of cultural production, from art and literature to mass media such as film, television and popular music. The importance of the culture industries, in a collection of writing on urban cultures, is threefold: first, most culture is produced and consumed in cities, and much of it references the city directly or indirectly. Second, ideas and memories of cities are mediated through the products of the culture industries; so, for instance, the depiction of Los Angeles as site of an imminent collapse in social values in the film *Strange Days* (Kathryn Bigelow, 1995, USA) conditions how we think of such mega-cities generally. Third, the culture industries are often in the vanguard of urban development. This includes the use of redundant industrial buildings for cultural purposes (such as the power station converted to become the Tate Gallery at Bankside, London), and the redefinition of whole quarters as cultural districts, such as Temple Bar in Dublin.

Research in the 1980s (for instance by John Myerscough for the Policy Studies Institute in London, published as *The Economic Importance of the Arts in Britain*, 1988, PSI) demonstrated the contribution of the culture industries to invisible earnings and the financial economies of cities such as London. And Sharon Zukin (in Section II) shows how the arts contribute to the symbolic economy of New York.

The centrality of the culture industries in mediating perceptions of cities means that any effort to change the future directions of urban development, for example towards a greater degree of sustainability or social justice, requires their interrogation. Culture is produced in specific historical conditions, and affirms or resists specific structures of power and value. In a capitalist society, the culture industries play a key role in the generation of surplus wealth, and in maintaining the equilibrium (however false) of the operations of capital. That is, they both contribute to and affirm the effects of market forces on everyday life. As Theodor Adorno and Max Horkheimer write in the extract from *Dialectic of Enlightenment* which opens this section: 'The whole world is made to pass through the filter of the culture industry' (1944 [1997], Verso).

Adorno and Horkheimer raise the problem that, as they put it, 'real life is becoming indistinguishable from the movies'. This introduces three difficulties: first, that a fictionalization of the world obscures contradictions such as that between bourgeois society's proclamation of liberty for all and its mechanisms of exchange which limit liberty to the owners of capital. Second, that the dominance of mass culture, acting like an opiate, restricts the ability of people to imagine a society other than that which is. This prefigures arguments made by Adorno and Herbert Marcuse elsewhere that the role of art in society is to rupture the surfaces of the dominant reality. And, third, that as the culture industries increasingly permeate everyday life, 'real life' becomes the movies and people's lives are lived out through commercialized cultural representations.

A particular example of another difficulty is presented by Patricia Phillips in her seminal article for *Artforum*: 'Out of Order: The Public Art Machine'. Written in the late 1980s, a time of rapid expansion in the commissioning of site-specific sculpture for urban spaces – when advocates and agencies were proclaiming the competitive edge offered by art in development schemes – Phillips' essay pricks the bubble. She argues that public art has little to do with democracy, the expression of which is a key function of public space, and that a definition of publicness by physical site is too restrictive. In place of that, Phillips sees a psychological domain of public issues and values. The meanings attached to spaces by different publics demonstrate the diversity of society and contrast with the narrow focus of art when it colonizes public space as aesthetic space.

Public monuments have, since the erection of Roman triumphal arches, imposed certain readings on public spaces. Some, like the Statue of Liberty or the Eiffel Tower, become signs for the cities in which they are sited. The Statue of Liberty acts also as a sign for a dominant idea of democracy. By an irony of history it was unveiled in the same year in which the first laws were passed to exclude certain categories, such as Chinese labourers, from entry to the land of the free. A notable case of public monuments used to reconstruct history is given by Samir al-Khalil in *The Monument* (1991, University of California Press). In this book, he draws attention to images of the Iraqi President backed by the remains of Babylon, to give a sense of the inevitability of history. The extract included here concerns the architectural redefinition of Baghdad as a modern world-class city, in which public monuments, such as one scaled up from a cast of the President's own arms, colour the city's meaning.

If Baghdad is like a political theme park, heritage districts may be no less political, though their message of consumption obscures it. Yet, whilst the past is turned into a commodity by heritage districts, such as New York's South Street Seaport, or Fisherman's Wharf in San Francisco, there is another aspect to heritage in museums of everyday life and the histories of toil and struggle. Raphael Samuel, in 'Theme Parks – Why Not?' (1995), celebrates the history of ordinary and suburban things and the role of collectors and enthusiasts in making the archives to which tomorrow's researchers will go for information and ideas.

Dolores Hayden takes a version of this argument further in her account of the work of The Power of Place, a multi-disciplinary team working to make monuments from the memories of place of members of minority communities in Los Angeles. The team's work spans lobbying for the designation of historical districts to limit destructive development, and the creation of site-specific visual narratives. In a way, this is a resistance to the dominant city and its narrative of money and corporate power.

An overview of urban policies for cultural development is provided, finally, by Joost Smiers in a chapter from *Rough Weather* (previously available only in Dutch and French). Smiers contrasts the Parisian emphasis on large projects with the more integrated approach characteristic of Barcelona. He notes the scope for cultural policies that promote events such as festivals, rather than monuments – a policy adopted in Budapest. Another case covered by Smiers is the *Estate Romana*, a series of summer projects in Rome between 1976 and 1985 which sought a democratization of culture throughout the city. Smiers cautions against assumptions that cultural marketing can turn a city without a sufficient critical mass of creativity into a cultural capital, and warns of the danger that neighbourhoods into which cultural producers move, attracted by low living costs, can thereby become gentrified, defeating their purpose and aiding the dominant form of city development as art which might once have been resistant becomes commodified.

14 THEODOR W. ADORNO AND MAX HORKHEIMER

'The Culture Industry: Enlightenment as Mass Deception'

from *Dialectic of Enlightenment* (1944)

[. . .]

The whole world is made to pass through the filter of the culture industry. The old experience of the movie-goer, who sees the world outside as an extension of the film he has just left (because the latter is intent upon reproducing the world of everyday perceptions), is now the producer's guideline. The more intensely and flawlessly his techniques duplicate empirical objects, the easier it is today for the illusion to prevail that the outside world is the straightforward continuation of that presented on the screen. This purpose has been furthered by mechanical reproduction since the lightning takeover by the sound film.

Real life is becoming indistinguishable from the movies. The sound film, far surpassing the theater of illusion, leaves no room for imagination or reflection on the part of the audience, who is unable to respond within the structure of the film, yet deviate from its precise detail without losing the thread of the story; hence the film forces its victims to equate it directly with reality. The stunting of the mass-media consumer's powers of imagination and spontaneity does not have to be traced back to any psychological mechanisms; he must ascribe the loss of those attributes to the objective nature of the products themselves, especially to the most characteristic of them, the sound film. They are so designed that quickness, powers of observation, and experience are undeniably needed to apprehend them at all; yet sustained thought is out of the question if the spectator is not to miss the relentless rush of facts. Even though the effort required for his response is semi-automatic, no scope is left for the imagination. Those who are so absorbed by the world of the movie – by its images, gestures, and words – that they are unable to supply what really makes it a world, do not have to dwell on particular points of its mechanics during a screening. All the other films and products of the entertainment industry which they have seen have taught them what to expect; they react automatically. The might of industrial society is lodged in men's minds. The entertainments manufacturers know that their products will be consumed with alertness even when the customer is distraught, for each of them is a model of the huge economic machinery which has always sustained the masses, whether at work or at leisure – which is akin to work. From every sound film and every broadcast program the social effect can be inferred which is exclusive to none but is shared by all alike. The culture industry as a whole has molded men as a type unfailingly reproduced in every product. All the agents of this process, from the producer to the women's clubs, take good care that the simple reproduction of this mental state is not nuanced or extended in any way.

15 PATRICIA PHILLIPS

'Out of Order: The Public Art Machine'

from *Artforum* (1988)

There's good reason to be wary these days when the signs of another specialization start emerging – when one small point is established at the sacrifice of the wide horizon. Contemporary society has become remarkably undisciplined in the ways that it spontaneously endorses new disciplines in almost unimaginable areas of expertise. Those involved in the art world are well accustomed to the coalescences and lightning-like dissipations of style, but a new speciality is not a common notion. In the past 25 years, traditional distinctions between sculpture, painting, drawing, photography, and installation – as well as the idea of art and architecture as independent, exclusive phenomena – have eroded, causing fused and hybrid forms and unusual intersections. And conceptual catholicity, openness, and negotiable categorization have provided the groundwork for the galvanization of a new art: the now very active and hierarchically complex world of public art. Within this arena, there are many players and many productions, some enlightened ideas and little criticism.

Public art – as it is normally understood and encountered today – is a nascent, and perhaps naive, idea. It bears so little resemblance to earlier manifestations – especially the most immediate precedent of civic, elegiac art of the 19th and early 20th centuries – that the idea of a historical progression of uninterrupted continuity seems spurious; there are few instructive models. And so, though public art in the late 20th century has emerged as a full-blown discipline, it is a field without clear definitions, without a constructive theory, and without coherent objectives. When the intentions have been apparent they are usually so modest (amenity) or so obvious (embellishment or camouflage) that

they seem to have little to do with art at all. In short, the making of public art has become a profession, whose practitioners are in the business of beautifying, or enlivening, or entertaining the citizens of, modern American and European cities. In effect, public art's mission has been reduced to making people feel good – about themselves and where they live. This may be an acceptable, and it certainly is an agreeable, intention, but it is a profoundly unambitious and often reactionary one. And even these small goals are infrequently satisfied; public art doesn't generally please or placate, or provide any insistent stimulation. Instead, public art today, for the most part, *occupies*. And just at the moment when so much apparatus has been assembled and oiled that might aid in the development of a rigorous critical foundation for public art, there is a growing feeling of – well, why bother? Indeed, an enterprise that emerged with such idealism now feels like a lost opportunity.

Yet many artists, art administrators, and bureaucrats worked hard to promote the current proliferation and professionalization of public art, and did so with the noblest of intentions. Some reflection on the past indicates that those involved had good reason to lobby for "official" policy and protections. For art that appears beyond the configurations and machinations of the gallery and museum encounters different forces and greater risks, and thus should be provided, they believed, with some fundamental assurances and safeguards – for the sake of the artist, as well as the community. And given the very real need for relief from, or challenge to, the loud monotony of the urban landscape, state and federal guidelines for "percent for art" programs were initiated; standards and criteria for selection and review drafted; and bureaucratic procedures codified. But this clarification of operations has ultimately led to a "minimum basic standard" mentality. Not unlike American housing reform in the late 19th century – which was not based on constructive legislation for a sound life, but on the absolute lowest standards of acceptability – the public art "machine" now often encourages mediocrity. To weave one's way through its labyrinthine network of proposal submissions to appropriate agencies, filings and refilings of budget estimates, presentations to juries, and negotiations with government or corporate sponsors, requires a variety of skills that are frequently antithetical to the production of a potent work of art. If the "machine" itself can be put to use as a conduit, rather than as a molder of the art that emerges, then there is still the potential for transforming methodology and materials into positive energy. But more often the result of this process has been what Gordon Matta-Clark, James Wines, and others have referred to as "the turd in the plaza."

Public art operates on a practical as well as a philosophical level, but the contemporary preoccupation has been with the pragmatic. Thus we can find abundant information on the strategies that initiate public art, but we can search far and wide for any compellingly articulated theory of public art. Can provocative art endure the democratic composition of the selection panel and process? Are art and ecumenicism in opposition? Can public art illuminate cultural ideas that other forms frequently cannot? What is it that public art can uniquely do? These are the kinds of questions, I would argue, that must be more vigorously explored. And I would further propose that this discourse will serve to overturn some knee-jerk assumptions about the very nature of the hybrid beast we call public art.

One basic assumption that has underwritten many of the contemporary manifestations of public art is the notion that this art derives its "publicness" from where it is located. But is this really a valid conception? The idea of the public is a difficult, mutable, and perhaps somewhat atrophied one, but the fact remains that the public dimension is a psychological, rather than a physical or environmental, construct. The concept of public spirit is part of every individual's psychic composition: it is that metaphysical site where personal needs and expression meet with collective aspirations and activity. The public is the sphere we share in common; wherever it occurs, it begins in the decidedly "somewhere" of individual consciousness and perception.

Therefore, the public is not only a spatial construct. And thus a truly public art will derive its "publicness" not from its location, but from the nature of its engagement with the congested, cacophonous intersections of personal interests, collective values, social issues, political events, and wider cultural patterns that mark out our civic life. Unfortunately, what we have traditionally seen is a facile definition that links those areas that cities (with private developers) designate as public spaces with the notion of public art. It is presumed that these sites, by virtue of their accessibility or prominence, are the ones where public art *can* and *should* appear. This is a questionable idea of many reasons – not the least of which is that public space, as it is emerging in our time, bears little kinship to the public space of the town square, plaza, or common in which the public art of the past traditionally found its home. Public space, as defined today, is, in fact, the socially acceptable euphemism used to describe the area that developers have "left over," the only "negotiable" space after all of their available commercial and residential space has been rented or sold. The City of New York, for example, has granted many developers the right to upscale the height or bulk of their buildings, contingent upon their agreement to provide a little more "public space" at ground level. But what qualities and characteristics these spaces must offer have been inconsistently interpreted. Thus public space has served as a great new incentive – not to be "public," however, but to satisfy far more profit-motivated market objectives. When public space and public art seem to appear spontaneously, it is usually because some savvy or enlightened developer has discovered that beauty can be profitable, and that offering something to the community (even if no one really understands the nature of the gift) can enhance the corporate image. In the same way that "good fences make good neighbors," the clear delineation of a public space has been packaged as a neighborly gesture, with public art the fence that identifies boundaries.

But a public art that truly explores the rich symbiotic topography of civic, social, and cultural forces can take place anywhere – and for any length of time. It would not have to conform to such formal parameters, for it would not find its meaning through its situation *in* a forum, but would *create* the forum for the poignant and potent dialogue between public ideals and private impulses, between obligation and desire, between being of a community and solitude. Wherever we might find that art, we would be inspired to extend its discourse into the variety of public and private domains we enter. Those two domains are different, of course, but they are interdependent. To define the public as merely that which exists outside the private is to deny the essential and complex relationship between the two.

A major exhibition in lower Manhattan this fall has helped to emblematize these and other disquieting questions about the relationship between so-called public space and public art. In and around a major and unfinished portion of ground-level space in the World Financial Center of the newly emerging Battery Park City complex, the real estate development firm of Olympia & York provided a space for invited artists and architects to install temporary, site-specific works. The Olympia & York assembly of art, entitled "The New Urban Landscape," was an extravaganza – in the best and worst senses of that word. This rich variety of projects announced loudly and emphatically that here lies another public space. And so here, once again, art was defined as public because of its location. Yet there was a particularly shrewd inversion at work. By dangling the bait of abundant and chewy art by some of the "hottest" accomplished and emerging artists from around the world, Olympia & York succeeded in appropriating the notion of public art to entice the public to a new site – that didn't, by any other definition, look or feel very public. And the lure for this consecration was both savory and spicey. The organizers and artists had the courage, and the developer the good sense (and beneficence), to endorse some politically loaded, controversial, and critical work in a corporate-sponsored setting. And yet "The New Urban Landscape" sends out troubling – and by now familiar – messages about public art's application. For "The New Urban Landscape"

was a fin-de-siècle enterprise – in some ways, the coda for fifteen years of fervor. And when it all ended, art had served as just one more ingredient in an elaborate coronation that attempted to transform nothing more than a low-ceilinged hallway into a dynamic public space, and a private developer into a public patron.

The involvement of corporations in the sponsorship and support of art is not a new thing. After many years of stimulating the production of private art, it seems quite natural that corporations would eventually find their way to public art, which can now not only boost a corporation's reputation as intelligent and concerned, but can also serve as the vehicle to demonstrate community spirit, a belief in the idea of place in an age of placeless architecture. With this project there was a generous and open sponsor, some very good art, and thoughtful, insightful organizers. So what is the problem? What is it that disturbs?

In fact, some of the answers to these questions will be found in other questions: those that address the implications of the temporary in public art. For in the bureaucratization of public art, there has been a tremendous emphasis on the installation of permanent projects. (Organizations such as the Public Art Fund Inc. and Creative Time, Inc. – dedicated to sponsoring short-lived exhibitions and installations in sites throughout New York – are two of the exceptions.) When evaluating proposals for art that will be commissioned to last "forever," it is not shocking that selection panels have often clammed up and chosen the safe, well-traveled path of caution. When faced with the expanses of eternity, it is not surprising that many artists themselves have tended to propose those cautious, evenhanded solutions. Therefore, the temporary is important because it represents a provocative opportunity to be maverick, or to be focused, or to be urgent about immediate issues in ways that can endure and resonate. But I would argue that the power of the temporary asserts itself productively and genuinely in situations where the pressure of the moment is implicit in the work. Seen in these terms, the temporary is not about an absence of long-term exhibition commitment on the part of any particular sponsor, but about a pledge of a different kind, with more compressed intensity, on the part of the artist.

The nucleated setting and agenda of Olympia & York's endeavour raises serious concerns about the potential for co-opting and institutionalizing even this radical fringe of public art. For what will be the lasting impact of this great event of Olympia & York's? In what significant ways has this exhibition marked this site, or furthered the idea of art as a critical public catalyst, once the gypsy encampment has packed up and moved on? In fact, wasn't this project just another schedule-driven exhibition that had little to do with the present or future of the public life at this site? If a succession of temporary exhibitions might, in fact, animate this public space (something the developers apparently desire) and begin to generate some meaningful dialogue (something the rest of us might like) about space, art, and contemporary urban life, such a possibility is entirely contingent upon some long-range vision as opposed to a shrewd public relations strategy, however magnificently or munificently that strategy is enacted. By "dressing up" (or disingenuously "dressing down" what would be considered even a poorly designed indoor sculpture garden in the garb and lingo of social conscience and inquiry ("The public spaces of The World Finance Center are an ideal context for public art," a four-color brochure tells us. "The works in this exhibition make unusual demands on the viewer," etc.), Olympia & York have, as much as anything, demonstrated to us just how subject to manipulation the concept of public art has become.

Perhaps another one of the great problems of public art today stems from its fundamentally ecumenical intentions. Artists striving to meet the needs of their public audience have too easily subscribed to the notion that these needs can best be met through an art of the widest possible relevance. The ideas of ecumenicism and relevance are not onerous, but they can have – and in the case of public art, often *do* have – insidious and oppressive dimensions. For broad-based appeal and the search for a universal common denominator are not a priori esthetic concepts, but a posteriori results. Reverse that order, and the art's in trouble, for art is an investigation, not an application. So it's disturbing when it looks as if artists are campaigning for public office – going for the majority consensus at all costs.

Not surprisingly, this goal of unanimity has also led to the establishment of what is considered a more democratic composition of public-art selection committees. There has been a generous and well-intentioned effort to include on these committees not only panelists with backgrounds in the arts, but also representatives from the local community in which the public installation will be situated. Yet if followed to its logical conclusion, the concept of "public" that this phenomenon implies reveals itself to be quite ludicrous. For public space is either communal – a part of the *collective* citizenry – or it is not. Somewhere along the line, our democratic process has presumed that the sentiments of one particular community, simply because of its members' propinquity to the prospective installation, should be granted greater significance. What this suggests is that we have arrived at some reliable formula for articulating the precise radius that distinguishes that community's interests from the larger field of public life. Thus the ideas of the local community and of the general public are put into an adversarial relationship, implying a fundamental conflict between those inside a particular neighborhood, area, city, etc., and those outside. This peculiar endorsement of community opinion, sometimes at the expense of larger public concerns, subtly yet effectively affirms a notion of what I would call "psychological ownership," at the same time that it refuses to ground that notion in any terms other than geographic. Thus, because I live, work, or relax near a certain site, I believe that, in a sense, I "have" that site, and am empowered to exert control, regulation, or power over those who are the "have-nots." We have seen the ramifications of this kind of thinking in a variety of controversies. For example, which is the community that should have the most say in "approving" the design of the Vietnam War Memorial in Washington, D.C.? Veterans? The family members of men killed or missing in action? The group of office workers and government bureaucrats who work nearby? The public at large, who might feel a sense of possession of this tragic, poignant space? And which is the community to be consulted when installations are contemplated for City Hall Park in Manhattan? The government employees who work in City Hall and cross the park each day?

Or New York City voters for whose civic authority and commitment the site speaks? Or the many homeless who spend their days and their nights in the park? And to which group should the artist throw his or her appeal?

Rather than digging in our heels to examine and analyze the implications of these questions, too many public-art sponsors and makers seem to be trying to sidestep them with a "minimum-risk" art; that is, an art that can be slipped quietly into space and somehow manage to engage everyone but seriously offend or disturb no one. But isn't it ironic that an enterprise aimed, even at the least, at enlivening public life is now running on gears designed to evade controversy? And that so many involved in public art express such dismay – even hurt – if and when controversy occurs? Curiously, Richard Serra's *Tilted Arc*, 1981, in Manhattan's Federal Plaza remains one of the great moments in contemporary public art, not despite but *because* of the conflict its installment generated. Is it offensive? Does it obstruct? Is it public if it does not please? Should the artist's personal vision of site-specificity be permitted to override the desires of the local (specifically professional) community most frequently exposed to the work? In fact, Serra's work achieved its most profound public resonance and significance precisely at the moment when its future seemed most threatened. And that inflexible, somewhat dogmatic object in a deplorable architectural context has been enriched by the color and texture of public debate that continues to surround it. *Tilted Arc* is an important symbol for public art because of the questions it has stimulated – and not because it should not be where it is.

Unfortunately, the avoidance of such controversy has generated an attitude about public art that constrains and segregates thinking. In the 1970s, when troubled cities felt a great vulnerability to the aggressive, often destructive gestures of disenfranchised citizens, the idea of "defensible space" became an important concept. It was Oscar Newman who first proposed that public space could and should be designed in a way that protected it from the onslaughts of graffitists, vandals, and other assaulters. We can see the influence of this proposition in the clunky, immovable concrete benches and barriers in our parks and city streets; the barbed grillwork appearing on ground-level heating ducts to stave off loiterers and the homeless seeking warmth. But we can also see the flip side of this proposition at work in the public-art mentality taking hold today. Public art may not be required to be physically "defensible" but it is, more and more, expected to be *defendable*. So every possible – and ludicrous – objection is raised at the early stages of the artist selection and proposal process, to anticipate and fend off any possibile community disfavor. With programs dependent on such tightly woven sieves, it's not surprising that plenty of hefty, powerful projects don't make their way through. And it's not surprising that, over the years, the artists who might propose such projects have turned their energies elsewhere, while the studios of the artists who have learned the appropriate formula have become minifactories for the churning out of elegant maquettes for current and future projects.

It is important to consider that the most public and civic space of many early American cities was the common. The common represented the site, the concept, and the enactment of democratic process. This public area, used for everything from the grazing of livestock to the drilling of militia, was the forum where information was shared and public debate occurred: a charged, dynamic coalescence. The common was not a place of absolute conformity, predictability, or acquiescence, but of spirited disagreement, of conflict, of only modest compromises – and of controversy. It was the place where the ongoing dialogue between desire and civility was constantly reenacted, rather than restrained or censored. If it's true that the actual space of the common does not exist as it did two hundred years ago, the idea is still vital. Its problematic shadow image, the idea of an enormous, happy cultural melting pot, was challenged and generally dismissed twenty-five years ago – except by a lot of people involved with public art. So if there is a tragedy here, it is that public art is in the unique position to reconstitute the idea of the common, and yet, by misconstruing the concept – by too often rewarding the timid, the proven, the assuaging – the public art machine has consistently sabotaged its own potential to do so.

Still, it seems, we're returned to the question of where that "common" might be today, or at least where – or how – we might look for the public art to create it. And though I've painted a bleak picture of the contemporary scene, it is not a hopeless one. In fact, if we take that step beyond conceiving of the new urban landscape as a geographic grid of buildings, spaces, and art, to view it instead as an ever-mutating organism sustained by multiple, interrelated vortices and networks and the private trajectories that complicate them, then the horizon line of public art expands to include the "invisible" operations of huge systems and the intimate stories of individual lives. Certainly the artists who choose to work with these polarities where the edges of the public are invented and realized and not the only ones whose projects provide significant stimulation. But their work, by pressing against calcified notions of public art, suggests some fresh visions of the common.

For example, artist Mierle Laderman Ukeles' work with the Department of Sanitation of the City of New York engages city residents in one of the most crucial, life-sustaining but maligned operations of urban life – garbage collection. Her most recent project, *Flow City* (scheduled for completion in 1990), will bring people into a cavernous marine transfer facility at 59th Street and the Hudson River for what is, in effect, a multimedia performance of trucks dumping their loads of household and commercial waste into barges destined for landfills. Ukeles' work proposes that the public in public art is defined by subject rather than object.

At the other end of the spectrum, some of the most fruitful investigations of public life and art are occurring in the most private, sequestered site of all – the home. For just as the public space has become diminished as a civic site, the home has become, in many senses, a more public, open forum. The public world comes into each home as it never has before, through television, radio, and personal computer. So that the rituals that were once shared conspicuously in a group are now still shared – but in isolation. An example of this ambiguous condition is the annual celebration of New Year's Eve in Times Square. Which is the more public event – the throng of people gathering at 42nd Street to watch a lighted apple drop, or the millions of people at home, each watching this congregation on TV? In other words, more and more, the home has become the site for the complex play of social

meanings. For this reason, it is a fruitful domain for dialogue about the public/private dialectic. Following in the footsteps of Gent's Museum van Hedendaagse Kunst's 1986 exhibition "*Chambres d'amis*," the Santa Barbara (California) Contemporary Arts Forum organized their 1988 "Home Show," with ten California residents welcoming ten artists into their homes to explore the region of interiority – as it relates to the external, public world.

For this project, the collaborative team of Kate Ericson and Mel Ziegler, with a work entitled *Picture out of Doors*, methodically removed all the doors in Pat and David Farmer's home, including doors from closets, cupboards, and cabinets, even from bedrooms and bathrooms. The tangible evidence of sanctioned voyeurism was stacked in the living room. In a sense, the team's project publicized intimacy by denying privacy. In many ways, the Santa Barbara installation was a tame project for Ericson and Ziegler. For the past ten years they have conducted their own investigations of the private/public dialectic, with much of their work occurring on their own instigation, that is, without the benefit or legitimacy of an arts organization. They have placed advertisements in local newspapers seeking homeowners willing to collaborate on projects. In one project in Hawley, Pennsylvania, for example, called *Half Slave, Half Free*, 1987, the team asked a homeowner to continue to cut only half of his lawn and leave the other portion unmaintained. *Half Slave, Half Free* suggests an expanded and provocative definition of public art, one that has sustained a commitment to independent "guerrilla" activity as an alternative to institutionalized commissioning, and that appeals to and enlists the support of the single vote (the homeowner/collaborator) as opposed to the majority rule in order to explore the half-slave, half-free relationship of personal to public, and vice versa.

Individual vision and independent thinking *are* possible in the realm of public art; what we've come to expect – or accept with a sigh – is not all we *need* expect. Two years ago, architects Donna Robertson and Robert McAnulty proposed a design of an apartment for an exhibition called "Room in the City." The proposal was spare yet complex. Within the space of a small Manhattan apartment, a procession of five video monitors, hooked up to a satellite communications disk placed outside the window, showed random images of the city, creating an ethereal glow of violated or collaged information. On the other side of a diagonal wall that slashed through the space, a single monitor, unattached to the communications dish and the surrounding city, sat at the foot of the bed. In the traditional site of domesticity and intimacy, this project stands as a metaphor for our new urban landscape; that site where private and public, the intimate and the shared, are fragmented and reconceptualized, where culture both originates and ends, and where the public is permitted to assert itself as an idea of ever-shifting focus and fruitful frustration.

16 SAMIR AL-KHALIL

'The Monument and the City'

from *The Monument: Art, Vulgarity and Responsibility in Iraq*
(1991)

[. . .]

Baghdad never lived up to the romance of its name. Before Saddam Husain's monument arrived, the streets were changing, growing wider and ever more impersonal, crowding in the mosques and the courtyard houses of the old city quarters. In the 1950s and 1960s, the dirty, picturesque 'old' Baghdad – with its sectarian and ethnically divided neighbourhoods, its colourful souks, its horizontal skyline punctuated with pretty vertical minarets, its inward-looking houses and shaded narrow alleyways – was being destroyed, we are constantly reminded, by modern 'International Style' architecture and that rampant engine of individual freedom: the automobile. Today even that much maligned, anarchic and crumblingly cosmopolitan city of the 1960s has shed its skin. Under the same name, a new assertive metropolis of five million people has taken the place of the city which housed a mere quarter of a million people in the 1920s, and which had grown to one and a half million by 1968 when the Ba'th first came to power. Hence, a large and growing number of people who today live in the new monument-filled city of Baghdad, have in fact never lived in any other kind of city.

Baghdad first began to metamorphose into its present form in the late 1970s. The regime's efforts had previously been taken up with eliminating domestic opposition, a Kurdish civil war, infrastructure development, nationalizations, eradicating illiteracy, industrialization, and military buildup. After a decade in power focused on these issues, the Iraqi political leadership felt secure, united and self-assured: all necessary preconditions for the President's decision to go to war with Iran in the spring of 1980. Moreover, the nonaligned nations conference was scheduled to be held in Baghdad in the summer of 1982. Saddam Husain, who had just hosted the anti-Camp David Arab Summit, was going to take over the mantle of Third World leadership from Fidel Castro. And he was going to have this leadership bestowed upon him in his own city. The desire to construct monuments and to assert political authority coincided therefore, as they have done since time immemorial in virtually every civilization.

Overnight Baghdad became a giant construction site: new and wider roads, redevelopment zones, forty-five shopping centres in different parts of the city opened to the public by 1982, parks (including a new tourist centre on an artificial island in the Tigris), a plethora of new buildings designed by Iraqi and world class architects, a crash programme for a subway station, and many new monuments – all were put in hand. The time had come to crystallize political facts and goals within the public realm in great architecture and lasting monumental art; to translate the collective force of the Iraqi people, as that leading theoretician of the modern movement Sigfried Giedion would have put it, into symbols.

The Mayor of Baghdad, Samir 'Abd al-Wahab al-Shaikhly, was fond of saying that twenty billion Iraqi dinars had been budgeted for this purpose. In particular six large areas in the central business and residential districts of Baghdad became in 1979 the focus of a massive accelerated urban redevelopment programme: Khulafa street, the Bab al-Sheikh area, the al-Kadhimmiyya shrine and vicinity, the al-Karkh area, Abu Nowas street and Haifa street. The Khulafa street redevelopment, for instance, consists of many new buildings along the street, two squares, a civic centre, a huge mosque extension and the rehabilitation of some old quarters.

The outcome of at least some of that expenditure can be experienced in Haifa street which was opened amid much fanfare in 1985. Nothing remotely like Haifa street has ever existed in Baghdad before. And insofar as the environment in which people live has any influence on their behaviour, or sense of who they are as a community, it can safely be said that an Iraqi nurtured in the bosom of Haifa street will not resemble one nurtured in 'old' Baghdad, however one wished to define that long lost city.

Among the big monuments of the new city special note must be made of the highly acclaimed *Shaheed* Monument of 1983, consisting of a circular platform, 190 metres in diameter, floating over an underground museum and carrying a 40-metre high split dome. The ensemble sits in the middle of a huge artificial lake. It cost the Iraqi exchequer a quarter of a billion dollars. Built by the Mitsubishi Corporation to the very exacting specifications of the engineering consulting firm of Ove Arup and Partners (of Sydney Opera House fame), the monument was conceived by the gifted Iraqi artist Ismail Fattah al-Turk and carried through the various stages of detail design and working drawings by a group of young Iraqi architects all formed in the Baghdad School of Architecture. Kenneth Armitage, the internationally renowned sculptor, is said to have been so overwhelmed by it during a visit in 1986, that he hugged the artist in a fit of emotion quite uncharacteristic of an Englishman.

The 'sculptural idea in the monument has been inspired by the principles of glorification of the "Martyr"', says the pamphlet issued on the opening day. But Saddam Husain's great war with Iran was in its first months when the monument was publicly announced, and had not yet enrolled all that many 'martyrs'. (The war began in September 1980 and on-site construction started in April 1981. The project would have been planned, drawn, managed and detailed many months before. Could it too, like the President's victory monument, actually have been conceived before the reality it commemorated, before the war had even started?)

[. . .]

A second enormous edifice of purely monumental import (obviously public buildings can also assume monumental significance but I am not here dealing with these) was built in tandem with the first and must have cost a great deal more. Unlike *Shaheed* which succeeds in creating an abstract, yet powerful and evocative symbolism by its use of the split dome, the 'Unknown Soldier' monument on July 14th street is simpleminded in the extreme. The tilted behemoth, which looks like a flying saucer made from reinforced concrete and frozen in mid flight, represents a traditional shield – a *dira'a* – dropping from the dying grasp of the archetypal Iraqi warrior. No one, Iraqi or otherwise, sees the imagery in quite this way. And even when you are told what to look for, credulity is strained. This second major edifice was conceived by Khalid al-Rahal, to whose experience Saddam Husain would later turn.

Monuments like *Shaheed* and the Unknown Soldier, along with the scale of the physical changes in the urban fabric, did not completely escape the attention of the world. In 1985, for instance, the *National Geographic* published a photographic essay entitled 'The New Face of Baghdad', and in 1988 the Arabic daily, *Al-Hayat*, ran an article on how excellent it was to see the Iraqi sculptor moving 'from the atelier to the street'. On the more sober professional front, a prominent Japanese architectural journal compared what was going on in Baghdad in the 1980s to Haussmann's late nineteenth-century reconstruction of Paris carried out under Napoleon III (in the relatively short span of twenty years, the journal stresses).

[. . .]

Shortly after the completion of *Shaheed* and the Unknown Soldier, the President conceived the third monument which is the focus of our study: the Victory Arch. The three monuments clearly form a unit. All refer to the gruelling eight-year war and the collective experience its pain and suffering forged in Iraq. But they do so in different ways. In Saddam Husain's final creation, I believe, the true conditions of monumentality as defined by Giedion, Leger and Sert are met. The President's monument is capable of haunting Iraqi memories and scratching at the psyche of all human beings in ways that even Fattah's overwhelming dome is not, much less Rahal's white elephant, although both are bigger, more deeply rooted in the ground (with concrete basements, exhibitions and all), and cost a great deal more to build. It exudes an ineffable special quality which is not shared with the two colossal monuments that preceded it, but is possessed by another curious and relatively modest scheme recently opened in the city of Basra.

In a green park on the banks of the corniche facing the Shatt Al Arab waterway, framed by a silhouette of construction cranes and set amidst a surreal landscape of black billboards which remind visitors how many Iraqis died fighting for Basra and Faw and how many shells landed on Iraqi soil, are eighty lifesize and lifelike sculptures of officers and commanders who fell in those battles. This was a priority project in the frenzied post-war rush to rebuild virtually from scratch the pulverized cities of Basra and Faw. Made from real family snapshots by a collective of Iraqi sculptors, they depict those heroes in a variety of forms of dress, in combat gear or without. Only one thing is common to all: every man's gaze is sternly fixed on the Iranian shore across the Shatt, and each has an arm accusingly stretched out pointing in the same direction.

If there is such a thing as a particular aesthetic 'mode' in Saddam Husain's city (and I think there is), then the latest monument in the Baghdad series and the eerie surreal landscape in Basra are its apogee. Such forms and images bestow meaning upon that mode in the very specific conditions of Iraqi culture, history and politics. The President's 'Baghdad Style', not the early modernism transplanted from the West, nor the latest 'post-modern' aesthetic gymnastics of all those who so enthusiastically participated in the redevelopment of Baghdad, truly celebrates the Ba'thist city.

Is monumentality an objective quality inherent in an urban community which is simply made manifest by the gifted artist, as the devotees of modernism have long assumed? Or is it merely an artist's eclectic interpretation of how the values and forms of his city might re-enact some tenets of culture or history, as the post-modernist would claim? Who comes first, the artist or the city? From the standpoint of this debate, Saddam Husain's monument is set apart from its precursors in being a concrete physical intervention in the city by the man who is behind all the other changes. (Likewise, although the Basra landscape is the pooled effort of many sculptors, it was inspired by one imagination, probably Saddam Husain's.) Hence, its creation momentarily dissolves a dichotomy which has plagued all those who have reflected on the city; it might even shed light on the general nature of monumentality itself.

The monument, Aldo Rossi wrote, 'is the sign upon which one reads something that cannot otherwise be said, for it belongs to the biography of the artist and the history of society'. Rossi thought of monuments as fixed references within the city, itself conceived as a collective human artefact changing through time and made of many parts. For him as for Sert, Leger and Giedion, those three masters of modernism, monuments are those parts of the city imbued with persistence and permanence, a result of their capacity to encapsulate the history, art and collective memory of a place. They are works of art in their own right and 'more than' art in that they are signs of a collective will. The often repeated organic metaphor that likens the city to a biological organism whose form is constituted by evolutionary incremental adaptations to function, is rejected by Rossi. Instead, the monument is a special kind of work of art within an even larger work of art: the city. It is both a public event and a form identified with a 'style', unique in time and place, yet linking the city's past with its future. Always, therefore, the essence of a monument, like that of a city, lies in its destiny.

Rossi was thinking of Rome, Paris, Granada and great urban artefacts like the Roman Forum and the Alhambra; he did not have the new face of Baghdad and Saddam Husain's monument in mind. His way of thinking, however, reminds us that cities and their most genuine monuments are always mirror images of one another; inside the belly of the one you can untangle the meaning of the other. Saddam Husain's new Baghdad is not merely a run-of-the-mill Third World capital, haphazardly stuffed with people and ugly or beautiful artefacts as the case may be; it is also a collective state of mind which has its gods. Thinking about the meaning of these gods must remain as incomplete as the unfulfilled destiny of the Victory Arch itself.

17 RAPHAEL SAMUEL

'Theme Parks — Why Not?'

from *Independent on Sunday* (1995)

Today, as on any other Sunday of the year, thousands of our fellow citizens will visit "historic" towns, ancient monuments, country parks, living history museums and theme parks. In the summer, thousands more will go to bird sanctuaries and wildlife reserves or join nature "mystery trails" and "historic" walks; last year, the steam railways alone carried some 50 million passengers. "Heritage" is as popular with the general public in the late 20th century as the wonders of science and invention were in the 1870s when, on a Whit Monday, some 70,000 visitors are said to have flocked to Liverpool to see the new warehouses.

All this, it might be thought, would be a matter for celebration. "Heritage" has given millions of people an active interest in the past and, as in the battle for Oxleas Wood, the defence of it has stimulated numerous environmental campaigns. It is one of the few areas of national life in which it is possible to invoke an idea of the common good. It has given a new lease of life – and a new visual form – to what used to be called, in the 1890s and 1900s when it found expression in municipal libraries, swimming baths and bandstands, the Civic Gospel.

We live in an expanding historical culture, one far more open to the stigmatised and hitherto excluded than those of the past. Family history societies, one of the true grass-roots movements of our time, working on the archives with an erudition which the professional scholar might imitate, have democratised the study of genealogy, just as the collecting manias of the last 30 years have made us all, in some sort, curators or memory keepers.

Yet heritage-baiting has become a favourite sport of the metropolitan intelligentsia. Barely a week goes by without it being targeted for abuse in one or other of the "quality" newspapers. The charge is that it wants to commodify the past and turn it into tourist kitsch, presenting a Disneyfied version of history in place of the real thing. "A loathsome collection of theme parks and dead values" is how Tom Paulin has described the British heritage industry. Heritage, according to the critics, is the mark of a sick society, one which, despairing of the future, has become besotted with an idealised vision of its past. It is a symbol of national decadence, a malignant growth which testifies at once to the strength of this country's *ancien regime* and to the weakness of radical alternatives.

For these reasons, denunciation of heritage has been as strong on the left as on the right. One of the most eloquent critics is Neal Ascherson, of this newspaper, who has written: "Where there were mines and mills, now there is Wigan Pier Heritage Centre, where you can pay to crawl through a model coal mine, watch dummies making nails, and be invited 'in' by actors dressed as 1900 proletarians. Britain, where these days a new museum opens every fortnight, is becoming a museum itself."

Ascherson uses the word "vulgar" and that may seem strange from the lips of a socialist, a republican and a democrat. But as moral aristocrats, waging war on the corruptions of capitalist society, socialists, like the radical nonconformists who preceded them, have often been at their fiercest when denouncing Vanity Fair, or what Aneurin Bevan called the "vulgar materialism" of capitalist society. And from the time of William Morris onwards they have been apt to rebuke the masses for what another great Labour leader, Ernest Bevin, called the "poverty" of their desires.

"Heritage" has many different histories. One

version is certainly aristocratic nostalgia, as exemplified by *Brideshead Revisited*. I am as indignant as others on the left at attempts to use heritage to promote a country house version of the national past. But I have come to think the dangers of this exaggerated; not the least attractive feature of Covent Garden is that it is a Sloane-free zone. Heritage is an idea which belongs at least as much to the left as to the right. Its French cousin, *patrimoine*, goes back to the Jacobin educational projects of the egalitarian priest, L'Abbe Gregoire. In the United States, the creation of black heritage centres and the rediscovery of African roots were central to the Black Power movement of the 1960s. In Britain, the National Trust was originally a progressive cause. Conservationism is now, arguably, the principal outlet for the reformist impulse in national life.

Today's open-air museums give pride of place to farmers, labourers and artisans and put a premium on the craftsman-retailer. Family, work and home are placed at the centre of history. The opponents of heritage charge it with imposing a conservative view of the past. On the contrary, it is part of a change in attitudes which has left any unified view of the past – liberal, radical or conservative – in tatters. Culturally it is pluralist. Everything is grist to its mill: the inter-war "semi", no less than the stately home. And, far from simply domesticating or sanitising the past, it often makes a great point of its strangeness, of the brute contrast between now and then.

The denigration of "heritage", though voiced in the name of radical politics, echoes some of the right-wing jeremiads directed against "new history" in the schools. It is accused of taking the mind out of history, offering a package-holiday view of the past as a substitute for the real thing. Still worse – a kind of ultimate profanity in the eyes of the purists – is the use of the performing arts, as with the ex-teachers who dress up in period costumes and act as demonstrators, interpreters and guides. Literary snobbery comes into play: the belief that only books are serious; perhaps, too, a suspicion of the visual, rooted in a Puritan or Protestant distrust of graven images.

There is social condescension here. It is a favourite conceit of the aesthete that the masses, if left to their own devices, are moronic; that their pleasures are unthinking and their tastes cheap and nasty. Theme parks – doubly offensive because they seem to us to come from America and because they link history to the holiday industry – are a particular bugbear for the critics. As engines of corruption, or seducers of the innocent, they seem to occupy the symbolic space of those earlier folk-devils of the literary imagination: jukeboxes and transistor radios, or candyfloss and milk bars. In contemporary left-wing demonolgy they have become the latest in a long line of opiates of the masses, on a par with Butlin's holiday camps and bingo halls in the 1950s; "canned entertainment" a "Hollywood films" in the 1930s, what JB Priestley feared was the "Blackpooling" of English life and leisure.

Historians are only too ready to join the chorus of disdain. Does envy play some part? Heritage has a large public following, mass membership organisations whose numbers run to hundreds of thousands, whereas our captive audiences in the lecture hall or the seminar room can sometimes be counted on the fingers of one hand. Heritage involves tens of thousands of volunteers. It can command substantial Exchequer subsidies, and raise large sums by appealing to the historically minded public. It fuels popular campaigns and is at the very centre of controversy about the shape of the built environment.

Yet historians are no less concerned than conservationists to make their subjects imaginatively appealing. We may not prettify the past in the manner of English Heritage or the National Trust, but we are no less adept than conservation officers and museum curators at tying up loose ends and removing unsightly excrescences. We use vivid detail and thick description to offer images far clearer than any reality could be. Is not the historical monograph, after its fashion, as much a packaging of the past as costume drama?

Education and entertainment need not be opposites; pleasure is not by definition mindless. There is no reason to assume that people are always more passive when looking at old photographs or film footage, handling a museum exhibit, following a local history trail, or even buying a historial souvenir, than when reading a book. People do not simply "consume" images in the way in which, say, they buy chocolate. As in any reading, they assimilate them as best they can to pre-existing images and narratives. The pleasures of the gaze are different in kind from those of the written word but not necessarily less taxing on historical reflection and thought.

And who are today's memory keepers? Museum curators? Clarice Cliff collectors? Professional historians? Vinyl freaks? One of my critics derides me for attempting to rescue the metal detectorists, the historical re-enactors and the steam engine freaks from the condescension of posterity. Yet today's vast army of collectors and detectors are not only creating an archive for the future but also pointing the way to what are likely to be some of its leading themes. When the history of the suburbs comes to be undertaken, the newly opened museum of the lawnmower in Southport will be as good a starting point as the British Museum newspaper library in north London.

18 DOLORES HAYDEN

'The Power of Place: Urban Landscapes as Public History'

from I. Borden, J. Kerr, A. Pivaro and J. Rendell (eds), *Strangely Familiar: Narratives of Architecture in the City* (1996)

The power of place – the power of ordinary urban landscapes to nurture citizens' public memory, to encompass shared time in the form of shared territory – remains untapped for most working people's neighborhoods in most cities, and for most ethnic history and women's history. The sense of civic identity that shared history can convey is missing. And even bitter experiences and fights communities have lost need to be remembered – so as not to diminish their importance.

To reverse the neglect of physical resources important to women's history and ethnic history is not a simple process, especially if we are to be true to the insights of a broad, inclusive social history encompassing gender, race and class. Restoring significant shared meanings for many neglected urban places first involves claiming the entire urban cultural landscape as an important part of history, not just its architectural monuments. This means emphasizing the buildings types – such as tenement, factory, union hall, or church – that have housed working people's everyday lives. Second, it involves finding creative ways to interpret modest buildings as part of the flow of contemporary city life. A politically conscious approach to urban preservation must go beyond the techniques of traditional architectural preservation (making preserved structures into museums or attractive commercial real estate) to reach broader audiences. It must emphasize public processes and public memory. This will involve reconsidering strategies for the representation of women's history and ethnic history in public places, as well as for the preservation of places themselves.

Some of the possible projects to emphasize gender and ethnic diversity include: identifying landmarks of women's history and ethnic history not yet seen as cultural resources; creating more balanced interpretations of existing landmarks to emphasize the diversity of the city; and recruiting nationally known artists and designers to collaborate with historians and planners to create new works of art celebrating women's history and ethnic history in public places. Often the subject of working women of color will draw the largest possible audience; in Los Angeles, for example, they are a majority of citizens but have long been neglected as a subject for public history.

The first public art project undertaken by The Power of Place took place in Los Angeles and focused on Biddy Mason. Mason was born in 1818 in the South and lived as the slave of a Mormon master in Mississippi who decided to take his entire household west to Salt Lake City, Utah and then on to the Mormon outpost of San Bernardino, California. Biddy Mason literally walked across the country behind her master's wagon train, herding the livestock and caring for her own three young daughters. She went to court to win freedom for herself and her master's other slaves in Los Angeles in 1855. After the trial in 1856 she settled in the city to earn a living as a midwife. Her urban homestead, purchased with ten years' earnings, eventually included a two-storey brick building, which provided a place for her grandsons' business enterprises as well as for the founding of the First African Methodist Episcopial Church (FAME).

When I first saw the site of Mason's homestead, it was a parking lot in Downtown Los Angeles between 3rd and 4th Streets, Spring and Broadway – an unlikely place for a history project. But a new building was being erected on the site, and the Los Angeles Community Redevelopment Agency (a city agency) was interested in some commemoration. I served as the project director as well as historian on the team, which included graphic designer Sheila de Bretteville, artists Betye Saar and Susan King, and curator Donna Graves. The first public event was a workshop in 1987 where historians, planners, community members, students and the project team members discussed the importance of the history of the African-American community in Los Angeles and women's history within it. Mason's role as a midwife and founder of the FAME church were stressed. Eventually, the Biddy Mason project included five parts. First, Betye Saar's assemblage, "Biddy Mason's House of the Open Hand", was installed in the elevator lobby. It includes motifs from vernacular buildings of the 1880s as well as a tribute to Mason's life. Second, Susan King's large format letterpress book, *HOME/Stead*, was published in an edition of 35. King incorporated rubbings from the Evergreen Cemetery in Boyle Heights where Mason is buried. These included vines, leaves and an image of the gate of heaven. The book weaves together historical text with King's meditations on the homestead becoming a ten-story building. Third, an inexpensive poster, "Grandma Mason's Place: a Midwife's Homestead", was designed by Sheila de Bretteville. Historical text I wrote for the poster included midwives' folk remedies. Fourth, "Biddy Mason: Time and Place", a black poured concrete wall with slate and granite inset panels, designed by Sheila de Bretteville, chronicles the story of

Biddy Mason and her life. The wall includes a midwife's bag, scissors, and spools of thread debossed into the concrete. De Bretteville also included a picket fence around the homestead, agave leaves, and wagon wheels representing Mason's walk to freedom from Mississippi to California. Both her "Freedom Papers" and the deed to her homestead are among the historic documents and photographs bonded to granite panels. And fifth, my article, "Biddy Mason's Los Angeles, 1856–1891", appeared in *California History*, Fall 1989.

The various pieces share some common imagery: gravestone rubbings in the book and carved letters in the wall; a picket fence, a medicine bottle and a midwife's bag in the lobby and the wall. One old photograph of Mason and her kin on the porch of the Owens family's house appears in four of the five pieces, as does the portrait.

Everyone who gets involved in a public history or public art project hopes for an expanded audience, beyond the classroom or the museum. The wall by de Bretteville, finished in 1989, has been especially successful in evoking the community spirit of claiming the place. Youngsters run their hands along the wagon wheels, elderly people decipher the historic maps and the Freedom Papers. People of all ages ask their friends to pose for snapshots in front of their favorite parts of the wall.

Excerpted from Dolores Hayden. The Power of Place: Urban Landscapes as Public History, *(MIT Press, 1995).*

'European Cities — First Sow, then Reap'

from *Rough Weather* (1997)

ARTS IN CITIES: BOLSTER PRESTIGE: DEVELOP IDEAS

For many centuries European cities were breeding grounds for artistic creation. There are two basic approaches which guide large European cities in their involvement with the arts. One puts strong emphasis on using the arts to bolster prestige, aid urban renewal, or promote tourism. The second understands that you must first sow before you may reap; that you will not have major works of art when creators, performers, and craftspeople do not have the opportunity to develop ideas that are not yet well accepted.

The reality of European cities is more complex than is implied in this simple dichotomy; the good cannot be separated from the bad along such clearcut lines. But we can make an inventory of aspects that distinguish municipal cultural policies.

'Sometimes superficial artistic events take place in Barcelona,' Ferran Mascarell, the chief of the cultural department of the Catalan capital, admits. 'But in general a well-balanced relationship exists between the stated purpose, the phenomena, money spent, the artistic result, and the number of visitors.' He does not find it difficult to list a whole variety of initiatives in all fields of the arts, including architecture, which are making Barcelona a very lively and inspiring city, all realised since Franco died.

The Parisian model, with its emphasis on great projects, does not inspire him at all. We must keep in mind that in a city as enormous as Barcelona there are risks: how does it create the possibility for a population of four million to be involved with artistic phenomena that are not

superficial? The 1992 Olympic Games did not rank high on the list of Ferran Mascarell's favourite projects because they sucked away energy and resources for improving the climate of developing the arts in 1993 and beyond. Eduardo Delgado, head of a study centre for municipal cultural policies, sounds an alarm for his city Barcelona and other cities that are presently much praised for their cultural climate: 'A fashionable city is an endangered one.'

Budapest has different problems. The former regime has neglected this magnificent metropolis. Economically, the country must be rebuilt and it is suffering from a lack of know-how and investment capital. The infrastructure, also in the arts, and previously one-sided support for the arts requires a complete overhaul. Thanks to the relatively liberal policies in place before 1989 Hungary is in a better position today both culturally speaking and in general when compared to other East and Central European countries. Its difficulties are still immense.

Miklós Marschall, alderman of culture until 1994, intended to put this beautiful old city by the river Danube back on the world cultural map. 'Budapest has the flavour to be a cosmopolitan city.' It is a city that was economically, scientifically, and culturally in contact with the rest of the world. 'Between the two world wars, Budapest was culturally full of life: every important artist came to Budapest. And we had a very well-educated audience.'

The policy today is to restore Budapest as a centre of exchange between Eastern and Western Europe. That means restoring an artistic climate of energy and vitality, in which scientists and artists will once again present their work in the city, and spend time there as their counterparts had earlier in the century. Miklós Marschall feels that for the sake of its own population and the viability of the city, it is necessary to maintain contact with many different artistic developments. He says, 'For this image cultural tourism, which is a strategic focus of the municipal development programme, must be attracted – avoiding, however, the suffocating congestion bedeviling Prague.' A rich artistic and cultural life may also give the international business community the sense that Budapest is a good place to locate. The alderman for culture knows that Budapest must hurry to improve its cultural climate. 'Now is the time when many enterprises are deciding where to take up residence.'

The task of refurbishing the image of the city 'means spectacular things indeed. For instance, we will finance – together with some sponsors – a whole new municipal orchestra, the Festival Orchestra, which will recruit only young musicians and Hungarian musicians who live abroad. Improving the old ensembles is a hopeless task. By the tradition of the Communist period they are lazy and no longer used to rehearsing six hours a day: they are not interested in developing artistic excellence. A whole generation of musicians has been lost.'

Miklós Marschall also says that, 'no high-quality culture can exist without innovation at the grassroots. Cultural organisations are needed to stimulate and accommodate new artistic developments. This is our outspoken political approach: we are very much aware of the necessity of supporting alternative and innovative groups. They were neglected over the last forty years, but since we have been in office – that is, since 1990 – we have supported, for instance, alternative theatres very heavily. This is a significant new element in our arts policy in Budapest.'

In the opinion of those I ask in the theatre world of Budapest, he does not exaggerate. Many small theatres have opened their doors and now flourish in pursuit of a variety of theatrical approaches. This is thanks to the presence of artists and arts leaders of tremendous creativity, whose efforts are sustained by a politician of great vision.

His approach also accommodates a variety of festivals: 'The traditional spring festival presents highbrow artistic events. It is supported by the government. We schedule some festivals during the summer to make the city attractive to tourists. And we plan to give the Budapest autumn festival a completely new character from 1992 on. It offers an opportunity to the whole spectrum of contemporary Hungarian arts. They will be heard and have the chance to meet each other. This autumn festival is not intended to make money.'

Miklós Marschall is sure that 'the broad range of cultural investments we are making will pay off.' This is a remarkable note to sound these days, when in Eastern and Central Europe all concentration is on expecting salvation from the free market, without any public support for the infrastructure of economic and cultural life.

Many cities try to distinguish themselves by investing in artistic events. Franco Bianchini, a reader in cultural planning and policy at Montfort University, Leicester, estimates that such policies make sense if the prospects are realistic: 'the direct impact of 1980s cultural policies on the regeneration of employment and wealth was relatively modest, in comparison with the role of culture in constructing positive urban images, developing the tourist industry, attracting inward investment, and strengthening the competitive position of cities.'

This list of results does not really convince Andreas Wiesand (director of the Institute of Cultural Research, in Bonn). 'The "hubbub approach", coming forth from those policy intentions, will eventually simmer down. At a certain moment enough museums will have been built and enough spectacular festivals will have taken place. By simply doing more of that, one will no longer gain a decisive advantage over competing cities!'

As an example he points to Glasgow: 'Marketing does not suddenly make Glasgow the cultural capital of Europe. In itself it is an interesting city, but it would be better if it would refrain from promoting itself with all that rubbish you are obliged to put out when you are selected to be "European cultural capital" for a year. It costs money and it helps nobody. Renato Nicolini, member of the Italian parliament, suggests that it is time we dispense with the outdated project of a yearly European cultural capital, which was initiated by Melina Mercouri when she was the minister of culture of Greece. It does not correspond at all with the real cultural and artistic movements through Europe.'

In Glasgow, according to Andreas Wiesand, one should delve back into the history of the city and the region. There you can find many sources of inspiration – in design, in furniture, for instance. Let the sources that are actually there express what role they would like to play in the artistic field.

PUBLIC SPACE

Cities have always been characterised by the existence of a variety of places where people could meet and congregate on their own terms: collectively we call this the public space. Although vital for freedom of communication, including artistic communication, this phenomenon is endangered, according to Franco Bianchini. 'Many local councils, tempted with promises of increased rateable values and income from land sales, have allowed the development of privately owned and controlled indoor shopping malls in the town or city centre. Management of the new shopping centres have the right to physically remove or deny access to anyone they consider "undesirable".' Arts that used to find a welcome place on streets and squares are thus crossed out of public life.

In *Culture Inc.*, Herbert Schiller, communication scholar from San Diego, argues that the proliferation of the shopping malls – more than 30,600 in the USA alone, greater than the number of secondary schools – threatens freedom of communication. The American First Amendment governs the rights of citizens with respect to their government and not with respect to private individuals and interests, and thus the owners of shopping malls have the right to forbid all forms of communication of which they do not approve. Apart from this formal, legal dimension, Herbert Schiller notes: 'The private ownership of the shopping centre – its primary characteristic – makes it also an inhospitable place for free circulation of ideas, to say nothing of social action. It is a locale where the likelihood of independent thinking, let alone action, is matter-of-factly excluded.'

In his study on Los Angeles *City of Quartz*, Mike Davis directs our attention to the precise and deliberate aims underlying the design and organisation of the pseudo-public spaces being created in cities in the new style: 'Ultimately the aims of contemporary architecture and the police converge most strikingly around the problem of crowd control. As we have seen, the designers of malls and pseudo-public space attack the crowd by homogenizing it. They set up architectural and semiotic barriers to filter the "undesirables." They enclose the mass that remains, directing its circulation with behaviourist ferocity. It is lured by visual stimuli of all kinds, dulled by musak, sometimes even scented by invisible aromatizers.'

NINE FRIVOLOUS SUMMERS: THE ESTATE ROMANA

Renato Nicolini cannot imagine harbouring such a hostile attitude toward the population of his own beloved city. At the time of this writing he was running for mayor of Rome, where he served as the alderman for culture from 1976 through 1985. During that period he was responsible for the remarkable initiative of the *Estate romana*, the *Roman Summer*.

The outdoor areas in the centre of the city, the historical Roman places, and the nights themselves became settings for artistic experiences that forged a collective feeling, contrary to the hermetic sensibility that ordinarily attends such events. Renato Nicolini prompted artists to break through conventional, elitist, and dull ways of working. He opened performance and exhibition venues at the Basilica di Massenzio at the Forum Romanum, the Villa Pamphili, the Castel Sant'Angelo, the Castel Porziano, and many other formerly sacrosanct places. Just two examples: everywhere in the city a concert featuring, for instance, the music of John Cage would suddenly burst forth, accompanied by watering carts of the urban cleaning department spouting water in all colours and fragrances. Achillo Bonito Oliva, an art critic and theoretician of the Italian *transavanguardia* organised a three kilometer open-air exposition of paintings on the Mura Aureliane.

For nine Roman summers Renato Nicolini broke with the traditional Italian system of *clientelismo*: 'I choose completely different artists, it changed every year, and I let them do as they liked. That is not the usual way here!' Giovanni Bechelloni, professor in the sociology of cultural processes at the University of Florence, judges that the social significance was even more important than the cultural one: 'Italy was in the grip of terrorism at the time, which created an antisocial climate. A myriad of strikes paralysed the country. People were afraid. The Estate romana represented the taking back of public space from the terrorists and ended the sense of public unsafety. It showed that the city was not a dangerous place.'

'Another purpose,' Renato Nicolini explains, 'was to give the common people the feeling that the urban centre is not a place only for the elite. We wanted to change the monocultural "destiny" of the city. It is awful to be imprisoned in a place consisting only of shops, shops, and more shops. A city flourishes when the centre serves a multitude of functions and is an open space for a mixture of people. Why shouldn't the old monuments of Rome be a part of current public life, rather than something remote and sacred?' The Estate romana was a nine year expression of the artistic vitality of the time, delivering musical, poetic, and dance events. It was full of images and films, everything bursting out any moment. It brought (a bit belatedly, to be sure) the beat generation to Rome and hosted the painters of the *transavanguardia*.

One of the much-discussed aspects of the Estate romana was its *effimero* (ephemeral) character. The film historian Lino Micciché explains the double meaning of *effimero*: 'transitory, passing, short-lived, fleeting, pro-visional – but also frivolous, not completely serious, superficial, not backed by any theory, pastiche. The Estate romana had an American face.' Culturally Italy is split in two segments, Giovanni Bechelloni explains: 'Italian intel-lectuals are looking to the cultures of France, Germany, or England, while the mass of the people feels attracted to Russia – although that has pretty much passed – or to the American way of life.' In this plenitude, spontaneity, and higgledy-piggledy quality Renato Nicolini feels most at home.

He reproaches Italian intellectuals for being unable to muster vigorous or effective response to the mass-produced entertainment presented night after night on television by Berlusconi's Fininvest. Renato Nicolini believes that 'the arts that are worthwhile must be announced and discussed in myriad ways, and presented in lively ambiences like a bright and entrancing cake-walk. We may be confident that people will not sit forever as slaves to the television. There is no reason to be pessimistic at all. But dull arts are not an alternative.'

The blithely ephemeral Estate romana proved to be *too* effimero – too weak, too nebulous. In 1985 Renato Nicolini had to give up his job as alderman for culture because his party, the then still-Communist PCI, lost the elections in the wave of growing conservatism. The Estate romana did not continue. Achille Bonito Oliva assesses it 'as a postmodernistic phenomenon, breaking open rigid habits. After nine summers, it was just enough, and afterwards everybody went their own way. At the same time as he created the Estate romana, Renato Nicolini did not address the fact that there are nearly no libraries in the neighbourhoods, no artistic infrastructure to maintain the arts, and nothing was done about the museums. His programme was not complete; it was only effimero, and that was a problem. What he accomplished could too easily be destroyed, which it was: his successor broke immediately and completely with the concept of cultural spontaneity.'

If Renato Nicolini had been elected mayor of Rome he would not have repeated the Estate romana. 'We have serious things to do like banning cars from the part of the city that is within the walls. In the peripheral quarters we must create a new concept of urban life. We must stop consumerism and pollution. The mass media were the *grand bouffe*, the big greedy pigs. That is enervating, but it will end. I would like to plan more, rather than depending so largely on spontaneous impulses. Obviously, the artists have a lot to do in those processes.'

This raises the question of whether artists are prepared to direct their attention to social issues and neighbourhoods. Renato Nicolini says: 'I'm not so pessimistic about the artists. When we start the enormous process of urban renewal they will be inspired again. They will get involved. When you give them chances and when you challenge them, the initiatives and the artistic impulses will follow.'

ARTS IN THE NEIGHBOURHOODS

One of the great hopes of the 1960s and '70s was that art would be disseminated throughout the neighbourhoods of big cities. Large segments of populations apparently did not feel at home in the traditional artistic facilities, so art had to be made available in the places people spent the bulk of their time. Schematically we may identify two tendencies in the great decentralising movement of the 1970s and early 1980s: one was based on the needs of alternative initiatives themselves; the other was directed at attracting the inhabitants of working-class neighbourhoods to the arts.

In the sixties the number of people who felt themselves attracted to artistic professions grew rapidly. They required places to rehearse, perform, exhibit, and live. In many nineteenth century neighbourhoods industrial buildings were free, and were squatted by artists. Later, prudent city councils gave such buildings to young entrepreneurs and to artistic initiatives at low costs. Some of those artists briefly felt that their work had to relate to people living in the neighbourhood where they worked, but this was not a deeply felt intention: most of their creations were too experimental to communicate in the neighbourhoods where they settled.

In many European cities the presence of alternative theatres and galleries changed the character of entire quarters. People in ancillary fields – art dealers, clothing designers, services such as packing, shipping, and framing of works of art – found cheap housing in such neighbourhoods. Pubs moved in, followed in Western Europe by alternative shops. Businesses began to flourish in these increasingly bohemian districts. Such conglomerations inspired further artistic initiatives.

Such by then fashionable neighbourhoods were discovered by developers and financial investors. This process started in the eighties and still takes place here and there. The cost of living in such neighbourhoods eventually becomes too high for artists, who must move on to find affordable living and working spaces elsewhere. This phenomenon is on the verge of taking place in Prenzlauer Berg, in Berlin-East. During the Communist period, many artists lived in this quarter and created a lively bohemian, oppositional ambience. Now Prenzlauer Berg in the centre of the unified Berlin is very attractive to business. Real estate prices are rising. Thomas Flierl, who is the leader of the cultural department of this quarter, is afraid that commerce will displace the artistic community.

'Of course we will increasingly rely on abstract modes of communication. But artistic communication still must have a place somewhere – not imaginary, but tangible and visible. Metropolises need such places, where the cultural discussion, workshop-style institutions, and people belonging to various social milieux and cultures can meet. During the DDR period, Prenzlauer Berg was such a niche with alternatives, experiments, and international contacts. Unified Berlin needs such a free place, where the social changes can be reflected and expressed.'

Thomas Flierl is not optimistic about the chances of Prenzlauer Berg surviving as an artist community. The same fate awaits its West Berlin counterpart, a neighbourhood called Kreuzberg, located in the centre of the city unified by the dismantling of the Wall.

A completely different movement in the sixties and seventies was to *disseminate* artistic awareness and practice in neighbourhoods, to encourage the active participation of inhabitants in artistic creation. The artists involved in those projects were politically motivated to make workers artistically competent, dependent neither on what was called bourgeois culture, nor on the output of cultural industries.

In one of the working-class quarters of Rome, Valentina Valentini presented arts groups and events that were not seen there in the normal course of events. The members of her artistic collective prompted the inhabitants to dramatise themes from their own history and their recent transition from the countryside to the city.

Valentina Valentini left the university where she had started work as a theatre historian to get involved in this project (she is now back at the university): 'It was large and complex, with a very high standard of quality. We had no prejudices about the capacity of people to deal with unfamiliar forms of artistic expression. Our cultural and political purpose was that ordinary people would escape the seduction to be the silent and obedient objects of mass media. We worked to help them find their own voices back again, and to discover their own relation with and choice for artistic expressions.'

All over Europe such experiments took place, and nearly everywhere these intentions were overwhelmed by the breakthrough of commercial television and the rising tide of yuppie values. Andrea Volo, a painter and the secretary general of the Italian artists' trade union (Sindicato Nazionale d'Artisti) looks back on the Estate romana and the work of artists in the neighbourhoods: 'We lost the battle in a fight we never even prepared for!'

Efforts in the name of cultural democratisation are still carried out today. The goals, however, are generally not as profound and radical as those just described. Rather than empowering the inhabitants of the neighbourhoods to develop their own critical competence and proficiency in the arts, the current efforts limit themselves to making accessible artistic and cultural productions beyond those created and distributed by the culture and entertainment industries, and to providing training and opportunities in artistic disciplines. The difference is that neither the appreciation nor the practices have the social criticality of the efforts undertaken during the sixties and seventies. One could call the present practice a process of *cultural extension* but not really cultural *empowerment*.

[. . .]

SMALL CITIES: EXCELLENT BREEDING GROUNDS

The high visibility conferred by both the attractiveness and the enormous problems facing the famous metropolises makes it easy to forget that a substantial part of the European population lives in the smaller cities and towns. Smaller metropolitan areas are where children and adults will have their only opportunity to encounter the arts. If no such opportunities exist, most of them will never attend a dance performance or a concert, will never see a film on wide screen, will never enter a bookstore or a gallery, and will not have their own artistic potential nurtured.

A journey to the big city is an exceptional occasion, generally reserved for attending the more spectacular events. To be meaningful, ongoing exposure to the work of artists past and present must take place where people live. As regards the arts, the mountain must go to Mohammed, whether living in a fishing village in the Shetland islands, or in a working-class neighbourhood in Marseilles.

Today a delicate question is whether the inhabitants will have the opportunity to be exposed to a diversity of arts that are not superficial. I discussed this with several people, and we came to one important conclusion. A serious threat to the cultural life of the community arises when some influential social force – perhaps the mayor or the alderman for culture – gives a community of only one or two hundred thousand inhabitants the shine of a full-blown city.

The actress Litten Hansen and the actor Troels Munk – respectively, the outgoing and incoming president of the Danish Actors Union – can recite by heart the list of small cities in their country that have made this mistake: 'They have built enormous theatres. But after the prestige-raising decision has been made and the building has opened its doors, three things were lacking: the money to programme the theatre on a regular basis; the audience to fill the hall; and sufficient artistic creativity for making such a project a success.' Meanwhile the financial burden prevents the city or town investing in more modestly scaled artistic developments that are commensurate with the actual size of the city.

This does not mean that spectacular artistic projects are categorically wrong for medium-sized cities; the suitability depends on the existence of a strong social and artistic base in the community to support activities extra-ordinary in scope or ambition. Hans Dona, a former alderman for culture in Den Bosch, a city with a population of one hundred thousand in the southern part of the Netherlands, follows a policy of looking 'at what naturally prospers in one artistic field or another.' He feels that it is senseless to promote events that are actually prestigious hobbies. To establish a good profile as a medium-sized city, there is only one responsible way to do it: stimulate what is already happening in the artistic field in the community.

In the opinion of Hans Dona it only makes sense to support eye-catching projects when they are embedded in a broader cultural infrastructure of the city or town and have the support of different layers of the arts community. In fact, cities of moderate size have two obligations: one is toward their own population; the other is related to further development of the arts.

Hans Dona regrets the fashionable attitude that focuses attention only on the arts that already have a recognised value. 'A still-developing, not-yet-arrived, inchoate but promising art exists. Those experiments are important to ensure a high quality art in the future. But abundance and variety of artistic work of more ordinary quality has a value in itself: it gives life colour and helps develop deeper understanding.' It is the traditional role of the medium-sized cities to support and to cherish artistic evolution.

To invite and to lodge such a diversity of artists has a function for the cultural life in the city or town. They may contribute to arts education in the schools and to developing the artistic skills of amateurs and neophytes. Hans Dona claims that artists are the backbone of the artistic infrastructure of the community. Therefore the city should tap their professional skills in raising the quality of public spaces and the built environment: 'You must provide artists with studio space, commission work from them, and provide an opportunity to exhibit. The city must subsidise and stimulate performances and festivals and maintain theatres, rock music venues, and small museums.'

This decentralised approach to artistic development is an important contribution to internationalism, Dragan Klaic, director of Theatre Institute Netherlands says: 'In the heads of many politicians internationalism is still for prestige. But if you give the local city politicians an understanding of European culture and convince them that they can be the factors in creating this European culture, it is possible to find admirable amount of support that you wouldn't otherwise have.'

Smaller cities can be more flexible and faster in making decisions. They can more easily recognise cultural and artistic resources in their area. Those resources may represent a precious artistic value which connects self-evidently with a city abroad. Such activities may help make a city recognizable on an international level. It may also provide extra political support because it can be said: 'Look what international culture we are bringing to you citizens; we are making you citizens of Europe, even when you live in a small town of a hundred thousand people!'

SECTION IV

CULTURE AND TECHNOLOGIES

INTRODUCTION

The cultures of modernity are products of technological change. That is, their forms reflect new possibilities of production and dissemination which new technologies both make possible and suggest might be possible. At the same time, new technologies are products of a cultural climate, and their use changes according to shifts of ideology. Hence, the distribution of the Bible in vernacular languages during the sixteenth and seventeenth centuries was enabled by the invention of printing, but the invention of printing itself sprang from the culture of an urban class of traders (the bourgeois class) and their identification of city life with freedom, including a freedom to hold faith without priestly mediation.

In the twentieth century, the spread of mechanical means of reproduction for image and sound, in recorded music and film, or postcards of works of art, have two consequences: the challenge of such reproductive technologies to traditional ideas of the aura of the work of art, and the development of cultural industries (dealt with in Section III). Walter Benjamin, in his essay 'The Work of Art in the Age of Mechanical Reproduction' (included in *Illuminations*, 1970, Cape), to which the reader is referred, writes that art's authenticity is lost when the work ceases to have a unique presence in time and space. This detaches art from the historical moment captured in its making, and allows a mass audience to participate in what becomes a new art form, so that, it could be argued, film takes on the social role of painting in a new and distinctive way.

Most technological advances take place in urban settlements, facilitated by a concentration of capital and labour. But not all begin there. Lewis Mumford, in the extract from *The Human Prospect* (1956, Beacon Press) included here, traces the derivation of the seemingly commonplace town clock to the regular marking of the hours of religious office in monasteries. What began as an aid to spiritual observance becomes, under capitalism, a means to regulate production. The clock is thus no longer commonplace, but a sign of quantitative thought and the mechanism of exchange. This demonstrates the shift in values which takes place when a technology is transferred from one ideological context to another. Mumford sees the clock rather than the steam engine as the central machine of modern times.

One of the assumptions on which much urban social theory is based is that there is something distinct about metropolitan culture and lifestyle, that it differs radically from that of the small town or rural village. This is expressed in concepts such as the twenty-four hour city where bars remain open and public transport runs all night, and, in the text by the Archigram group, the instant city. This plug-in city, as much polemic as project, is conceived as taking metropolitan life on tour to the disadvantaged provinces. This mobile cultural feast, which defines metro-politanism largely in terms of new technologies of image reproduction, such as television, video and audio-visual display, was tested out in cultural sites such as the Kassel Documenta, Germany's most prominent art fair, in 1972.

The theme of electronic and digital media is taken up by Nigel Thrift in his essay on a 'Hyperactive World' (1995). Thrift mentions the twenty-four hour dealing rooms of world financial markets. In these virtual spaces, a world is constructed with neither horizon nor limits to speculation. It might be seen as the ultimate Cartesian realm, in which a self (as subject) not only observes an objectified world, but more or less invents it. But Thrift reminds us that even those of us 'waiting for the bus' rather than working in the financial fast lane are directly affected by the technology of plastic money and its networks of telecommunications. Thrift ends with a paradox: that the increasing weight of information which electronic technology can transfer leads not to legibility but its opposite, as the corresponding burden of interpretation grows beyond capacity. Even this world of superhighways, then, gets traffic jams.

Most writing on new technologies locates them in public domains such as commerce or the state, from cash machines to surveillance cameras and public broadcasting. But technologies affect the domestic sphere, too. It is possible, and the equipment is being piloted in Finland, for someone to control all the electric gadgets in a house, from entryphone to microwave, through a mobile phone. Whilst the technology which changed city life in the early years of the century was evident in new forms of transport, such as the tram and the motor car, today it is in the invisible form of electronic communications.

Stephen Graham and Simon Marvin write about the changing realities of the home, such as teleworking, in an extract from *Telecommunications and the City* (1996, Routledge). If the city of the nineteenth century, or twentieth-century Fordism, grew as workforces expanded, now those accumulations of human resource are dispersing, perhaps regaining some control over their working pattern and environment. Graham and Marvin note that in France the Minitel videotex system serves a third of households. They foresee homes in which all appliances will work on integrated electronic command systems, and where shopping is done in virtual reality.

Finally, how does contemporary technology affect the production and reception of architecture? Will new-technology-driven lifestyles require new approaches to environmental design? Paul Virilio, in his essay 'The Overexposed City' from the collection *Rethinking Architecture* (1997, Routledge) edited by Neil Leach (see suggestions for further reading), writes that 'From here on, urban architecture has to work with the opening of a new "technological time-space".' His statement echoes that of a mayor of Philadelphia, whom he quotes in his opening sentence: 'From here on in, the frontiers of the state pass into the interior of cities.'

Virilio is, in the end, pessimistic (or participates in an erosion of value linked to deconstructive postmodernism). Yet his analysis of the over-exposure of the city in the surfaces of television seems also like an extension to absurdity of Benjamin's view of cinematography. In the filmic close-up, Benjamin saw a deeper knowledge of what orders everyday life, and an opening of a field of action. On one hand, otherwise invisible moments of action, such as the exact position of a walker in mid-stride, are made visible; on the other, the manipulation of images in (what were then) new technologies enables new freedoms of representation.

20 LEWIS MUMFORD

'The Monastery and the Clock'

from *The Human Prospect* (1956)

One is not straining the facts when one suggests that the monasteries ... helped to give human enterprise the regular collective beat and rhythm of the machine; for the clock is not merely a means of keeping track of the hours, but of synchronizing the actions of men.

Where did the machine first take form in modern civilization? There was plainly more than one point of origin. Our mechanical civilization represents the convergence of numerous habits, ideas, and modes of living, as well as technical instruments; and some of these were, in the beginning, directly opposed to the civilization they helped to create. But the first manifestation of the new order took place in the general picture of the world: during the first seven centuries of the machine's existence the categories of time and space underwent an extraordinary change, and no aspect of life was left untouched by this transformation. The application of quantitative methods of thought to the study of nature had its first manifestation in the regular measurement of time; and the new mechanical conception of time arose in part out of the routine of the monastery. Alfred Whitehead has emphasized the importance of the scholastic belief in a universe ordered by God as one of the foundations of modern physics: but behind that belief was the presence of order in the institutions of the Church itself.

The technics of the ancient world were still carried on from Constantinople and Baghdad to Sicily and Cordova: hence the early lead taken by Salerno in the scientific and medical advances of the Middle Age. It was, however, in the monasteries of the West that the desire for order and power, other than that expressed in the military domination of weaker men, first manifested itself after the long uncertainty and bloody confusion that attended the breakdown of the Roman Empire. Within the walls of the monastery was sanctuary: under the rule of the order surprise and doubt and caprice and irregularity were put at bay. Opposed to the erratic fluctuations and pulsations of the worldly life was the iron discipline of the rule. Benedict added a seventh period to the devotions of the day, and in the seventh century, by a bull of Pope Sabinianus, it was decreed that the bells of the monastery be rung seven times in the twenty-four hours. These punctuation marks in the day were known as the canonical hours, and some means of keeping count of them and ensuring their regular repetition became necessary.

According to a now discredited legend, the first modern mechanical clock, worked by falling weights, was invented by the monk named Gerbert who afterwards became Pope Sylvester II, near the close of the tenth century. This clock was probably only a water clock, one of those bequests of the ancient world either left over directly from the days of the Romans, like the water-wheel itself, or coming back again into the West through the Arabs. But the legend, as so often happens, is accurate in its implications if not in its facts. The monastery was the seat of a regular life, and an instrument for striking the hours at intervals or for reminding the bell-ringer that it was time to strike the bells, was an almost inevitable product of this life. If the mechanical clock did not appear until the cities of the thirteenth century demanded an orderly routine, the habit of order itself and the earnest regulation of time-sequences had become almost second nature in the monastery. Coulton agrees with Sombart in looking upon the Benedictines, the great working order, as perhaps the original founders of modern capitalism: their rule certainly took the curse off work and their vigorous engineering enterprises may even have robbed warfare of some of its glamor. So one is not straining the facts when one suggests that the monasteries – at one time there were 40,000 under the Benedictine rule – helped to give human enterprise the regular collective beat and rhythm of the machine; for the clock is not merely a means of keeping track of the hours, but of synchronizing the actions of men.

Was it by reason of the collective Christian desire to provide for the welfare of souls in eternity by regular prayers and devotions that time-keeping and the habits of temporal order took hold of men's minds: habits that capitalist civilization presently turned to good account? One must perhaps accept the irony of this paradox. At all events, by the thirteenth century there are definite records of mechanical clocks, and by 1370 a well-designed "modern" clock had been built by Heinrich von Wyck at Paris. Meanwhile, bell towers had come into existence, and the new clocks, if they did not have, till the fourteenth century, a dial and a hand that translated the movement of time into a movement through space, at all events struck the hours. The clouds that could paralyze the sundial, the freezing that could stop the water clock on a winter night, were no longer obstacles to time-keeping: summer or winter, day or night, one was aware of the measured clank of the clock. The instrument presently spread outside the monastery; and the regular striking of the bells brought a new regularity into the life of the workman and the merchant. The bells of the clock tower almost defined urban existence. Time-keeping passed into time-serving and time-accounting and time-rationing. As this took place, Eternity ceased gradually to serve as the measure and focus of human actions.

The clock, not the steam-engine, is the key-machine of the modern industrial age. For every phase of its development the clock is both the outstanding fact and the typical symbol of the machine: even today no other machine is so ubiquitous. Here, at the very beginning of modern technics, appeared prophetically the accurate automatic machine which, only after centuries of further effort, was also to prove the final consummation of this technics in every department of industrial activity. There had been power-machines, such as the water-mill, before the clock; and there had also been various kinds of automata, to awaken the wonder of the populace in the temple, or to please the idle fancy of some Moslem caliph: machines one finds illustrated in Hero and Al-Jazari. But here was a new kind of power-machine, in which the source of power and the transmission were of such a nature as to ensure the even flow of energy throughout the works and to make possible regular production and a standardized product. In its relationship to determinable quantities of energy, to standardization, to automatic action, and finally to its own special product, accurate timing, the clock has been the foremost machine in modern technics: and at each period it has remained in the lead: it marks a perfection toward which other machines aspire. The clock, moreover, served as a model for many other kinds of mechanical works, and the analysis of motion that accompanied the perfection of the clock, with the various types of gearing and transmission that were elaborated, contributed to the success of quite different kinds of machine. Smiths could have hammered thousands of suits of armor or thousands of iron cannon, wheelwrights could have shaped thousands of great water-wheels or crude gears, without inventing any of the special types of movement developed in clockwork, and without any of the accuracy of measurement and fineness of articulation that finally produced the accurate eighteenth-century chronometer.

The clock, moreover, is a piece of power-machinery whose "product" is seconds and minutes: by its essential nature it dissociated time from human events and helped to create the belief in an independent world of mathematically measurable sequences: the special world of science. There is relatively little foundation for this belief in common human experience: throughout the year the days are of uneven duration, and not merely does the relation between day and night steadily change, but a slight journey from East to West alters astronomical time by a certain number of minutes. In terms of the human organism itself, mechanical time is even more foreign: while human life has regularities of its own, the beat of the pulse, the breathing of the lungs, these change from hour to hour with mood and action, and in the longer span of days, time is measured not by the calendar but by the events that occupy it. The shepherd measures from the time the ewes lambed; the farmer measures back to the day of sowing or forward to the harvest: if growth has its own duration and regularities, behind it are not simply matter and motion but the facts of development: in short, history. And while mechanical time is strung out in a succession of mathematically isolated instants, organic time – what Bergson calls duration – is cumulative in its effects. Though mechanical time can, in a sense, be speeded up or run backward, like the hands of a clock or the images of a moving picture, organic time moves in only one direction – through the cycle of birth, growth, development, decay, and death – and the past that is already dead remains present in the future that has still to be born.

Around 1345, according to Thorndike, the division of hours into sixty minutes and of minutes into sixty seconds became common: it was this abstract framework of divided time that became more and more the point of reference for both action and thought, and in the effort to arrive at accuracy in this department, the astronomical exploration of the sky focused attention further upon the regular, implacable movements of the heavenly bodies through space. Early in the sixteenth century a young Nuremberg mechanic, Peter Henlein, is supposed to have created "many-wheeled watches out of small bits of iron" and by the end of the century the small domestic clock had been introduced in England and Holland. As with the motor car and the airplane, the richer classes first took over the new mechanism and popularized it: partly because they alone could afford it, partly because the new bourgeoisie were the first to discover that, as Franklin later put it, "time is money." To become "as regular as clockwork" was the bourgeois ideal, and to own a watch was for long a definite symbol of success. The increasing tempo of civilization led to a demand for greater power: and in turn power quickened the tempo.

Now, the orderly punctual life that first took shape in the monasteries is not native to mankind, although by now Western peoples are so thoroughly regimented by the clock that it is "second nature" and they look upon its observance as a fact of nature. Many Eastern civilizations have flourished on a loose basis in time: the Hindus have in fact been so indifferent to time that they lack even an authentic chronology of the years. Only yesterday, in the midst of the industrializations of Soviet Russia, did a society come into existence to further the carrying of watches there and to propagandize the benefits of punctuality. The popularization of time-keeping, which followed the production of the cheap standardized watch, first in Geneva, then in America around the middle of the last century, was essential to a well-articulated system of transportation and production.

To keep time was once a peculiar attribute of music: it gave industrial value to the workshop song or the tattoo or the chantey of the sailors tugging at a rope. But the effect of the mechanical clock is more pervasive and strict: it presides over the day from the hour of rising to the hour of rest. When one thinks of the day as an abstract span of time, one does not go to bed with the chickens on a winter's night: one invents wicks, chimneys, lamps, gaslights, electric lamps, so as to use all the hours belonging to the day. When one thinks of time, not as a sequence of experiences, but as a collection of hours, minutes, and seconds, the habits of adding time and saving time come into existence. Time took on the character of an enclosed space: it could be divided, it could be filled up, it could even be expanded by the invention of labor-saving instruments.

Abstract time became the new medium of existence. Organic functions themselves were regulated by it: one ate, not upon feeling hungry, but when prompted by the clock: one slept, not when one was tired, but when the clock sanctioned it. A generalized time-consciousness accompanied the wider use of clocks: dissociating time from organic sequences, it became easier for the men of the Renaissance to indulge the fantasy of reviving the classic past or of reliving the splendors of antique Roman civilization: the cult of history, appearing first in daily ritual, finally abstracted itself as a special discipline. In the seventeenth century journalism and periodic literature made their appearance: even in dress, following the lead of Venice as fashion-center, people altered styles every year rather than every generation.

The gain in mechanical efficiency through co-ordination and through the closer articulation of the day's events cannot be overestimated: while this increase cannot be measured in mere horsepower, one has only to imagine its absence today to foresee the speedy disruption and eventual collapse of our entire society. The modern industrial régime could do without coal and iron and steam more easily than it could do without the clock.

21 ARCHIGRAM

'Instant City'

from Peter Cook et al. (eds), *Archigram* (1972)

THE NOTION

In most civilized countries, localities and their local cultures remain slow moving, often undernourished and sometimes resentful of the more favoured metropolitan regions (such as New York, the West Coast of the United States, London and Paris). Whilst much is spoken about cultural links and about the effect of television as a window on the world (and the inevitable global village) people still feel frustrated. Younger people even have a suspicion that they are missing out on things that could widen their horizons. They would like to be involved in aspects of life where their own experiences can be seen as part of what is happening.

Against this is the reaction to the physical nature of the metropolis: and somehow there is this paradox – if only we could enjoy it but stay where we are.

The Instant City project reacts to this with the idea of a 'travelling metropolis', a package that comes to a community, giving it a taste of the metropolitan dynamic – which is temporarily grafted on to the local centre – and whilst the community is still recovering from the shock, uses this catalyst as the first stage of a national hook-up. A network of information – education – entertainment – 'play-and-know yourself' facilities.

In England the feeling of being left out of things has for a long time affected the psychology of the provinces, so that people become either over-protective about local things, or carry in their minds a ridiculous inferiority complex about the metropolis. But we are nearing a time when the leisure period of the day is becoming really significant; and with the effect of television and better education people are realizing that they could do things and know things, they could express themselves (or enjoy themselves in a freer way) and they are becoming dissatisfied with the television set, the youth club or the pub.

A BACKGROUND FROM ARCHIGRAM WORK

The old Plug-in City programme of 1964 pulled together a series of seemingly disconnected notions and small projects (a throwaway unit here, an automatic shop there, or even an idea about a megastructure), reinforcing and qualifying the theme and eventually suggesting a total project – a portmanteau for the rest. Later on, the work that Archigram had done with the Hornsey Light-Sound Workshop and on several exhibitions (with the actual techniques of audience participation and control of a responsive audio-visual system) began to form a working laboratory for the techniques of Instant City.

The Instant City is both collective and coercive: by definition there is no perfect set of components. On the drawings which have been made over a period of two years there are often quotes from other pieces of work (for instance, Oslo Soft Scene Monitor as the parent of Audio-Visual Juke Box). In such a machine people tune in to their environment by choosing and making from a range of audio-visual programmes; the Oslo machine is its progenitor but it really implies something where the participant plays a completely open-ended creative game. Around 1966, at an exhibition on Brighton pier, we made an experiment of putting a man in a circular drum, spinning him round and then bombarding him (two feet from his face) with wild coloured slides and bombarding him (one foot from his ears) with wild sounds. A typical instance of a first-stage experiment, unsubtle and without feedback, unable to provide the man inside with a button to press to say 'stop'. Later, the Oslo machine moved on from this and the limit of choice was one of cost rather than of concept.

With our notion of the robot (the symbol of the responsive machine that collects many services in one appliance), we begin to play with the notion that the environment could be conditioned not only by the set piece assembly but by infinite variables determined by your wish, and the robot reappears in the Instant City in several of the assemblies.

The Instant City has planning implications the force of which have emerged more and more strongly as the project has developed, so that by the time we are making the sequence describing 'the airship's effect upon the sleeping town' it is the infiltrationary dynamic of the town itself that is as fascinating as the technical dynamic of the airship. Again we have to reflect on the psychology of a country such as England where there is a historical suggestion that vast upheaval is unlikely. We are likely to capitalize on existing institutions and existing facilities whilst complaining about their inefficiency – but a country such as England must now live by its wits or perish, and for its wits it needs its culture.

A PROGRAMME BACKGROUND

The likely components are audio-visual display systems, projection television, trailered units, pneumatic and lightweight structures and entertainments facilities, exhibits, gantries and electric lights.

This involves the theoretical territory between the 'hardware' (or the design of buildings and places) and 'software' (or the effect of information and programmation of the environment). Theoretically it also involves the notions of urban dispersal and the territory between entertainment and learning. The Instant City could be made a practical reality since at every stage it is based upon existing techniques and their application to real situations. There is a combination of several different artefacts and systems which have hitherto remained as separate machines, enclosures or experiments. The programme involved gathering information about an itinerary of communities and the available utilities that exist (clubs, local radio, universities, etc.) so that the 'City' package is always complementary rather than alien. We then tested this proposition against particular samples.

The first stage programme consisted of assemblies carried by approximately twenty vehicles, operable in most weathers and carrying a complete programme. These were applied to localities in England and in the Los Angeles area of California. Later, having become interested in the versatility of the airship, we came to propose this as another means of transporting the Instant City assembly (a great and silent bringer of the whole conglomeration).

Later we applied the method of the Instant City to proposals for servicing the Documenta exhibition at Kassel in Germany. By this time also there had developed a feedback of ideas and techniques between this project and our Monte Carlo entertainments facility.

A TYPICAL SEQUENCE OF OPERATIONS (TRUCK-BORNE VERSION)

1   The components of the 'City' are loaded on to the trucks and trailers at base.
2   A series of 'tent' units are floated from balloons which are towed to the destination by aircraft.
3   Prior to the visit of the 'City' a team of surveyors, electricians, etc. have converted a disused building in the chosen community into a collection, information and relay station. Landline links have been made to local schools and to one or more major (permanent) cities.
4   The 'City' arrives. It is assembled according to site and local characteristics. Not all components will necessarily be used. It may infiltrate into local buildings and streets, it may fragment.
5   Events, displays and educational programmes are partly supplied by the local community and partly by the 'City' agency. In addition major use is made of local fringe elements: fares, festivals, markets, societies, with trailers, stalls, displays and personnel accumulating often on an *ad hoc* basis. The event of the Instant City might be a bringing together of

events that would otherwise occur separately in the district.

6   The overhead tent, inflatable windbreaks and other shelters are erected. Many units of the 'City' have their own tailored enclosure.

7   The 'City' stays for a limited period.

8   It then moves on to the next location.

9   After a number of places have been visited the local relay stations are linked together by landline. Community (1) is now feeding part of the programme to be enjoyed by Community (20).

10   Eventually by this combination of physical and electronic, perceptual and programmatic events and the establishment of local display centres, a 'City' of communication might exist, the metropolis of the national network.

11   Almost certainly, travelling elements would modify over a period of time. It is even likely that after two to three years they would phase out and let the network take over.

As the Instant City study developed, certain items emerged in particular. First, the idea of a 'soft-scene monitor' – a combination of teaching-machine, audio-visual juke box, environmental simulator, and from a theoretical point of view, a realization of the 'Hardware/Software' debate (which is still going on in our Monte Carlo work, as the notion of an electronically-aided responsive environment). Next, the dissolve of the original large, trucked, circus-like show into a smaller and very mobile element backed by a wonderful, magical dream descending from the skies. The model of the small unit suggests two trucks and a helicopter as the carriage, with quick-folding arenas and apparatus that can quickly be fitted into the village hall. Another stimulus was the invitation to design the 'event-structure' for the 1972 Kassel Documenta – an elaborate art/event/theatre scene requiring a high level of servicing but a minimum of interference with the 'open-air creative act'. 'The Kassel-Kit' of apparatus can therefore be considered as a direct extension of the original IC Kit.

Are we back to heroics then, with a giant, pretty and evocative object? The Blimp: the airship: beauty, disaster and history. On the one hand we were designing a totally unseen and *underground* building at Monte Carlo, and on the other hand flirting with the airborne will-o-the-wisp. The Instant City as a series of trucks rushing round like ants might be practical and immediate, but we could not escape the loveliness of the idea of Instant City appearing from nowhere, and after the 'event' stage, lifting up its skirts and vanishing. In fact, the primary interest was spontaneity, and the remaining aim to knit into any locale as *effectively* as possible. For Archigram, the airship is a device: a giant skyhook.

Operationally, there were two possibilities. A simple airship with apparatus carried in the belly and able to drop down as required. Otherwise, a more sophisticated notion of a 'megastructure of the skies'. Ron Herron's drawings suggest that the 'ship' can fragment, and the audio/visual elements are scattered around a patch of sky. Once again, the project work of the group has picked up a dream of its own past – the 'Story of a Thing' made (almost) real.

We then built a model, which could hang out its entrails in a number of ways. This was the simpler 'ship' which reads with the scenario of a small town transformation. In the drawing with airship 'Rupert', a major shift in Instant City was first articulated: the increasing feeling for change-by-infiltration. The 'city' is creeping into half-finished buildings, using the local draper's store, gas showrooms and kerbside, as well as the more sophisticated setup. And there is a mysterious creeping animal: the 'leech' truck, which is able to climb up any structure and service from it: with the resulting possibility of 'bugging' the whole town as necessary. Gently then, the project dissolves from the simple mechanics or hierarchies of 'structuring' and like-objects. Just as did the Plug-in City: it sowed the seeds of its own fragmentation into investigations of a gentler, more subtle environmental tuning.

IDEAS CIRCUS

The notion of the Ideas circus came after the experience of several Archigram lectures and seminars where common characteristics of college and exhibition facilities could be experienced.

There is little interchange of ideas between one institution and the next, and display or documentation facilities have to be erected from scratch.

SCHEME

To institute a standard package of five or six vehicles that contain all the equipment necessary to set up a seminar, conference, exhibition, teach-in or display. The package can be attached to an existing building, plugging-into such facilities as are there and using the shelter of existing rooms for Circus equipment. The Circus can also be completely autonomous: set up in a field, if necessary. The idea would be to circulate between major provincial centres, tapping local universities, bleeding-off from them personalities, documentation and such things as film of laboratory experiments: then carrying on to the next town. Weekend visits to smaller places could be made. Some vehicles could hive off for an afternoon teach-in at the local Women's Institute. The Circus would be programmed with basic film and slide material. The feedback facility is most important: verbatim documentation of seminars, documents, films, etc., would be printed-off and left behind. Static educational facilities need topping-up. Mobile educational facilities could so easily be a nine-day wonder. The Ideas Circus is offered as a tool for the interim phase: until we have a really working all-way information network.

In the four weeks, the Circus is first programmed from London with tapes, filmstrips, etc., on the tour subject. These are prepared with help from institutions available.

The centres visited are geographically fairly close so that little time is spent actually on the road. In the multi-vehicle version there can be a programmed echeloning of the constituent parts so as to make best use of time and resources. In this version a single unit (vehicle simplified programme) can be sent to small towns nearby for a one-night stand or 'appetiser' demonstration. The instigation of a national information network such as that shown here between universities is important but not absolutely essential. Special 'personal boost' lectures are suggested and these plus the landlined information instance a meshing of the Circus network with any other available network.

The Circus is here shown as involved with an 'academic' tour such as 'Microbiology for All', 'New Maths', 'Modern Architecture', or whatever.

It could equally well cover similar territory with a commercial promotion or other non-academic tour.

22 NIGEL THRIFT

'A Hyperactive World'

from R.J. Johnston et al. (eds), *Geographies of Global Change: Remapping the World in the Late Twentieth Century* **(1995)**

INTRODUCTION

That light overhead, the one gliding slowly through the night sky, is a telecommunications satellite. In miniature, it contains the three main themes of this chapter. Through it, millions of messages are being passed back and forth. Because of it, money capital seems to have become an elemental force, blowing backwards and forwards across the globe. As a result of innovations like it, the world is shrinking – many places seem closer together than they once did.

The satellite is itself a sign of a world whose economies, societies and cultures are becoming ever more closely intertwined – a process which usually goes under the name of *globalization* (Giddens 1991). But what sense can we make of this process of globalization? Again, the satellite provides some clues. Those millions of messages signify a fundamental problem of *representation*. Simply put, the world is becoming so complexly interconnected that some have begun to doubt its very *legibility*. The swash of money capital registering in the circuits of the satellite comes to signify the "hypermobility" of a new *space of flows*. In this space of flows, money capital has become like a hyperactive child, unable to keep still even for a second. Finally, the shrinking world that innovations like the satellite have helped to bring about is signified by *time–space compression*. Places are moving closer together in electronic space and, because of transport innovations, in physical space too. These three simple themes of legibility, the space of flows, and time–space compression can therefore be seen as "barometers of modernity" (Descombes 1993), big ideas about what makes our modern world "modern."

[. . .]

[However] I hope to show, through a double-take on the world of international money, just how partial these accounts are, even in that sphere of the world's economy which might be expected most closely to approximate to them. Then in the last part of the chapter I want to propose the beginnings of a more moderate account of the signs of the times.

[. . .]

CRITIQUE 2: HOOKED ON SPEED

Take 1: masters of the universe?

An image which has become a cliché. A foreign exchange dealing room of a major bank in London, New York, or Tokyo. The mainly young men and women who inhabit these rooms for ten or eleven hours at a time are under pressure. They are under pressure to make profits. They are under pressure from their fellow traders – they don't want to lose face by screwing up a deal. They are under the pressure of constant surveillance – from managers, from video cameras, from tape recorders capturing all their calls. Above all, they are under pressure of time. Dealing itself is largely a matter of timing and dealers are expected not only to make profits from their deals but to make them quickly. You are only as good as your last deal.

To cap it all, these dealers are at the sharp end of time–space compression. Their world is a world where telecommunications have become more and more sophisticated and, as a result, space has virtually been annihilated by time. A dealer's world consists of a few immediate colleagues, the electronic screens which are the termini of electronic networks that reach round the world to other colleagues in other cities, and the electronic texts that can be read off the battery of screens. If there is a space of flows, then this is it.

Certainly, what these dealers can do to the world is, in its own way, quite extraordinary. In any day's work of foreign exchange dealing, currencies are being transmitted backwards and forwards across the world by dealers to the tune of $100 billion in a day. This money is increasingly able to gainsay governments. As one dealer put it, "it's a huge big global casino. If a government steps out of line they get their currency whacked." There are plenty of governments that can attest to this, as the example of the European Exchange Rate System (the ERM) shows only too well.

The ERM, introduced as a means of transforming the European Union into a zone of monetary stability in which the exchange rates of the different European national currencies would vary in an ordered and predictable way, came under an intense series of speculative assaults from foreign exchange traders over the course of 1992 and 1993.

[. . .]

The results of these speculative assaults were devastating. All but one ERM currency was devalued against the deutschmark, the anchor currency of the system; two currencies – sterling and the lira – were forced to leave the system altogether, and the rules of ERM membership were relaxed to such a degree that it became a pale shadow of its former self.

An example like this might be used to suggest that international money can flow where it will without let or hindrance – in other words, we see here a true space of flows. But like those apparently fraught young men and women in the dealing room . . . this is something of an exaggeration. First of all, the "phantom state" of international money is a nomad state. That is, it has no permanent spaces to call its own, only a series of transient sites in a few global cities. This constant mobility has its advantages. In particular, the world of international money is

difficult to tie down. But it also has its disadvantages. The phantom state is always in danger of being trapped by nation states which control territories, and are able to regulate what goes on within them. Thus this space of flows can be choked off by the rules nation states impose – like capital adequacy ratios, which force banks to set aside a certain portion of their capital, or rules on how or what financial investments to use. Secondly, the phantom state has to be constantly in motion, chasing into all the nooks and crannies of the world economy that might produce a profit. Such a task requires an enormous investment of not only money but also communication. Nowadays money is essentially information and getting that information, interpreting it, and using it at the right time, requires constant human interaction. The result is that the hypermobile world of international money is actually a hypersocial world, a world of constant interchange between people, whether over electronic networks, or in face-to-face meetings, or at the end of often lengthy journeys. In this sense this world of flows is not abstract at all – it is the product of and it is produced by people communicating about what is going on.

So the barometers of illegibility, of a space of flows, and of time–space compression can clearly be seen to be partial, even when an example is chosen which should cast them in a favorable light. But there is one last point that needs to be made . . . for the denizens of the world of international money these barometers do not represent some new condition. They are practiced in living with them. Since the international financial system has been in operation, its practitioners have had to live with uncertainty, using only limited information to assess the risks they run in investing money. Since international financial markets started to coalesce in the late nineteenth century, because of the telegraph and then the telephone, their practitioners have become well versed in living with time–space compression. It is all part of the game they play every day and it is a game they are good at.

Take 2: networks and ghettoes

The dealers in the international financial system live life in the fast lane. But what about those of us waiting for the bus or cursing the late train?

For us too, monetary transaction is speeding up. The installation of credit cards, automated teller machines (ATMs), and the like means that life in the slow lane is moving faster.

> I'm in Paris, it's late evening, and I need money quickly. The bank I go to is closed ... but outside is an ATM ... I insert my ATM card from my branch in Washington DC and punch in my identification number and the amount of 500 francs, roughly equivalent to $300. The French bank's computers detect that it is not their card, so my request goes to the Cirrus system's inter-European switching centre in Belgium which detects that it is not a European card. The electronic message is then transmitted to the global switching centre in Detroit, which recognises that there's more than $300 in my account in Washington and deducts $300 plus a fee of $1.50. Then it's back to Detroit, to Belgium and to the Paris bank and its ATM and out comes $300 in French francs. Total elapsed time: 16 seconds.

Increasingly, we are all dependent on the speed and processing power of telecommunications that examples like this illustrate. But the new space of telecommunications is not, in reality, a smooth global space over which messages can flow without friction. It is a skein of *networks* which are "neither local nor global but are more or less long and more or less connected" (Latour 1993). To think otherwise is to mistake length or connection for differences in scale level, to believe that some things (like people or ideas or situations) are "local" whilst others (like organizations or laws or rules) are "global."

The late twentieth century is covered by a lattice of networks. Public and private, civil and military, open and closed, the networks carry an unimaginable volume of messages, conversations, images and commands. By the early 1990s, the world's population of 600 million telephones and 600 million television sets will have been joined by over 100 million computer workstations, tens of millions of home computers, fax machines, cellular phones and pagers (Mulgan 1991).

But when we see this world of telecommunications as indeed a world of networks, we can also see other things. First of all, telecommunications networks still rely on many hundreds of thousands of people and machines all around the world who build, monitor, repair, and use them, just like the navvies and the telegraph engineers did of old. Second, modern telecommunications networks may be hybrid systems, in which machines and people are increasingly mixed together in queer combinations, but this does not make them more abstract or more abstracted. They break down. They stutter. They pause. They make errors. Whether it is a case of a vandalized ATM, a line fault or an atmospheric disturbance, these networks are not self-sustaining. Third, these networks can be organized in different ways which are more or less effective. Fourth, not everyone is connected to these networks. The new telecommunications networks have produced "electronic ghettoes" (Davis 1992) in which the only signs of globalization that can be found flicker across television screens, endlessly mocking their viewers by producing in front of them the lives and possessions that they will never be able to obtain. This is not even life in the slow lane. It is life on the hard shoulder.

In the electronic ghettoes the space of flows comes to a full stop. Time–space compression means time to spare and the space to go nowhere. It is all horribly legible – as the example of the South Central area of Los Angeles shows only too well. There, access to normal monetary transactions and credit, represented by a network of facilities like ATMs, bank branches and the like is in decline as these facilities have been shut down by banks whose bottom line is under pressure, further fuelling South Central's economic decline. Its inhabitants are forced back on an informal system of cheque-chasing services, mortgage brokers, credit unions, and cash. Yet just a few miles from South Central are the recently constructed corporate towers of Los Angeles's financial district, a place with all the necessary connections with telecommunications networks that are long and interconnected all around the world. It is a district that is nearer in this network time and space to New York, or London, or Tokyo than it is to South Central.

CONCLUSIONS

When we actually look at the space of flows, instead of taking it as a given, we find nothing very abstract or abstracted. What we find, as the example of the world of money shows only too well, is a new topology that makes it possible to go almost anywhere that networks reach (but without occupying anything but very narrow lines of force). But what we also find is a system where people are often in interaction with only four or five other people at a time – on a trading floor, or in a bank branch. What we find, in other words, are networks that are always both "global" and "local".

The problem is that writers like Jameson and Castells and Harvey, with their ideas of a global capitalist order typified by barometers such as an illegible globalization, a space of flows and time–space compression, are in constant danger of simply reproducing modernist views of an increasingly frantic commodified world filled with decontextualized rationalities like capitalism, various kinds of organizational bureaucracy and markets – rationalities which are soulless and relentless and shiveringly impersonal. But things, as they say, ain't necessarily like that;

> An organisation, a market, an institution are not supralunar objects made of a different matter from our poor local sublunar relations. The only difference stems from the fact that they are hybrid and have to mobilise a greater number of objects for their descriptions. The capitalism of Karl Marx or Fernand Braudel is not the total capitalism of (some) Marxists. It is a skein of longer networks that rather inadequately embrace a world on the basis of points that become centres of profit and calculation. In following it step by step, one never crosses the mysterious lines that should divide the local from the global. The organisation of American big business described by Alfred Chandler is not the organisation described by Kafka. It is a braid of networks materialised in order slips and flow charts, local procedures and special arrangements which permit it to spread to an entire continent so long as it does not cover that continent. One can follow the growth of an organisation in its entirety without ever changing (scale) levels and without ever discovering "decontextualised" rationality. . . . The markets . . . are indeed regulated and global, even though none of the causes of that regulation and that aggregation is itself either global or total. The aggregates are not made from some substance different from what they are aggregating. No visible or invisible hand suddenly descends to bring order to dispersed and chaotic individual atoms.
>
> (Latour 1993)

In other words, what we need to produce

geographies of global change is less exaggeration and more moderation.

But what would the more modest accounts look like? There are three closely related ways in which we need to ring the changes. First of all, we need to change the way that we do theory. In particular, that means recognizing that large-scale changes are always complex and contingent. They must be seen as:

> a multiplicity of often minor processes, of different origin and scattered location, which overlap, repeat, or imitate one another, support one another, distinguish themselves from one another according to their domain of application, converge and gradually produce the blueprint of a general method.
>
> (Foucault 1977)

Secondly, we need to recognize that all networks of social relation, whether we are talking about capitalism, or firms, or markets, or any other institutions, are incomplete, tentative, and approximate. They are constantly in the process of ordering a somewhat intractable geography of different and often very diverse geographical contexts. "And ordering extends only so far into that geography. The very powerful learn this quickly" (Law 1994). Accounts of these networks therefore need to recognize that it is a struggle to keep them at a particular size. There is no such thing as a scale. Rather, size is an uncertain effect generated by a network and its modes of interaction. A network of social relations has no natural tendency to be a particular size or operate at a particular scale; "some network configurations generate effects which, so long as everything is equal, last longer than others. So the tactics of ordering have to do, in general, with the construction of network arrangements that might last a little longer" (Law 1994). Usually, this trick depends upon the invention and use of materials that can be easily carried about and retain their shape, "immutable mobiles" like writing, print, paper, money, a postal system, cartography, navigation, ocean-going vessels, cannons, gunpowder, xerographics, computing, and telephony. Electronic telecommunications can be interpreted as the latest of these mobiles, yet another means of allowing certain networks of social relations to retain their integrity by ordering distant events. Accounts of these networks also need to recognize that new forms of connection produce new forms of disconnection. We will never reach a totally connected world. As the example of the new electronic ghettoes shows, new peripheries are constantly being created. Thirdly, and finally, we need to beware of confusing our theoretical ambitions with the reality that we are located *in* the world. Too often, the works of authors like Jameson, Castells and Harvey seem to assume that there is a place, like a satellite, from which it is possible to get an overview of the whole world. Yet, as numerous feminist commentators, have made clear, this assumption now looks increasingly like a classic masculine fantasy, a dream of being able to find a vantage point from which everything will become clear and a way of refusing to recognize that the world is a mixed, joint, commotion. Once we see that this assumption is a fantasy then we are also able to take account of all the subjects which in the past were often considered to be, somehow, "local" – like gender, or sexuality, or ethnicity – which recent work has shown are crucial determinants of global capitalism. In other words, one could understand "global capitalism" the better and, at the same time, realize that not everything can be explained by studying "global capitalism" (Walker 1993).

So now how might we see barometers like illegibility, the space of flows and time–space compression? To begin with, we can recognize that a globalizing world offers new forms of legibility which in turn can produce new forms of illegibility. For example, electronic tele-communications provide more and new kinds of information for firms and markets to work with but the sheer weight of information makes inter-pretation of this information an even more pressing and difficult task. The space of flows is revealed as a partial and contingent affair, just like all other human enterprises, which is not abstract or abstracted but consists of social networks, often of a quite limited size even though they might span the globe. Finally, time–space compression is shown to be something that we have learned to live with, and are constantly finding new ways of living with (for example, through new forms of subjectivity). It might be more accurately thought of as a part of a long history of immutable mobiles that we have learnt to live through and with. After all, each one of us is constructed by these "props", visible and invisible, present and past, as much as we con-struct them.

For those looking for big answers to the big questions that challenge us now in a globalizing world, this level of provisionality may all seem a bit frustrating. Yet the history of the last one hundred years or so suggests that big answers founder when they come up against the messy, contingent world that we are actually landed with. Worse than this, the big answers sometimes become a part of the problem, as their proponents force order on the world by applying ordered force. In other words, the same history suggests that, if we are going to try to clean up the uglier bits of our messy, contingent world, big answers are not the solution. Of course, this means that our actions are likely to be modest and sometimes mistaken, but the fact of recognizing this state of affairs is in itself empowering; it means that the future is open and we can all do something to shape it. There is positive value to be gained in "striving to incorporate the problems of coping with that openness into the practice of politics" (Gilroy 1993). To put it another way, there is nothing definite about the geographies of global change, and we do not have to be definite to change them.

REFERENCES

Davis, M. (1992) *Beyond Blade Runner: Urban Control. The Ecology of Fear*, Open Magazine Pamphlet 23, New Jersey, Westfield

Descombes, V. (1993) *The Barometer of Modern Reason: On the Philosophies of Current Events*, Oxford, Oxford University Press

Foucault, M. (1977) *The Archaeology of Knowledge*, London, Tavistock

Giddens, A. (1991) *Modernity and Self Identity: Self and Society in the Late Modern Age*, Cambridge, Polity Press

Gilroy, P. (1993) *The Black Atlantic: Modernity and Double Consciousness*, London, Verso

Latour, B. (1993) *We Have Never Been Modern*, Brighton, Harvester Wheatsheaf

Law, J. (1994) *Organizing Modernity*, Oxford, Blackwell

Mulgan, G. J. (1991) *Communication and Control: Networks and the New Economics of Communication*, Cambridge, Polity Press

Walker, R. B. J. (1993) *Inside/Outside: International Relations as Political Theory*, Cambridge, Cambridge University Press

'The Social and Cultural Life of the City'

from *Telecommunications and the City: Electronic Spaces, Urban Places* (1996)

THE HOME AS A DOMESTIC 'NETWORK TERMINAL'

In the modern city, the home emerged as 'the last reserve space'. It was, to quote Helga Nowotny, 'a social space of special significance which has come to signify for us the last sanctuary in a bewildering outside world'. But this sense that the home is isolated from the rest of the social world is changing rapidly. Many homes, as we have seen, are being incorporated into more and more networks on increasingly global scales. These blur the dividing line between what is public and what is private. To Putnam [in his essay 'Beyond the Home' (1993)], this means that the best way to consider the home is as a 'terminal'. He writes,

> in speaking about the modern home, we are talking about more than technological comforts. The modern home is inconceivable except as a terminal, according the benefits of, but also providing legitimate support to a vast infrastructure facilitating flows of energy, goods, people and messages. The most obvious aspect has been a qualitative transformation of the technical specification of houses and their redefinition as terminals of networks.

This is not the result of some simple technological 'logic' however. Nor can we attribute this process entirely to broad political and economic forces. Rather, Roger Silverstone has shown that new communications and information technologies are entering homes through complex and diverse processes of social construction and 'domestication'. [He writes, in his essay 'Domesticating the revolution' (1994) that the] consumption of these new technologies, and the services that are accessible through them, are

pursued by consumers who seek to manage and control their own electronic spaces, to make mass-produced objects and meanings meaningful, useful and intelligible to them. This is a process of 'domestication' because what is involved is quite literally a taming of the wild and a cultivation of the tame. In this process new technologies and services, unfamiliar, exciting, but also threatening, are brought (or not) under control by domestic users.

Often, though, these technologies create conflicts – particularly gender conflicts – over who has control and access within the household.

But these processes operate within the context of broader political, social and economic changes. The shift towards information labour, home and telecentre-based teleworking, flexible labour markets and the cultural shifts towards globalisation provide the context for the changes underway. [Geoff Mulgan found] ... however, that beyond the hype, there remains 'deep resistance to electronic interactivity beyond a small number of enthusiasts'. Most of the development of technology in the home still centres on stand-alone media and entertainment systems such as CD players, personal computers and TVs. Most cable services are not interactive – they merely pump more channels to be passively consumed than terrestrial broadcasting. Telephones remain by far the most important interactive home communications system.

We must also be conscious of the varied prospects for genuine tele-based services to the home. In the United States, home tele-services seem to have a greater chance of reaching 'critical mass'. In Europe, resistance seems to be much deeper. A recent study by the Inteco consultancy of 12,000 people across Europe, funded by 30 retailers and telecom companies, found that there is deep consumer resistance to interactive shopping; entertainment and security services were marginally more attractive.

The main problem, then, is that, in mass markets 'the audience does not yet provide the revenue which supports the services'. Thus the focus . . . is on 'demassification' and the 'cherry picking' of markets by offering higher value-added services for more lucrative consumers and those small markets where genuine prospects exist. The Inteco study also concluded that the only attractive markets at present comprise 'active, relatively high income families where time is at a premium, where someone in the family will be learning to play a musical instrument or be keen on gardening or some form of home study. [These] will be the fertile ground for multimedia seeds to root'.

Despite uncertainties in demand, commercial efforts to enhance the role of homes as terminals for many technical networks, offering 'pay-per' services (at least for these affluent groups), are intensifying. These are driven by the view that there are huge commercial rewards that may be in store from commercial services offered over the much-vaunted prospect of multimedia, interactive television or the information super-highway. Because of this, the still rather marginal shifts towards home banking, home- and centre-based teleworking, and tele-based access to services may eventually grow, as new technological capabilities come into the social mainstream – or, perhaps more likely, the financial reach of socially affluent groups.

Already, 'narrowcasted' electronic newspapers have been developed in the United States, accessible by computer and geared to the particular programmed interests of consumers. 'Video on demand' technologies are about to enter the market, offering selected videos played down phone lines as compressed signals. In a few places, cable networks are beginning to offer value added services rather than just TV. A growing range of consumer services such as 'electronic bookstores' are developing, based on the Internet (which is emerging as a domestic service for technology enthusiasts). 'Smart', 'interactive' or 'high definition' TVs have been developed which, when combined with signal compressions technologies, offer the technological potential for user-friendly and interactive services.

A range of full-blown commercial trials are now in the offing for interactive, sophisticated home telematics systems. Time Warner and US West are already operating an interactive TV trial in Orlando, Florida, for 4,000 homes. A system called 'stargazer', developed by Bell Atlantic, will soon offer education, entertainment, information and shopping on a trial basis to 2,000 homes in Northern Virginia. But the commercial and commodified nature of these emerging systems makes the widely used analogy of the electronic highway profoundly misleading.

If the dream is of consumers who 'spend at the touch of a button', those without money are likely to have little to do with these developments. Brody writes that 'what is emerging has less the character of a highway than a strip mall, focusing on services with proven demand that can rapidly pay for themselves . . . cable operators are seeking to establish a presence in the lush, unregulated territory of enhanced services'. They also seem to be supply-driven – often crude attempts by technology and telecommunications companies to increase the use of existing networks and technologies and so improve profitability.

'ELECTRONIC COTTAGE OR NEO-MEDIEVAL MINI-FORTRESS?'

While there are uncertainties over the viability of these systems, there does seem to be an overall trend towards the mediation of more work, travel and consumption via home-based telematics which intensifies the home-centredness of urban society. These trends suggest that homes may be progressively disembedding from their immediate social environment within urban places through their linkage into electronic spaces.

[. . .]

In this context, Barry Smart asks the important question: will home-centredness emerge as benign and positive development as in the Tofflerian 'electronic cottage' scenario or, conversely, will it 'encourage a retreat from public life and public space and serve thereby to increase the sense of insecurity which has become an increasing feature of the "post-modern" urban environment'. The trends are complex and varied and belie easy generalisation. But there seems to be a shift, particularly in the middle classes, towards 'cocooning' – their withdrawal from public spaces in cities and the use of home-centred and self-service technologies and network access points in their place. Fear of crime and social alienation with urban life are key supports to this trend. A recent Channel 4 television programme commented that access to telematics networks for shopping, leisure and work 'looks safe indeed compared to urban decay. Paranoia, violence and pollution are eating away at the soul of America, driving it inward – to the protection of the home, private security, entry codes, and video-surveillance-controlled gated fortresses'. In this vein, Manuel Castells warns of a dystopian (American) urban future where 'secluded individualistic homes across an endless suburban sprawl turn inward to preserve their own logic and values, closing their doors to the immediate surrounding environment and opening their antennas to the sounds and images of the entire galaxy'.

There are two key issues here. First, it is becoming increasingly clear that there are dangers that these home-based technological innovations will simply exacerbate existing trends towards individualisation and polarisation within western cities. Relying in the telematics field on 'a totally individualised society completely ruled by market mechanisms' [as Kubieck recognises] seems to threaten to undermine the public, civic sense of cities as physical and cultural spaces of social interaction. Certainly, the reliance on technical networks for social interaction and entertainment and the withdrawal and cocooning that goes along with this, encourages fear of crime and a shift toward the 'tribalisation' of socioeconomic groups in cities.

Second, as we have seen, these trends threaten to exclude and marginalise already disadvantaged groups who are unable to afford participation in these technological futures. Women, for example, may become ensnared in domestic space by a pervasive shift towards home-based teleshopping.

PART 2

*L*iving Urban Culture

SECTION V

EVERYDAY LIFE

INTRODUCTION

The city is not just a place of famous people, spectacular events, great works of art, powerful politicians, super-rich business and grand architectural projects. It is also the place of all people, common occurrences, various creative acts, community workers, small shops and quotidian buildings. This is the city of everyday life – the city that occurs normally, routinely, without fuss and bother. It is the city that is at once boring and banal, and profoundly connected to the culture of the city that everyone experiences every day of their lives.

As such, everyday life has been treated ambiguously by cultural commentators. In his seminal publications on everyday life – including *Everyday Life in the Modern World* (1984, Transaction Publishers) and *Critique of Everyday Life* (1991, Verso) (see suggestions for further reading) – Henri Lefebvre described the 'bureaucratic society of controlled consumption' as a 'terrorism' in which every aspect of everyday life is invaded, organized and exploited by capitalism. On the other hand, everyday life also offers the chance to enact simple loves and desires, to be creative, to undertake 'revolutionary' acts out of the reach of the state.

Although not themselves included here (Lefebvre's related thoughts on space are instead included in Section VII, Representations of the City), Lefebvre's various thoughts on the everyday are echoed in many of the contributions in this section. For example, the essay 'Ludwigshafen–Mannheim' (1928) by Ernst Bloch provides a remarkable precursor to many of Lefebvre's formulations on the subject. Bloch, a fellow unorthodox Marxist and friend of cultural theorist Walter Benjamin, here describes the development of his birthplace, Ludwigshafen, initially in terms of the various streets and urban space which make up the grimy physical fabric. As Bloch puts it, this is 'the factory dirt which had been compelled to become a city', and little else. However, everyday life is more than objects; it is the routinized proletarian worker 'placed on the conveyor belt'; it is the absence of high culture mixed with a 'shoddy-bold cinema glamour in the sad streets'. But Bloch's secularized Judaic Messianism also allows him to see signs of redemption here: he ends by viewing Ludwigshafen as a town of 'bursting places', a loose 'sea-port on the land' that holds clues not to the past but to the present and to the future.

In what is perhaps one of the most famous and influential essays ever written on the city, Georg Simmel provided another ambiguous perspective. Writing like Bloch about his home city (Berlin), Simmel (and here unlike Bloch) focused in 'The Metropolis and Mental Life' (1903) (included in David Frisby and Mike Featherstone (eds) *Simmel on Culture*, 1997, Sage, pp. 174–85) on the interaction of metropolitan dwellers, seeing that the size, density and sheer quantitative excess of the city causes an 'intensification of nervous stimuli' among its residents, leading them to adopt an indifferent blasé attitude to each other, combined with a more overt, precious individualistic personality. Their everyday exchanges are then like the exchanges of money commodities – impersonal, distant and cold – while other aspects of city life (notably time and punctuality) are rendered into matters of quantity rather than quality.

Similar themes are picked up in Siegfried Kracauer's 'The Hotel Lobby' from *The Mass Ornament* (1963 [1995], Harvard), which considers modernity in Berlin. Influenced heavily by his teacher Georg Simmel and by Benjamin, and himself a teacher of Theodor Adorno (see Section III, The Culture Industries), Kracauer here uses the spatial type of the hotel lobby not so much to explain this kind of internalized architecture in itself but as a means to understand society at large. Hence the lobby appears as a place of silence, avoided eye contact, distant relations, unreality, anonymity and insubstantial aesthetics – all aspects that are paralleled in Kracauer's view of the meaningless existence of modern life, governed by a dominant 'ratio' or rationality.

Other authors have, however, interpreted the hotel lobby in more positive terms. The feminist cultural theorist and historian Elizabeth Wilson, for example, has seen it as a semi-public, semi-private, amoral realm that allows women to enter into the spaces of the city. From another chapter of her book *The Sphinx in the City* (1991, University of California Press), the extract reprinted here continues the theme of gender but in different geographic locations: Latin America and Africa. Wilson notes that severe infrastructural and architectural deficiencies of, for example, shanty towns, may also be countered by easy social contacts and supportive friendships and familial relations. In terms of gender, and arguing that the everyday life of cities provides both dangers and exhilaration for women, Wilson seeks in places opportunities to break out of constricting family stereotypes, work patterns, community activities and architectural design processes. As she concludes, 'the gulf between what is and what might be may appear to widen; on the other hand, the city both raises aspirations and gives more chance of their realisation'.

The idea that everyday practices offer emancipation in the city is also prevalent in Michel de Certeau's *The Practice of Everyday Life* (1984, University of California Press) (see suggestions for further reading), a book dedicated to 'the ordinary man' and to 'a common hero, an ubiquitous character, walking in countless thousands on the streets'. As this dedication and other famous sections of the book suggest, it is the everyday spatial actions of commonplace people that provide their experience and knowledge of the city. De Certeau differentiates between the stability of places and the operations and actions of spaces, between abstract maps based on vision and operational tours that involve bodily movement through the city. Everyday life is for him an action, a language that is not remote from but embedded in common practices in the city.

Influenced greatly by the kinds of writing and theories in this section, the architect Bernard Tschumi, in 'Spaces and Events' from *Architecture and Disjunction* (1994, MIT), proposes that for architecture to be truly architecture it must contain events and programmes. In opposition to the concentration on styles and form in much postmodern architecture of the 1970s and early 1980s, Tschumi suggests that it is the everyday actions of people that architects should consider in designing their buildings. This does not mean, however, reducing such activities to 'functions', or seeing that buildings would determine the actions of their users (as many twentieth-century modernist architects have suggested). Rather, architects should promote disjunctions between form and use ('pole vaulting in the chapel, bicycling in the laundromat' or, alternatively, 'the most intricate and perverse organization of spaces' occupied by 'average suburban families'). Other tactics open to the architect include, suggests Tschumi, the disjunction of building and urban context, different notational techniques, and an assemblage of urban codes.

Not all architecture, however, provides such opportunities for celebratory and creative everyday life. In their account of the Meadowhall shopping mall in South Yorkshire, taken from their book *A Tale of Two Cities* (1996, Routledge), Karen Evans, Ian Taylor and Penny Fraser show how this, the most everyday of late twentieth-century building types, is also a microcosm of everyday life as a whole. Generic in form, replicated in cities all across the world, the shopping mall is a standardized provision of standardized experiences: consumption over production, vision over other senses, surveillance over spontaneity, convenience over difficulty, passivity over activity, homogeneity over heterogeneity. Despite the oppressive character of their object of study, the authors are careful to reproduce the views of individual shoppers and nearby residents, showing how everyday life is a matter of perception, choice and individual action.

24 ERNST BLOCH

'Ludwigshafen—Mannheim'

from *Heritage of our Times* (1928)

Otherwise the oil smokes more to itself. Or the shavings only lie where there is planing. The planers live in rented holes, the streets are cheerless. Far away, however, the masters live with the money that has been earned by others. Had houses like knick-knacks, dressed in the old-fashioned style. No sound of daily work penetrated them.

Thus the quarters were previously already divided where the work and the devouring is done. But of course technology sprang up even in older refined parts of towns, destroyed the picture. The approach road from the station had usually become a different one to that from the country road, shifted the old axis. But even so the traditional urban culture did not die completely; the rampart, the ring was planted with trees or even became a residential area. The new water-tower was embarrassed to be one in the eighties, was built like a tankard. Socially, too, the bourgeoisie moved into courtly pleasure, had its good concerts, chatted in the boxes.

But new cities came off badly here with nothing to guide their steps. Particularly if they lie next to an old cultural city, like Ludwigshafen next to Mannheim, on either side of the Rhine. If the river did not divide them enough already, the Bavarian – Baden border made sure any equality was prevented. Ludwigshafen was consequently obliged to become a town in its own right, not just a suburb into which the sewage of industry flows. At its foundation seventy-five years ago it was definitely intended as competition against Mannheim; so it continued to manage on its own resources in a highly current way. Here is a place therefore, thoroughly typical of the capitalist. Now, where the planers live in the town itself, where no beautiful houses far away, and certainly no previous urban cultures lie giddily above the Now. The Baden Aniline and Soda Factory, the nucleus of IG Farben (moved here so that the smoke and proletariat did not drift over Mannheim), became the literal true emblem of the town. Over there lay the chessboard of the old royal Residence, a cheerful and friendly building, as if in the time of Hermann and Dorothea; had instead of the biggest factory the biggest château in Germany, perhaps less of a true emblem, in the nineteenth century, but still a beautiful ornament which gave the bourgeoisie standing. Whenever it was time for coffee and cigars and higher things. Ludwigshafen, on the other hand, remained the factory dirt which had been compelled to become a city: random and helpless, cut in two by the arc of the embankment, a Zwickau without inhibitions, after the false dawn of the Biedermeier period which occurred at the time of its foundation, an extremely wet day. The initial attempts at 'Art and Science' proved ridiculous, were all intercepted by Mannheim; today there is still no theatre in the numerically long since perfect big city. In the marketplace stands a 'Monumental Fountain' (that is what it is called); it is grey, yellow, white, red, because it is supposed to contain all the various kinds of Rhineland Palatinate sandstone. Men's heads, heraldic mottoes, columns, niches, urns, wreaths, little ships, crowns, bronze, basin, obelisk, all on the puniest scale – the whole thing is perhaps the most beautiful Renaissance monument of the nineteenth century. A thousand parlours look down on us from these stones; here is 1896 in a nutshell and in the provinces. And from the embankment a weeping willow waves across to the 'Jubilee Fountain' (again that is what it is called); over there cast iron stands on tuff, the goddess Bavaria presents the goddess

Ludwigshafenia with the city crown, below at an angle Father Rhine leans grotto-like, pours a trickle of water from his cornucopia. At the station there is a bust of Schiller, and the big dipper sings the words to it, the brandy bars are called 'The Parisian Hour' and the theatre club are performing 'The Executioner of Augsburg': such is or was until recently this petit-bourgeois Wild West on the Rhine. On the most solemn river in Germany halfway between Speyer and Worms, in the midst of the Nibelungenlied as it were, right next to the Jesuit church, the rococo library, Schiller's Court and National Theatre in Mannheim. Seldom did one have the realities and the ideals of the industrial age so close together, the dirt and the royally built-in money.

Why are we writing about it with such a long preamble? Precisely because something turned around here, because where the age is marching emerges here. Because Ludwigshafen, which stands for various things, has suddenly become more important, in the new air, than Mannheim. There lies, no, there now *sails* the ugly city, but it kicks up such a crude row, money circulates and IG Farben streams. Something there has become Front which brings everything to light and is no longer educatedly embarrassed. Now even the city goddess Ludwigshafenia and even Father Rhine have had nooses put around their necks, at the Jubilee Fountain, and have been pulled down, which is at least as symbolic as master builder Solness. And they are going to put an amusement machine in the romantic place, and a theatre as well with visiting groups who are currently prominent; in short, all the mixing noise which the bourgeoisie now allows and which is at any rate more concrete than Schiller and Ibsen played by a stock of people entitled to a pension. The boys of Ludwigshafen have cranes in front of their eyes, fair and Karl May, the middle class reads its Rudolf Herzog even here of course, but without believing it, most read nothing at all, but their world looks like Sinclair, sometimes even like Jack London. Wassermann and Thomas Mann, elegant bourgeois problems of an older stratum, have no place in this. Here there is only the stage for factories and what goes with them, there is rawness and stench but without stuffy air. IG Farben, which founded the city to begin with, now gives it more than ever the pure, raw-cold, fantastic face of late capitalism.

Cities of this kind should therefore be especially weighed up. There are many such places in the Ruhr area, though without such sharp contrast so near. They still have enough reactionary mustiness, stupidity of lasting petit bourgeois, a dreadful provincial press. Yet Ludwigshafen has a more honest face when compared with the Mannheim type; its industry did not first destroy natural, cultural connections, but is alien to them ab ovo. Here there is the most genuine hollow space of capitalism: this dirt, this raw and dead-tired proletariat, craftily paid, craftily placed on the conveyor belt, this project-making of ice-cold masters, this profit-business without remnants of legends and clichés, this shoddy-bold cinema glamour in the sad streets. This is what it now looks like in the German soul, a proletarian–capitalist mixed reality without a mask. And round about Ludwigshafen the hazy plain with swamp-holes and ponds, a kind of prairie which knows no little estates and no idylls, to which factory walls and fiery chimneys are significantly suited; the telegraph pole sings along. This is a good standpoint from which to see current reality, to grasp even better the tendency which it is and which it will resolve. Older, more comfortable cities, plush cities as it were, also have this tendency, but not in such a traditionless vacuum. In fifty years a real swine of a city could stand on the crude earth which has not even cleaned up its swine, but is the most direct growth from ship-building, silos, elevators, factory buildings. The coming age has more to topple here, but less to ignite than in the old culture, which gives more to plunder in return.

In the current dirt hardly anything still blooms worth talking about. Nothing but miserable dross from Berlin, at best, here and there, reaches the most advanced places. In return, however, there are bursting places which Berlin does not yet have, and in which improvisations can nest which no 'cultural will' anticipates. Places like Ludwigshafen are the first sea-ports on the land, fluctuating, loosened up, on the sea of an unstatic future. The cosy Rhineland Palatinate vineyards, half an hour away, Court and National Theatre, the nearby cathedrals of Worms and Speyer, move into the distance for the time being. The international station-boundness fuses everything together, has neither the earlier muse nor can content itself with the receptive enjoyment of handed-down pictures. The zero which screams, the chaos that displaces itself coldly and currently, is probably closer to the origin which created the cultural pictures than the merely 'educated' bourgeoisie which hangs them up in the dining room. Even plundering the old, creating a new montage from it, succeeds best from the standpoint of such cities. They are themselves a knotting place; workers and entrepreneurs tie the knot clearly, contemporaneously, objectively, between themselves and what is coming in the future.

25 SIEGFRIED KRACAUER

'The Hotel Lobby'

from *The Mass Ornament* (1963)

The community of the higher realms that is fixated upon God is secure in the knowledge that as an oriented community – both in time and for all eternity – it lives within the law and beyond the law, occupying the perpetually untenable middle ground between the natural and the supernatural. It not only presents itself in this paradoxical situation but also experiences and names it as well. In spheres of lesser reality, consciousness of existence and of the authentic conditions dwindles away in the existential stream, and clouded sense becomes lost in the labyrinth of distorted events whose distortion it no longer perceives.

The *aesthetic* rendering of such a life bereft of reality, a life that has lost the power of self-observation, may be able to restore to it a sort of language; for even if the artist does not force all that has become mute and illusory directly up into reality, he does express his directed self by giving form to this life. The more life is submerged, the more it needs the artwork, which unseals its withdrawnness and puts its pieces back in place in such a way that these, which were lying strewn about, become organized in a meaningful way. The unity of the aesthetic construct, the manner in which it distributes the emphases and consolidates the event, gives a voice to the inexpressive world, gives meaning to the themes broached within it. Just what these themes mean, however, must still be brought out through translation and depends to no small extent on the level of reality evinced by their creator. Thus, while in the higher spheres the artist confirms a reality that grasps itself, in the lower regions his work becomes a harbinger of a manifold that utterly lacks any revelatory word. His tasks multiply in proportion to the world's loss of reality, and the cocoon-like spirit [*Geist*]

that lacks access to reality ultimately imposes upon him the role of educator, of the observer who not only sees but also prophetically foresees and makes connections. Although this overloading of the aesthetic may well accord the artist a mistaken position, it is understandable, because the life that remains untouched by authentic things recognizes that it has been captured in the mirror of the artistic construct, and thereby gains consciousness, albeit negative, of its distance from reality and of its illusory status. For no matter how insignificant the existential power that gives rise to the artistic formation may be, it always infuses the muddled material with intentions that help it become transparent.

Without being an artwork, the *detective novel* still shows civilized society its own face in a purer way than society is usually accustomed to seeing it. In the detective novel, proponents of that society and their functions give an account of themselves and divulge their hidden significance. But the detective novel can coerce the self-shrouding world into revealing itself in this manner only because it is created by a consciousness that is not circumscribed by that world. Sustained by this consciousness, the detective novel really thinks through to the end the society dominated by autonomous *Ratio* – a society that exists only as a concept – and develops the initial moments it proposes in such a way that the idea is fully realized in actions and figures. Once the stylization of the one-dimensional unreality has been completed, the detective novel integrates the individual elements – now adequate to the constitutive pre-suppositions – into a self-contained coherence of meaning, an integration it effects through the power of its existentiality, the latter transformed not into critique and exigency but into principles of aesthetic composition. It is only this entwinement into a unity that really makes possible the interpretation of the presented findings. For, like the philosophical system, the aesthetic organism aims at a totality that remains veiled to the proponents of civilized society, a totality that in some way disfigures the entirety of experienced reality and thereby enables one to see it afresh. Thus, the true meaning of these findings can be found only in the way in which they combine into an aesthetic totality. This is the minimum achievement of the artistic entity: to construct a whole out of the blindly scattered elements of a disintegrated world – a whole that, even if it seems only to mirror this world, nevertheless does capture it in its wholeness and thereby allows for the projection of its elements onto real conditions. The fact that the structure of the life presented in the detective novel is so typical indicates that the consciousness producing it is not an individual, coincidental one; at the same time, it shows that what has been singled out are the seemingly metaphysical characteristics. Just as the detective discovers the secret that people have concealed, the detective novel discloses in the aesthetic medium the secret of a society bereft of reality, as well as the secret of its insubstantial marionettes. The composition of the detective novel transforms an ungraspable life into a translatable analogue of actual reality.

In the *house of God*, which presupposes an already extant community, the congregation accomplishes the task of making connections. Once the members of the congregation have abandoned the relation on which the place is founded, the house of God retains only a decorative significance. Even if it sinks into oblivion, civilized society at the height of its development still maintains privileged sites that testify to its own nonexistence, just as the house of God testifies to the existence of the community united in reality. Admittedly society is unaware of this, for it cannot see beyond its own sphere; only the aesthetic construct, whose form renders the manifold as a projection, makes it possible to demonstrate this correspondence. The typical characteristics of the *hotel lobby*, which appears repeatedly in detective novels, indicate that it is conceived as the inverted image of the house of God. It is a negative church, and

can be transformed into a church so long as one observes the conditions that govern the different spheres.

In both places people appear there as *guests*. But whereas the house of God is dedicated to the service of the one whom people have gone there to encounter, the hotel lobby accommodates all who go there to meet no one. It is the setting for those who neither seek nor find the one who is always sought, and who are therefore guests in space as such – a space that encompasses them and has no function other than to encompass them. The impersonal nothing represented by the hotel manager here occupies the position of the unknown one in whose name the church congregation gathers. And whereas the congregation invokes the name and dedicates itself to the service in order to fulfill the relation, the people dispersed in the lobby accept their host's incognito without question. Lacking any and all relation, they drip down into the vacuum with the same necessity that compels those striving in and for reality to lift themselves out of the nowhere toward their destination.

The congregation, which gathers in the house of God for prayer and worship, outgrows the imperfection of communal life in order not to overcome it but to bear it in mind and to reinsert it constantly into the tension. Its gathering is a *collectedness* and a unification of this directed life of the community, which belongs to two realms: the realm covered by law and the realm beyond law. At the site of the church – but of course not only here – these separate currents encounter each other; the law is broached here without being breached, and the paradoxical split is accorded legitimacy by the sporadic suspension of its languid continuity. Through the edification of the congregation, the community is always reconstructing itself, and this elevation above the everyday prevents the everyday itself from going under. The fact that such a returning of the community to its point of origin must submit to spatial and temporal limitations, that it steers away from worldly community, and that it is brought about through special celebrations – this is only a sign of man's dubious position between above and below, one that constantly forces him to establish on his own what is given or what has been conquered in the tension.

Since the determining characteristic of the lower region is its lack of tension, the togetherness in the hotel lobby has no meaning. While here, too, people certainly do become detached from everyday life, this detachment does not lead the community to assure itself of its existence as a congregation. Instead it merely displaces people from the unreality of the daily hustle and bustle to a place where they would encounter the void only if they were more than just reference points. The lobby, in which people find themselves *vis-à-vis de rien*, is a mere gap that does not even serve a purpose dictated by *Ratio* (like the conference room of a corporation), a purpose which at the very least could mask the directive that had been perceived in the relation. But if a sojourn in a hotel offers neither a perspective on nor an escape from the everyday, it does provide a groundless distance from it which can be exploited, if at all, *aesthetically* – the aesthetic being understood here as a category of the nonexistent type of person, the residue of that positive aesthetic which makes it possible to put this nonexistence into relief in the detective novel. The person sitting around idly is overcome by a disinterested satisfaction in the contemplation of a world creating itself, whose purposiveness is felt without being associated with any representation of a purpose. The Kantian definition of the beautiful is instantiated here in a way that takes seriously its isolation of the aesthetic and its lack of content. For in the emptied-out individuals of the detective novel – who, as rationally constructed complexes, are comparable to the transcendental subject – the aesthetic faculty is indeed detached from the existential stream of the total person. It is reduced to an unreal, purely formal relation that manifests the same indifference to the self as it does to matter. Kant himself was able to overlook this horrible last-minute sprint of the transcendental subject, since he still believed there was a seamless transition from the transcendental to the performed subject-object world. The fact that he does not completely give up the total person even in the aesthetic realm is confirmed by his definition of the "sublime," which takes the ethical into account and thereby attempts to reassemble the remaining pieces of the fractured whole. In the hotel lobby, admittedly, the aesthetic – lacking all qualities of sublimity – is presented without any regard for these upward-striving intentions, and the formula "purposiveness without purpose"

also exhausts its content. Just as the lobby is the space that does not refer beyond itself, the aesthetic condition corresponding to it constitutes itself as its own limit. It is forbidden to go beyond this limit, so long as the tension that would propel the breakthrough is repressed and the marionettes of *Ratio* – who are not human beings – isolate themselves from their bustling activity. But the aesthetic that has become an end in itself pulls up its own roots; it obscures the higher level toward which it should refer and signifies only its own emptiness, which, according to the literal meaning of the Kantian definition, is a mere relation of faculties. It rises above a meaningless formal harmony only when it is in the service of something, when instead of making claims to autonomy it inserts itself into the tension that does not concern it in particular. If human beings orient themselves beyond the form, then a kind of beauty may also mature that is a fulfilled beauty, because it is the consequence and not the aim – but where beauty is chosen as an aim without further consequences, all that remains is its empty shell. Both the hotel lobby and the house of God respond to the aesthetic sense that articulates its legitimate demands in them. But whereas in the latter the beautiful employs a language with which it also testifies against itself, in the former it is involuted in its muteness, incapable of finding the other. In tasteful lounge chairs a civilization intent on rationalization comes to an end, whereas the decorations of the church pews are born from the tension that accords them a revelatory meaning. As a result, the chorales that are the expression of the divine service turn into medleys whose strains encourage pure triviality, and devotion congeals into erotic desire that roams about without an object.

The *equality* of those who pray is likewise reflected in distorted form in the hotel lobby. When a congregation forms, the differences between people disappear, because these beings all have one and the same destiny, and because, in the encounter with the spirit that determines this destiny, anything that does not determine that spirit simply ceases to exist – namely, the limit of necessity, posited by man, and the separation, which is the work of nature. The provisional status of communal life is experienced as such in the house of God, and so the sinner enters into the "we" in the same way as does the upright person whose assurance is here disturbed. This – the fact that everything human is oriented toward its own contingency – is what creates the equality of the contingent. The great pales next to the small, and good and evil remain suspended when the congregation relates itself to that which no scale can measure. Such a relativization of qualities does not lead to their confusion but instead elevates them to the status of reality, since the relation to the last things demands that the penultimate things be convulsed without being destroyed. This equality is positive and essential, not a reduction and foreground; it is the fulfillment of what has been differentiated, which must renounce its independent singular existence in order to save what is most singular. This singularity is awaited and sought in the house of God. Relegated to the shadows so long as merely human limits are imposed, it throws its own shadow over those distinctions when man approaches the absolute limit.

[. . .]

The observance of *silence*, no less obligatory in the hotel lobby than in the house of God, indicates that in both places people consider themselves essentially as equals. In "Death in Venice" Thomas Mann formulates this as follows: "A solemn stillness reigned in the room, of the sort that is the pride of all large hotels. The attentive waiters moved about on noiseless feet. A rattling of the tea service, a half-whispered word was all that one could hear." The contentless solemnity of this conventionally imposed silence does not arise out of mutual courtesy, of the sort one encounters everywhere, but rather serves to eliminate differences. It is a silence that abstracts from the differentiating word and compels one downward into the equality of the encounter with the nothing, an equality that a voice resounding through space would disturb. In the house of God, by contrast, silence signifies the individual collecting himself as firmly directed self, and the

word addressed to human beings is effaced solely in order to release another word, which, whether uttered or not, sits in judgment over human beings.

Since what counts here is not the dialogue of those who speak, the members of the congregation are anonymous. They outgrow their names because the very empirical being which these names designate disappears in prayer; thus, they do not know one another as particular beings whose multiply determined existences enmesh them in the world. If the proper name reveals its bearer, it also separates him from those whose names have been called; it simultaneously discloses and obscures, and it is with good reason that lovers want to destroy it, as if it were the final wall separating them. It is only the relinquishing of the name – which abolishes the semisolidarity of the intermediate spheres – that allows for the extensive solidarity of those who step out of the bright obscurity of reciprocal contact and into the night and the light of the higher mystery. Now that they do not know who the person closest to them is, their neighbor becomes the closest, for out of his disintegrating appearance arises a creation whose traits are also theirs. It is true that only those who stand before God are sufficiently estranged from one another to discover they are brothers; only they are exposed to such an extent that they can love one another without knowing one another and without using names. At the limit of the human they rid themselves of their naming, so that the world might be bestowed upon them – a word that strikes them more directly than any human law. And in the seclusion to which such a relativization of form generally pushes them, they inquire about their form. Having been initiated into the mystery that provides the name, and having become transparent to one another in their relation to God, they enter into the "we" signifying a commonality of creatures that suspends and grounds all those distinctions and associations adhering to the proper name.

26 ELIZABETH WILSON

'World Cities'

from 'The Sphinx in the City: Urban Life, the Control of Disorder, and Women' (1991)

In Latin America cities grew up in a number of different ways. There, the strong planning tradition of Spanish and Portuguese colonialism determined the form of the older cities. São Paulo, Brazil, for example, was founded in 1554 by the Jesuits as a bridgehead to the hinterland to further their 'domestication' of indigenous Indians. After independence in 1822 it became a provincial administrative and political centre. By the mid nineteenth century coffee-growing was substantial; this, and the related growth of a railroad network, transformed São Paulo by 1900 into a city of a quarter of a million inhabitants (as against 65,000 in 1890) and an important industrial centre.

An infrastructure of sewage, roads, water supplies and lighting always lagged behind the needs of the raggedly growing city. By the 1930s, when Claude Lévi-Strauss was living there, the city was dealing with its housing problem by a variety of hand-to-mouth solutions, and 'in 1935 the citizens of São Paulo boasted that, on an average, one house per hour was built in their town . . . The town is developing so fast that it is impossible to obtain a map of it; a new edition would be required every week.' Everything was chaotic, and 'vast roadworks' were being built adjacent to old crafts quarters like Syrian bazaars. A new public avenue cut through a once exclusive neighbourhood, 'where painted wooden villas were falling to pieces in gardens full of eucalyptus and mango trees'; then came a working-class neighbourhood, 'along which lay a red light district, consisting of hovels with raised entresols, from the windows of which the prostitutes hailed their clients. Finally, on the outskirts of the town, the lower-middle-class residential areas . . . were making headway.'

Today, many city dwellers live in shanty towns, and these have become the most telling and guilt-inducing images of 'third-world poverty', inviting the voyeuristic horror of the westerner. In the 1960s about one quarter of the population of cities such as Manila and Djakarta, one third of the population of Mexico City and half that of Lima lived in shanty towns. In the late 1970s it was estimated that by 1990 three-quarters of Lima's poor would be living in such conditions.

The literature on shanty towns usually emphasises their squalor and poverty. Dwellings are constructed of wattle, cardboard or corrugated iron, and placed in close proximity, with narrow lanes running between them which also serve as open drains. Seven or eight members of a household sleep together in one room, and animals often share the human living space. Whether located in Venezuela or Calcutta, Cairo or Djakarta, the descriptions of the 'barrios' are extraordinarily reminiscent of the Victorians' lurid depictions of nineteenth-century slums. Like the London or Paris poor of that time, women of the barrios queue for water, which is supplied – often intermittently – from standpipes. There are no proper sewage systems in the 'typical' shanty town, which in general lacks all basic amenities.

It is often assumed that the people living in such conditions must somehow themselves be inadequate, just as the Victorian reformers assumed for the most part that the poor were locked into the slums because they were lacking in moral fibre. For example, an American academic, M.H. Ross, described the Nairobi shanty towns as follows:

Downtown Nairobi is beautiful, with its tall buildings, modern architecture and flowering trees . . . Four miles [away] . . . live some 10,000 to 20,000 urban squatters . . . The houses, crammed together in an apparently haphazard fashion dictated by the uneven terrain of the valley's walls, are built of mud and wattle and have roofs made of cardboard, flattened-out tin cans, or even sheet metal. A visitor entering the area is struck by the lack of social services; the roads are makeshift, garbage is piled high in open areas, and children play in the dust.

He dismissed the inhabitants of Mathare as 'generally urban misfits and rural outcasts . . . [who] lack the skills necessary to find jobs in the modern economy' – yet went on to reveal that the community organised nursery schools, a co-op and social events, and that 'the most striking aspect of Mathare is that it is highly organised and politically integrated . . . There is a clearly identifiable group of community leaders.'

In the 1950s and 1960s governments in many countries responded to spontaneous settlements with harassment and evictions. Since the 1970s, a different strategy has been more often used: a 'site and services' policy. This recognises the impossibility of providing better housing, and also that the squatters will not, indeed cannot, return to their former rural homes. The trend has therefore been to legalise the occupation of the shanty towns. Where possible sewage, electricity and water may be provided. The residents are encouraged to build their own homes (which they were already doing). National government and international agencies recognised that this self-help alternative could be a low-cost housing policy, and since 1974 self-help has been officially endorsed by the World Bank.

Some researchers and planners have given these policies enthusiastic support. Critics, on the other hand, have argued that they simply let governments off the hook and act as a justification for low wages and frightful living conditions.

Running through many of these debates, which for the most part originate in western institutions, has been the continuing theme of the planners' contempt for the poor. Individual studies of African or other non-western cities have testified to their ebullience, diversity and variety. There is friendliness, and social contacts are established with ease. Most of those who come to the towns do not come as strangers, but already have members of the family established there. This family will cushion the shock of the transition to urban life, and provide much-needed help and support in the initial months. The new arrival relies precisely on ties of kinship to see her or him through the bewildering early days of life in the town.

Yet many who have written of third-world city life have emphasised poverty, the breakdown of family life, prostitution, crime and psychological maladjustment, and have blamed the impersonal nature of city life with its alienation and anomie. The work of Oscar Lewis, who described an alleged 'culture of poverty', was extremely influential in the 1950s and 1960s in the development of this view, which, consciously or not, built on the assumptions of Georg Simmel and Louis Wirth, who had emphasised the impersonal factors in urban life.

More recently, there has been a greater recognition of the persistence of family obligations in the city. In the cities of the third world as in nineteenth-century New York or Chicago, or for that matter in the new industrial towns of nineteenth-century Lancashire, family connections play an important role in the process by which immigrants and rural workers are transformed into urban dwellers.

Yet the negative view of the effects of city life on the individual and the family has persisted. In addition, writers, planners and officials have then imposed the weight of the theories and assumptions of postwar western planning with its emphasis on zoning, segregation and surveillance. The result has been that in third-world cities, as in the cities of postwar Britain and the United States, redevelopment, zoning, skyscraper business districts and dormitory suburbs are to be the answers to the 'chaos' and 'moral breakdown' of the unplanned industrial city. The same exaggerated faith in the ability of neutral, scientific planning to solve the urban crisis has led to a reproduction of the same mistakes as in western cities, but with far more extreme results.

Western assumptions about the normality and universality of the nuclear family have been extended to the shanty towns and urban populations of the third world. The view that the slums of Latin America fostered unmarried motherhood, delinquency, prostitution and a psychology of apathy and living for the moment has ceded to that of writers who were more likely to emphasise the 'normality' and stability in western terms of family life and social existence in the shanty towns. 'Normality', however, has meant the stereotype of family life in which the husband works and the woman remains at home, engaged in full-time domestic work and child rearing. This is not the actual experience even of the majority of western families. It has been used to marginalise women in the paid workforce, and has also perpetuated inequalities generally between men and women inside and outside the family.

It is even more inappropriate as a model of the family in non-western urban settlements. Its basic assumptions do not take into account the situation and crucial importance of women in and to these housing settlements. National and international investment in self-help housing projects has failed to answer the needs of women and, on the contrary, has excluded them. Governmental and financial policies tend to be based on the assumption that the nuclear family with a male breadwinner is the usual and indeed natural family form, whereas in fact different household forms coexist. Recent figures have suggested that one in three of the world's households are now headed by women. In urban areas, however, approximately 50 per cent of households are headed by women, and in the refugee camps of Central America the figure is closer to 90 per cent.

Women-headed households are placed in a kind of double jeopardy. Their very existence is denied, minimised or not taken account of. Then, in addition, it is often assumed that only male-headed households are sufficiently stable to merit inclusion in housing schemes. In one project in São Paulo, funded by the Brazilian Housing Bank Profilurb programme, women who headed households were excluded by the criteria of eligibility. Yet some observers have argued that families headed by women may be more stable, show more responsibility in paying back loans, in paying rent and in improving properties.

Governments have also failed to recognise that women play an especially important role in 'community management'. Because women are more directly concerned than men with the welfare of the household, and with 'community' issues such as water supply and safety, they are normally more aware than men of the needs of a housing project, and more committed to its success.

A neglect of the needs of women has often militated against the success of self-help housing projects. For example, a failure to recognise that women as well as men engage in paid employment leads to a male bias when 'squatters' are relocated, often miles from the centre of the city. This bias arises either when rehousing is near factories or other sites of male employment, or simply because it is assumed that men can travel long distances to work. Planners pay little heed to the fact that women both need work, and need to be near their work if, as they invariably have to, they are successfully to combine it with domestic duties. For example, many women take in laundry or have jobs as maids, and for this they need to live near to their employers. Even if it is acknowledged that the relocation has created transport difficulties, these may be interpreted as difficulties for men. In Belo Horizonte, Brazil, for example, public transport was laid on at peak hours, so that men could travel to work, but was withdrawn during the day when women needed to use the service, either to take children to school, to go shopping, or get to their own part-time jobs.

Another way in which women are ignored and excluded is that they are not consulted on issues such as the design of houses. For example, the object of the Tanzanian 'Better Housing Campaign' was to persuade people to build more durable houses, built of imported materials, to replace traditional ones built with local materials. One unintended effect of the higher costs of building with imported materials was that it impeded the tradition of building separate accommodation for male and female members of families. As a result, women were redefined as dependants, and their traditional autonomy was undermined. In two housing projects in Tunis, houses were designed with a much smaller courtyard than in traditional design. Because Muslim women, spending most of their time in the home, needed this internal space, its absence caused depression and even suicide.

In spite of all the obstacles and prejudice, women have managed to involve themselves in many local housing projects. They have organised locally against the fearful hardships of the majority of the *barrios*. In one settlement in

Guayaquil, capital of Ecuador, initial conditions were terrible, as the settlement consisted of swampland. Settlers bought plots on which they then had to build their own houses. To begin with, the settlement had no services, not even water. It was approached by perilous catwalks above the swamps. These were so dangerous that two children drowned, while two women died as a result of wading through the swamp, and a man was electrocuted while trying to fix a light connection.

Women, given that they less frequently went out to work than did their male partners, were particularly isolated. Many had only just left their parents' home, others had given up paid work, and for those who tried to carry on with their laundry work the shortage of water was a disaster. They became more economically dependent than hitherto on their husbands or partners. So, while to settle on the swamp had economic advantages – they no longer had to pay rent and therefore had more money for education and consumer goods – the disadvantages of the area bore particularly upon women. Soon, however, they established mutual-aid networks, and this led to the formation of organised committees to agitate for change. To take a leading role in these did involve a challenge to traditional views of how women should behave (submissively, with the emphasis on her role as mother and homemaker); on the other hand, the traditional role itself validated membership of groups seeking to improve domestic living conditions.

In Nicaragua, the Sandinista Revolution in 1979 resulted in a political commitment to tackle problems such as these, but changes were brought about only with great difficulty. In one project in Managua, men objected to women sharing in the actual work of house construction, but, after considerable discord, some women did manage to learn the basic building skills. The women in the project were more radical than the men in that they wanted the finished houses allocated according to need, while the men wanted them to be allocated according to the amount of work put in (advantageous to them, since on the whole the men had fewer alternative calls on their time). During a dispute over the allocation of one particular house, the issues of *machismo* (also referred to as *somocismo*, i.e. behaviour worthy of the Somoza regime,

overthrown by the Sandinistas) was raised. This was possible because the revolution had initiated a reassessment of the roles of men and women, and was committed in principle to the equality of women. Increasingly throughout the 1980s, however, the Nicaraguan experience was distorted by the incursions of the Contras and American destabilisation, so that towards the end of the decade, the mounting economic crisis was testing the survival strategies of urban women to the limit.

Throughout Latin America, the position of single mothers is ambiguous. Confined to low-paid work, they, with their children, are among the poorest in the community, and still tend to be stigmatised, blamed for their plight, although many have been irresponsibly deserted by husband or lover. Some, however, have made the choice to bring up the children on their own, preferring the hardships of this way of life to the domineering behaviour of their menfolk, many of whom refuse to allow their women freedom of movement, abuse them physically, and spend the family income on heavy drinking. An Ecuador study showed that participation in the organisation of the *barrio* had also enabled some of the cohabiting women involved to achieve economic independence so that they were in a much more powerful position in relation to their husbands/partners.

In countries such as Brazil, where feminism influenced radical middle-class women in the 1970s, the general turbulence of the political situation made possible, for a time at least, alliances between the feminists and women's groups from the working class and the poorest sections of society, including women in the shanty towns who were fighting for the provision of basic needs. In this way, the impact of feminist ideas on an educated middle class affected a much broader group of women, although in different ways. Female domestic servants, for example, became a militant and well-organised force – a far cry from the usual picture of maids as unorganised, and, indeed, impossible to organise. For these women, life in the *barrios* and settlements was not a life of apathy and despair. Most of those living there neither came from nor wished to return to a rural life.

Life in the city provides the preconditions for continuing struggle, since in the city the poor, although 'excluded from the comforts of the city, are exposed to its modernity'. The existence of the benefits of urban life – even though they are excluded from them – justifies their demands for *inclusion*. The gulf between what is and what might be may appear to widen; on the other hand, the city both raises aspirations and gives more chance of their realisation.

Although cities in the non-western world differ both from one another and from the cities of Europe and North America, it has become customary to refer to the 'third-world city' or the 'world city' as a separate and recognisable entity.

A global capitalism must surely be creating a global city: this seems to be the assumption upon which such a generalisation is based, although it is also recognised that global capital may act to differentiate one city, and one economy, from another. A fear remains that the world, and its cities, are becoming homogenised: difference is ironed out and everything is the same. At the same time, within every city a growing distance between rich and poor makes for another kind of unreality, and a gulf in experience that cannot be bridged. We move, then, from world city to postmodern city.

27 BERNARD TSCHUMI

'Spaces and Events'

from *Architecture and Disjunction* (1994)

Can one attempt to make a contribution to architectural discourse by relentlessly stating that there is no space without event, no architecture without program? This seems to be our mandate at a time that has witnessed the revival of historicism or, alternatively, of formalism in almost every architectural circle. Our work argues that architecture – its social relevance and formal invention – cannot be dissociated from the events that "happen" in it. Recent projects insist constantly on issues of program and notation. They stress a critical attitude that observes, analyzes, and interprets some of the most controversial positions of past and present architectural ideologies.

Yet this work often took place against the mainstream of the prevalent architectural discourse. For throughout the 1970s there was an exacerbation of stylistic concerns at the expense of programmatic ones and a reduction of architecture as a form of knowledge to architecture as knowledge of form. From modernism to postmodernism, the history of architecture was surreptitiously turned into a history of styles. This perverted form of history borrowed from semiotics the ability to "read" layers of interpretation but reduced architecture to a system of surface signs at the expense of the reciprocal, indifferent, or even conflictive relationship of spaces and events.

This is not the place for an extensive analysis of the situation that engulfed the critical establishment. However, it should be stressed that it is no accident that this emphasis on stylistic issues corresponded to a double and wider phenomenon: on the one hand, the increasing role of the developer in planning large buildings, encouraging many architects to become mere decorators, and on the other, the tendency of many architectural critics to concentrate on surface readings, signs, metaphors, and other modes of presentation, often to the exclusion of spatial or programmatic concerns. These are two faces of a single coin, typical of an increasing desertion by the architectural profession of its responsibilities vis-à-vis the events and activities that take place in the spaces it designs.

At the start of the 1980s, the notion of program was still forbidden territory. Programmatic concerns were rejected as leftovers from obsolete functionalist doctrines by those polemicists who saw programs as mere pretexts for stylistic experimentation. Few dared to explore the relation between the formal elaboration of spaces and the invention of programs, between the abstraction of architectural thought and the representation of events. The popular dissemination of architectural images through eye-catching reproductions in magazines often turned architecture into a passive object of contemplation instead of the *place* that confronts spaces and actions. Most exhibitions of architecture in art galleries and museums encouraged "surface" practice and presented the architect's work as a form of decorative painting. Walls and bodies, abstract planes and figures were rarely seen as part of a single signifying system. History may one day look upon this period as the moment of the loss of innocence in twentieth-century architecture: the moment when it became clear that neither super-technology, expressionist functionalism, nor neo-Corbusianism could solve society's ills and that architecture was not ideologically neutral. A strong political upheaval, a rebirth of critical thought in architecture, and new developments in history and theory all triggered a phenomenon whose consequences are still unmeasured. This general loss of innocence resulted in a variety of moves by architects according to their political or ideological leanings. In the early 1970s, some denounced architecture altogether, arguing that its practice, in the current socioeconomic context, could only be reactionary and reinforce the status quo. Others, influenced by structural linguistics, talked of "constants" and the rational autonomy of an architecture that transcended all social forms. Others reintroduced political discourse and advocated a return to preindustrial forms of society. And still others cynically took the analyses of style and ideology by Barthes, Eco, or Baudrillard and diverted them from their critical aims, turning them over like a glove. Instead of using them to question the distorted, mediated nature of architectural practice, these architects injected meaning into their buildings artificially, through a collage of historicist or metaphorical elements. The restricted notion of postmodernism that ensued – a notion diminished by comparison with literature or art – completely and uncritically reinserted architecture into the cycle of consumption.

At the Architectural Association (AA) in London, I devised a program entitled "Theory, Language, Attitudes." Exploiting the structure of the AA, which encouraged autonomous research and independent lecture courses, it played on an opposition between political and theoretical concerns about the city (those of Baudrillard, Lefèbvre, Adorno, Lukács, and Benjamin, for example) and an art sensibility informed by photography, conceptual art, and performance. This opposition between a verbal critical discourse and a visual one suggested that the two were complementary. Students' projects explored that overlapping sensibility, often in a manner sufficiently obscure to generate initial hostility through the school. Of course the codes used in the students' work differed sharply from those seen in schools and architectural offices at the time. At the end-of-year exhibition texts, tapes, films, manifestos, rows of storyboards, and photographs of ghostlike figures, each with their own specific conventions, intruded in a space arranged according to codes disparate from those of the profession.

Photography was used obsessively: as "live" insert, as artificial documentation, as a hint of reality interposed in architectural drawing – a reality nevertheless distanced and often manipulated, filled with skilful staging, with characters and sets in their complementary relations. Students enacted fictitious programs inside carefully selected "real" spaces and then shot entire photographic sequences as evidence of their architectural endeavors. Any new attitude to architecture *had* to question its mode of representation.

Other works dealing with a critical analysis of urban life were generally in written form. They were turned into a book, edited, designed, printed, and published by the unit; hence, "the words of architecture became the work of architecture," as we said. Entitled *A Chronicle of Urban Politics*, the book attempted to analyze what distinguished our period from the preceding one. Texts on fragmentation, cultural dequalification, and the "intermediate city" analyzed consumerism, totems, and representationalism. Some of the texts announced, several years in advance, preoccupations now common to the cultural sphere: dislocated imagery, artificiality, representational reality versus experienced reality.

The mixing of genres and disciplines in this work was widely attacked by the academic establishment, still obsessed with concepts of disciplinary autonomy and self-referentiality. But the significance of such events is not a matter of historical precedence or provocation. In superimposing ideas and perceptions, words and spaces, these events underlined the importance of a certain kind of relationship between abstraction and narrative – a complex juxtaposition of abstract concepts and immediate experiences, contradictions, superimpositions of mutually exclusive sensibilities. This dialectic between the verbal and the visual culminated in 1974 in a series of "literary" projects organized in the studio, in which texts provided programs or events on which students were to develop architectural works. The role of the text was fundamental in that it underlined some aspect of the complementing (or, occasionally, lack of complementing) of events and spaces. Some texts, like Italo Calvino's metaphorical descriptions of "Invisible Cities," were so "architectural" as to require going far beyond the mere illustration of the author's already powerful descriptions; Franz Kafka's *Burrow* challenged conventional architectural perceptions and modes of representation; Edgar Allan Poe's *Masque of the Red Death* (done during my term as Visiting Critic at Princeton University) suggested parallels between narrative and spatial sequences. Such explorations of the intricacies of language and space naturally had to touch on James Joyce's discoveries. During one of my trips from the United States I gave extracts from *Finnegans Wake* as the program. The site was London's Covent Garden and the architecture was derived, by analogy or opposition, from Joyce's text. The effect of such research was invaluable in providing a framework for the analysis of the relations between events and spaces, beyond functionalist notions.

The unfolding of events in a literary context inevitably suggested parallels to the unfolding of events in architecture.

SPACE VERSUS PROGRAM

To what extent could the literary narrative shed light on the organization of events in buildings, whether called "use," "functions," "activities," or "programs"? If writers could manipulate the structure of stories in the same way as they twist vocabulary and grammar, couldn't architects do the same, organizing the program in a similarly objective, detached, or imaginative way? For if architects could self-consciously use such devices as repetition, distortion, or juxtaposition in the formal elaboration of walls, couldn't they do the same thing in terms of the activities that occurred within those very walls? Pole vaulting in the chapel, bicycling in the laundromat, sky diving in the elevator shaft? Raising these questions proved increasingly stimulating: conventional organizations of spaces could be matched to the most surrealistically absurd sets of activities. Or vice versa: the most intricate and perverse organization of spaces could accommodate the everyday life of an average suburban family.

Such research was obviously not aimed at providing immediate answers, whether ideological or practical. Far more important was the understanding that the relation between program and building could be either highly sympathetic or contrived and artificial. The latter, of course, fascinated us more, as it rejected all functionalist

leanings. It was a time when most architects were questioning, attacking, or outright rejecting modern movement orthodoxy. We simply refused to enter these polemics, viewing them as stylistic or semantic battles. Moreover, if this orthodoxy was often attacked for its reduction to minimalist formal manipulations, we refused to enrich it with witty metaphors. Issues of intertextuality, multiple readings and dual codings had to integrate the notion of program. To use a Palladian arch for an athletic club alters both Palladio and the nature of the athletic event.

As an exploration of the disjunction between expected form and expected use, we began a series of projects opposing specific programs with particular, often conflicting spaces. Programatic context versus urban typology, urban typology versus spatial experience, spatial experience versus procedure, and so on, provided a dialectical framework for research. We consciously suggested programs that were impossible on the sites that were to house them: a stadium in Soho, a prison near Wardour Street, a ballroom in a churchyard. At the same time, issues of *notation* became fundamental: if the reading of architecture was to include the events that took place in it, it would be necessary to devise modes of notating such activities. Several modes of notation were invented to supplement the limitations of plans, sections, or axonometrics. Movement notation derived from choreography, and simultaneous scores derived from music notation were elaborated for architectural purposes.

If movement notation usually proceeded from our desire to map the actual movement of bodies in spaces, it increasingly became a sign that did not necessarily refer to these movements but rather to the *idea* of movement – a form of notation that was there to *recall* that architecture was also about the movement of bodies in space, that their language and the language of walls were ultimately complementary. Using movement notation as a means of recalling issues was an attempt to include new and stereotypical codes in architectural drawing and, by extension, in its perception; layerings, juxtaposition, and superimposition of images purposefully blurred the conventional relationship between plan, graphic conventions and their meaning in the built realm. Increasingly the drawings became both the notation of a complex architectural reality *and* drawings (art works) in their own right, with their own frame of reference, deliberately set apart from the conventions of architectural plans and sections.

The fascination with the dramatic, either in the program (murder, sexuality, violence) or in the mode of representation (strongly outlined images, distorted angles of vision – as if seen from a diving airforce bomber), is there to force a response. Architecture ceases to be a backdrop for actions, becoming the action itself.

All this suggests that "shock" must be manufactured by the architect if architecture is to communicate. Influence from the mass media, from fashion and popular magazines, informed the choice of programs: the lunatic asylum, the fashion institute, the Falklands war. It also influenced the graphic techniques, from the straight black and white photography for the early days to the overcharged grease-pencil illustration of later years, stressing the inevitable "mediatization" of architectural activity. With the dramatic sense that pervades much of the work, cinematic devices replace conventional description. Architecture becomes the discourse of events as much as the discourse of spaces.

From our work in the early days, when event, movement, and spaces were analytically juxtaposed in mutual tension, the work moved toward an increasingly synthetic attitude. We had begun with a critique of the city, had gone back to basics: to simple and pure spaces, to barren landscapes, a room; to simple body movements, walking in a straight line, dancing; to short scenarios. And we gradually increased the complexity by introducing literary parallels and sequences of events, placing these programs within existing urban contexts. Within the worldwide megalopolis, new programs are placed in new urban situations. The process has gone full circle: it started by deconstructing the city, today it explores new codes of *assemblage*.

28 K. EVANS, I. TAYLOR AND P. FRASER

'Shop 'Til You Drop'

from *A Tale of Two Cities: Global Change, Local Feeling and Everyday Life in the North of England* (1996)

[. . .]

Meadowhall

The defining moment in the post-war history of shopping in this city [Sheffield] was without question, however, the opening of the Meadowhall Shopping Centre in September 1990. Meadowhall is built directly on the site of a major local steelworks (Hadfield's), which in 1980 had been the site of the biggest mass picket in the steel strike of that year, but which closed three years later. The site is directly adjacent to the M1 motorway, to the north-east of the city, convenient in the past for deliveries to and from the works, and now for the car-travelling shoppers of the North of England. On opening, it was proclaimed as 'the largest shopping mall in Europe'. The brainchild of the local entrepreneur E.D. Healey and what is now known (given its more recent activities in the Don Valley) as the Stadium Group, Meadowhall houses 223 stores on two levels in 1.5 million square feet of consumer space and provides free parking and its own rail and bus station. In 1994, Meadowhall was connected directly with the city centre and the council estates to the east by light rapid transit link, the South Yorkshire Supertram. As the publicity produced by Meadowhall's own management company proclaims, the mall is located in one of the most densely populated areas of England, with fully 9 million people living within one hour's drive of its car parks. 'Average weekly patronage' at Meadowhall in 1994 was measured at 480,000 people. Apart from its massive collection of shops, Meadowhall also features a giant 'videowall' constantly playing in front of the

'Oasis' eating zone, where there is an array of take-away options with foods 'from all over the world'. A further leisure development is planned in the form of a 'Tivoli Gardens' pleasure park and cinema on a site immediately adjacent to the mall.

There is no question that the effect of Meadowhall on the city centre of Sheffield has been devastating, with an estimated 20 per cent reduction in city-centre trade at the end of the second year of Meadowhall's trading, although the effects of the national recession may also have contributed to declining trade figures in the city centre.

[. . .]

MALL LIFE IN THE NORTH OF ENGLAND: MEADOWHALL

The focus of contemporary writing on the city, by architects and cultural theorists alike, is the shopping mall, particularly the large, out-of-town shopping malls like the MetroCentre in Gateshead and Meadowhall in Sheffield. In this writing, the mood is either one of uncritical celebration of the new sphere of freedom these malls are seen to produce, or of warning as to the new forms of social discipline the malls encourage.

[. . .]

However, in a situation where the 'undisputed aim' of such shopping centres is to encourage spending, there are many who simply 'play at being shoppers' in the mall – the ultimate subversion of the planners' and owners' objectives:

> The users, both young and old, are not just resigned victims, but actively subvert the ambitions of the mall developers by developing the insulation value of the stance of the jaded world-weary flâneur; asserting their independence in a multitude of ways apart from consuming. It is this practice, as opposed to the only too modern centralising ambitions of the mall builder, which is the heart of postmodern experience of the mall.
>
> (Shields, *Places on the Margin*, London, Routledge, 1989: 160)

Meaghan Morris, reacting to both Australian and American exemplars, notes how malls are presented to us, in their own promotional literature, in a very idealised, static and abstract fashion. She also observes that those who develop plans eventually have to confront the social composition of their local customer population. At the same time, malls must attempt to evolve over time, particularly in respect of how they position themselves in the local consumption markets. Crystal Peaks in South Sheffield was a local instance of this type of mall: indeed Meadowhall itself commissioned a statue, 'Teeming' (albeit cast in an artist's studio in Tuscany), for its main mall, depicting three Sheffield forge workers pouring molten steel into a crucible.

Morris's commentary is helpful in encouraging our recognition of the kind of local-spatial and temporal considerations that some writing about the mall ignores. Simply because the malls are ready-made shopping and socialising environments without any history, unlike the traditional High Street, this does not preclude the possibility of their being endowed with particular kinds of local meaning – as the sites of culturally specific practices which, we will argue later, are also put to use in individual shops in neighbourhoods.

Research in Southampton in the early 1990s suggested that the opening of a new shopping mall involved no necessary or final disturbance to established patterns of use of local city centres. Individual use of the mall did not automatically indicate approval of it, especially of its impact on the local economy or city centre, but did indicate that the range of goods in the mall was an improvement on the city centre and that the mall was convenient to use.

The striking feature of Meadowhall, of course, and of the MetroCentre in Gateshead, is the fact of their location in an old industrial area of the North of England, on the edge of cities (Newcastle and Sheffield) that have not defined themselves as metropolitan centres of consumption. Meadowhall itself, as we have already indicated, is built directly on the site of a major Sheffield steelworks, closed in the early 1980's on the outer edge of a city which has historically seen itself as decidedly non-metropolitan, and even as 'cut off' geographically and culturally from mainstream metropolitan practices in culture and trade. The response of Sheffielders to Meadowhall in that sense is a measure not just abstractly of shoppers' aesthetic response to a new building or even of their psychological appreciation of the convenience of its facilities; it is also a measure of the response of local citizens to a specifically metropolitan icon, a 'temple of consumption'. The majority of Sheffield people we met seemed, at first blush, to exhibit very similar, mixed responses to Meadowhall to the responses of people in Southampton to their new mall. Fear of the mall's effects on the city centre, set against an appreciation of its convenience were very much apparent, although there was also a range of other reactions. We identified five principal arguments, amongst which one, that of 'principled objection', was the most resonant of a local, Sheffield, reaction. We can illustrate each of these responses in turn.

Discourse of convenience

The convenience of malls such as Meadowhall for many people is undisputable: the parking is free, the opening hours are later than most shops in the city centre, there is the possibility of hiring an electric scooter or wheelchair for those with restricted mobility, and it provides a widely praised crèche. Furthermore, the atmosphere is clean and dry, any litter is swept up immediately and there is a sophisticated surveillance system in operation. The 'Videowall' and the Oasis eatery further extend the concept of the mall as an ideal day out for all the family.

A group of disabled Sheffielders were generally quite enthusiastic about Meadowhall, although some voiced criticism about the slipperiness of the floors, which was perceived to be a product of constant brushing up of litter and polishing, making it dangerous for those who are unsteady on their feet. Other groups also referred to the mall's convenience. Two unemployed people from Sheffield, Hazel Lowe and Steve O'Donnell, agreed on its attractions:

> It's pleasant and everything is in the one area. If it's raining then you're sheltered from the rain.

> Meadowhall, it's laid out nicely, I'll agree with that. They've done a good job with it.

Claire Moran, a clerical worker from Sheffield enthused:

> It's comfortable to shop in ... there's plenty of eating places.

Jacqui Partridge, an 'up-market shopper' thought that:

> If you're disabled or if you've got babies, then the answer really is to go to Meadowhall. Everything is laid on for you.

Principled objection

Many Sheffield residents expressed concern that Meadowhall was killing city-centre trade and atmosphere, and indicated that they would tend not to visit it on principle. Nat Feingold told us:

> I don't go there. We don't go there on principle. I want to support the shops in the centre of town. We don't want to go because we feel that it's unenvironmental and it's just killing the city centre ... the chains [stores] that are in Meadowhall are now putting more lines on. Say Marks and Spencers will have far more out there than in the city centre. With the result that nobody shops in M&S in the city centre. If you go in there wanting a suit or something, people say 'It's not here, you'll have to go to Meadowhall.'

Scott Trueman explained:

> I lived in North America for three years, I saw what happened to the city centres there. As soon as shopping malls came on the scene, on the outskirts of town, the city centre became unsafe at night, unstable. All the shopping was finished. You only went down town for work ... Houston, New Orleans. New Orleans not so much because it's got a tourist quarter. Where you get shopping malls ringing the town, the shopping dies in the centre of town. Take the small shops away, take the town centre shops away and you've got nothing but a dead American city. This is what's happening in Sheffield.

Sheffield and Meadowhall are two different places

A significant number of our Sheffielders, who we sensed were typical of a large body of local opinion, felt quite strongly that Meadowhall and Sheffield were two separate places. Kirstin Lloyd worked in a Sheffield hotel and found herself directing visitors to Meadowhall. She told us:

> [I tell them that] it's two miles down that road. Then they'll ask where the city centre is and I'll tell them that it's just down that way. It's then when it hits you that you really are directing people out of the city centre ... they always ask if they can walk to Meadowhall. You can see them looking and thinking why's that there in that area? It just brings them [visitors] to Meadowhall. It doesn't bring them to the city centre. It doesn't bring them to the things that we've got to offer, like the theatres, the cultural parts. It just brings them to Meadowhall.

This feeling is reinforced by the findings of our survey of Meadowhall users. Of 184 people surveyed there over one week, 58 per cent came from towns or cities outside Sheffield, and most of these people claimed not to know the city centre at all. Others were concerned that Meadowhall had done little for Sheffield people. Some Sheffield police officers were among them:

> It's done nothing for the local population at all. It causes traffic problems.

> When you consider how many Edgar Allen's [steelworks on whose derelict site Meadowhall was built] employed. The majority of people who use it are probably not even from Sheffield.

However, its appeal remains insidious: Josephine Armitage, a young woman not originally from Sheffield, explained:

> There's things that I sort of like about it. I go to the cinema. But it's quite frightening, the way it's like in America. What's sort of weird about it is that there's almost sort of nothing to buy. Everything is exactly the same ... what's frightening about it is that it overtakes. It's frightening how much you can get sucked into that feeling of ... I don't know, it's just ... nothingness. It's like watching television that you know is not very good. You sort of enjoy it at the time, but you know that it's not good for you. That it's not stimulating your brain.

Non-plussed by Meadowhall

Many Sheffielders simply seemed non-plussed by Meadowhall, and had perhaps been once or twice, just to have a look, or not at all. Imran Akbar had never been:

I've not been but I've heard about it. They say everything's dear.

Helen-Anne Goode was also unenthusiastic about Meadowhall:

We went once and had a look around and that was it. I just don't shop there.

Heather Moody visited it primarily for its restaurants:

I've been to Meadowhall, not to shop, because I'm the sort of person who'll shop in town. I've been down there on my birthday.

The celebration of Meadowhall

There are those, however, who expressed unreserved admiration for Meadowhall, and interestingly these often tended, in our research, to be older residents who had lived through many changes in the city already. Geoff Helliwell and Geoffrey Bundy opined:

I think it's a marvellous structure. I know it's done a lot of harm to the city centre, but it was derelict steel works land. It has improved the environment round there . . . most of the work was done by Italians. It's great. The marble floors and the pillars. It's all come from Italy, a lot of this. It's really a superb building, there's no question about that. The building itself is absolutely marvellous – the structure and everything about it – it's a palace.

I just wanted to say, in my opinion, for anyone with a car, Meadowhall is the best thing since sliced bread.

Edith Boardman was delighted with Meadowhall:

There's lots of forms to sit down on . . . there's a lot of seats, it's lovely . . . you go down these little side streets – there's all shops, then you get into the Oasis where all the food is . . . it's the best thing that's ever come to Sheffield.

The prevailing local folk wisdom in Sheffield, however, was that goods at Meadowhall are more expensive that they are in town. For many, this was deterrence enough, particularly on the part of those who have never been or been only once or twice. The group of retired men and women in the Afro-Caribbean Centre in Sheffield told us they considered it 'too dear', as did members of our group of unemployed and retired manual workers. The Oasis eating area, particularly, was considered 'a bit on the expensive side' by many. What a focus group conversation does not reveal, however, in and of itself, is the extent to which these kinds of account (discourses of cost and economy) are also an expression of the antipathy to metropolitan growth and 'development' which is an important defining element in the 'structure of local feeling' in old industrial enclave cities like Sheffield.

A group of mainly unemployed Pakistani women from the Nether Edge area of Sheffield were nevertheless very keen on Meadowhall, principally because they could go there with their husbands and families during the evening and parking was free. They were also impressed by the fact that there was now a Pakistani restaurant in the mall. Mabuba Razhar indicated he would like to see a fresh produce market there:

The only thing I miss is having a market. If they had that then I would never go to town.

However, a 'real' street market would, of course, contradict the principles of ordered, sedate consumption on which the new malls claim to be founded. Certainly, the only market in Meadowhall at present is a simulated Mediterranean street market with stalls of pre-packaged trinkets. In the imagination of the Meadowhall management, a 'real' market would risk attracting precisely the sorts of people associated with the unruly city-centre markets whom a significant number of 'up-market' mall shoppers go to Meadowhall to avoid.

Meadowhall had been in existence for some five years, at the time of writing, and had established itself as the leading shopping centre in the UK in terms of overall turnover, ahead of the MetroCentre in Gateshead and Merry Hill outside Dudley in Warwickshire (*Manchester Evening News* 24 January 1995). The Meadowhall management and the Sheffield Development Corporation have been at pains to emphasise the importance of Meadowhall in bringing visitors and 'spending power' to

Sheffield: in the summer of 1993, they gave great publicity to a study from Sheffield Hallam University showing that 20.2 million people had visited the city in 1992, with 16.9 million aiming for Meadowhall. The local paper's coverage of this report argued that this made Meadowhall a more popular destination than Blackpool. In the meantime, the city centre of Sheffield continued to show evidence of decline.

It may be more accurate, in fact, to identify the emergence of what Dr Gwyn Rowley (in the *Independent* 22 February 1992) calls a 'geographic dichotomy', or what we would want to see as a new and fundamental social division in the city – between those who can afford to shop at Meadowhall and come to feel a sense of belonging in the mall, and those who cannot, at least with any degree of regularity, and do not feel any such sense of belonging. For the latter population, we should remember, the city centre was often a necessary destination for other reasons (the Social Security Offices, the Town Hall), but the personal commitment to the city centre, albeit in its dilapidated state in the early 1990s, was often also underwritten by protests of local pride and loyalty. What has yet to be seen is whether the arrival of the Supertram in Sheffield, bringing passengers from outlying estates directly into the city centre and also into Meadowhall, further reconstructs the recently reorganised local stratification of the shopping population.

MEMORY, IMAGINATION AND IDENTITY

INTRODUCTION

The texts in this section investigate the role of memory and imagination in the perception of built environments, and the contribution of cultural frameworks to the formation of urban, social and individual identities.

Memories are often thought of as personal, almost private experiences. Like the meanings attached to spaces, they are subjective and mutable. Conditioned by cultural contexts and specific to the circumstances of their making, they are constantly remade as responses to those circumstances shift in retrospect. But, if memories are dialogues between past and present, self and other, reserved in a space of privacy, they remain dependent on a life in the world and have a collective dimension.

Gaston Bachelard, in *The Poetics of Space* (1969, Beacon Press), roots spatial experience in solitude. Here, memories accumulate to colour present perceptions. Although Bachelard cites lines from poetry which refer to roads, these, too, are part of an inner world articulated by thresholds and passages, in which actuality is internalized and transformed. Out of the forms of external appearances, Bachelard makes a palimpsest of memories which contain desires – for the safety of containment, the joy of (perhaps womblike) occupation. This might seem a Romantic attitude, a longing for the not-quite attainable or the joy just-gone – the kind of emotion found in paintings by Bonnard. A more careful consideration, however, suggests a re-articulation of the Cartesian paradigm which is, in many ways, a foundation of modernity. The world of a self constituted as Subject is necessarily objectified, in turn constituting the realm of the subject as a kind of solitude. The metaphor of a room, and a world seen through a window, framed as outside, acts appropriately to figure the space of Cartesian thought. In this realm of modern subjectivity, from which emerged the notion of privacy, revery is possible when the mind rests. In revery, the experiences and spatial imaginings of solitude are indelible. Bachelard also asserts in this extract the interplay of inside and outside. In revery, he writes, 'we cover the universe with drawings we have lived', and in these spaces of intimacy is well-being, a defining characteristic of shelter.

In Patrick Wright's *A Journey Through Ruins* (1993, Flamingo) another kind of imaginative world is investigated, closer to the collective and where only traces remain of celebration. Wright's account is located in the decay of London in the 1980s. His book is dedicated to 'Lady Margaret Thatcher', and charts a mood of disjunction and putrefaction. Litter drifts in the air, and people look on despondently, parodies of the nice figures in architects' models. Signs of damage appear on all sides of the intrepid narrator. Signs of occupation take forms such as racist graffiti in tower blocks designed as utopian social engineering. In such circumstances, memories of better times, often hard to place exactly, take on the rosy tints of an irretrievable world. Whilst Bachelard's space of revery has the comforting feeling of childhood safety, Wright's account of parts of East London is fixed in an uncomfortable present.

In the collective memory, monuments, such as war memorials, manipulate remembrance for the construction of national identity (see the text by al-Khalil in Section III). Aldo Rossi, in *The Architecture of the City* (1982, MIT), asks, in the passage preceding that reprinted here, how a city comes to be identified with an idea, for instance of what makes a capital city. For Rossi, a city is the collective memory of its citizens, and that memory is stored in specific objects and places. These are given special meaning, as signs of the city as a particular type of city; this process of lending value, which emphasizes the mutability rather than form of the city, is the articulation of urban space by urban publics. Rossi sees a rational intentionality in the gradual moulding of a city in the shape of its collective conceptualization, an identity out of memory and the interpretation of history.

The other three extracts foreground the formation of identities, though the final text links this again to memories of place. Kenneth Frampton (1985) reverses the modernist idea of an avant-garde, coining the term 'arriere-garde' to denote a critical distance from both the Enlightenment myth of progress and nostalgia for a pre-industrial world. Critical regionalism, taking region in the larger (geographical) sense of a part of the world rather than part of a country, he argues, must deconstruct the baggage of the global and universal (which often means the dominant, or western), and mediate both industrial and post-industrial technologies. From this position, that which is other than the dominant style of architecture no longer appears a deviation, in the senses of primitive or exotic, but becomes part of a plurality. Frampton's view, then, is explicitly not a call for the ersatz vernacular of tudorbethan supermarkets or the half-timbered semi-detached house, but it makes possible inventive solutions specific to context, in the absence of a mainstream which limits that invention. He gives the Bagsvaerd Church near Copenhagen, by Jorn Utzon, as an example, its inner vault derived from the Chinese pagoda and, without contradiction, referencing the western tradition of sacred spaces in which the vault acts as a metaphor for heaven.

Whilst Rossi sees a city's form as constructed incrementally, as the collective memory of its dwellers articulates a concept to which it increasingly conforms, Frank Mort emphasizes the construction of place identity in the mass media and style magazines. Looking at Soho in London in an extract from *Cultures of Consumption* (1996, Routledge), he notes the coincidence of economic and cultural factors. When rents in Soho, in the 1980s, were well below those of adjoining neighbourhoods, advertising, public relations, style publishing and film companies – all image makers or builders – moved in. Many small, young companies, and magazines for the style-conscious young, identified themselves with an area which, reciprocally, became identified as stylish. Here, too, an environment is nudged towards conformity with a concept, its existing built form co-opted to the agenda, often with only surface changes but deep shifts of meaning and added value. However, the interiors and streetscapes of Soho became a backdrop to what remained a partial and selective mythicization of urban space.

For the inhabitants of Cairo's working-class quarters, style is the subject-matter only of day-dreaming. They do not, as Farha Ghannam points out, have faxes or satellite dishes. Neither do they have many choices about where or how they live. In this chapter from *Space, Culture and Power* (1997, Zed Books), edited by Ayse Oncu and Petra Weyland, Ghannam tells the story of a community uprooted from an inner city area near the Ramses Hilton. As this area was cleared for more lucrative and, in the eyes of the state, prestigious, purposes, its inhabitants were rehoused on the periphery. Given the Egyptian state's lack of use for democratic process when its poor citizens are concerned, opposition to the move was fruitless. What Ghannam brings out is the persistence of resistance in the form of residual social forms, and the interpretation of the global as the suppression of the idea of locality which such forms represent. The community continues to be stigmatized, its public housing spatially and culturally differentiated from private housing schemes, yet in song and conversation, frequent reference is made to the old site and people retain their geographical identification. One factor unifies the contending strains of identity determined by place within this complex oppression: the inclusive space of faith. In the mosque, people from village, inner city and suburb find commonality, trust and safety – 'connected selves rather than separated and isolated others', as Ghannam writes.

29 GASTON BACHELARD

'The House. From Cellar to Garret. The Significance of the Hut'

from *The Poetics of Space* (1969)

[. . .]

I pointed out earlier that the unconscious is housed. It should be added that it is well and happily housed, in the space of its happiness. The normal unconscious knows how to make itself at home everywhere, and psychoanalysis comes to the assistance of the ousted unconscious, of the unconscious that has been roughly or insidiously dislodged. But psychoanalysis sets the human being in motion, rather than at rest. It calls on him to live outside the abodes of his unconscious, to enter into life's adventures, to come out of himself. And naturally, its action is a salutary one. Because we must also give an exterior destiny to the interior being. To accompany psychoanalysis in this salutary action, we should have to undertake a topoanalysis of all the space that has invited us to come out of ourselves.

Emmenez-moi, chemins! . . .

(Carry me along, oh roads . . .)

wrote Marceline Desbordes-Valmore, recalling her native Flanders (*Un Ruisseau de la Scarpe*).

And what a dynamic, handsome object is a path! How precise the familiar hill paths remain for our muscular consciousness! A poet has expressed all this dynamism in one single line:

O, mes chemins et leur cadence

    Jean Caubère, *Déserts*

(Oh, my roads and their cadence.)

When I relive dynamically the road that "climbed" the hill, I am quite sure that the road itself had muscles, or rather, counter-muscles. In my room in Paris, it is a good exercise for me to think of the road in this way. As I write this page, I feel freed of my duty to take a walk: I am sure of having gone out of my house.

And indeed we should find countless intermediaries between reality and symbols if we gave things all the movements they suggest. George Sand, dreaming beside a path of yellow sand, saw life flowing by. "What is more beautiful than a road?" she wrote. "It is the symbol and the image of an active, varied life." (*Consuelo*, vol. II, p. 116).

Each one of us, then, should speak of his roads, his crossroads, his roadside benches; each one of us should make a surveyor's map of his lost fields and meadows. Thoreau said that he had the map of his fields engraved in his soul. And Jean Wahl once wrote:

Le moutonnement des haies
C'est en moi que je l'ai.
    (*Poème*, p. 46)

(The frothing of the hedges
I keep deep inside me.)

Thus we cover the universe with drawings we have lived. These drawings need not be exact. They need only to be tonalized on the mode of our inner space. But what a book would have to be written to decide all these problems! Space calls for action, and before action, the imagination is at work. It mows and ploughs. We should have to speak of the benefits of all these imaginary actions. Psychoanalysis has made numerous observations on the subject of projective behavior, on the willingness of extroverted persons to exteriorize their intimate impressions. An exteriorist topoanalysis would perhaps give added precision to this projective behavior by defining our daydreams of objects.

However, in this present work, I shall not be able to undertake, as should be done, the two-fold imaginary geometrical and physical problem of extroversion and introversion. Moreover, I do not believe that these two branches of physics have the same psychic weight. My research is devoted to the domain of intimacy, to the domain in which psychic weight is dominant.

I shall therefore put my trust in the power of attraction of all the domains of intimacy. There does not exist a real intimacy that is repellent. All the spaces of intimacy are designated by an attraction. Their being is well-being. In these conditions, topoanalysis bears the stamp of a topophilia, and shelters and rooms will be studied in the sense of this valorization.

30 KENNETH FRAMPTON

'Towards a Critical Regionalism'

from Hal Foster (ed.), *Postmodern Culture* (1985)

[. . .]

Architecture can only be sustained today as a critical practice if it assumes an *arrière-garde* position, that is to say, one which distances itself equally from the Enlightenment myth of progress and from a reactionary, unrealistic impulse to return to the architectonic forms of the preindustrial past. A critical arrière-garde has to remove itself from both the optimization of advanced technology and the ever present tendency to regress into nostalgic historicism or the glibly decorative. It is my contention that only an arrière-garde has the capacity to cultivate a resistant, identity-giving culture while at the same time having discreet recourse to universal technique.

It is necessary to qualify the term arrière-garde so as to diminish its critical scope from such conservative policies as Populism or sentimental Regionalism with which it has often been associated. In order to ground arrière-garde in a rooted yet critical strategy, it is helpful to appropriate the term Critical Regionalism as coined by Alex Tzonis and Liliane Lefaivre in "The Grid and the Pathway" (1981); in this essay they caution against the ambiguity of regional reformism, as this has become occasionally manifest since the last quarter of the 19th century:

> Regionalism has dominated architecture in almost all countries at some time during the past two centuries and a half. By way of general definition we can say that it upholds the individual and local architectonic features against more universal and abstract ones. In addition, however, regionalism bears the hallmark of ambiguity. On the one hand, it has been associated with movements of reform and liberation; . . . on the other, it has proved a powerful tool of repression and chauvinism . . . Certainly, critical regionalism has its limitations. The upheaval of the populist movement – a more developed form of regionalism . . . has brought to light these weak points. No new architecture can emerge without a new kind of relations between designer and user, without new kinds of programs . . . Despite these limitations critical regionalism is a bridge over which any humanistic architecture of the future must pass.

The fundamental strategy of Critical Regionalism is to mediate the impact of universal civilization with elements derived *indirectly* from the peculiarities of a particular place. It is clear from the above that Critical Regionalism depends upon maintaining a high level of critical self-consciousness. It may find its governing inspiration in such things as the range and quality of the local light, or in a *tectonic* derived from a peculiar structural mode, or in the topography of a given site.

But it is necessary, as I have already suggested, to distinguish between Critical Regionalism and simple-minded attempts to revive the hypothetical forms of a lost vernacular. In contradistinction to Critical Regionalism, the primary vehicle of Populism is the *communicative* or *instrumental* sign. Such a sign seeks to evoke not a critical perception of reality, but rather the sublimation of a desire for direct experience through the provision of information. Its tactical aim is to attain, as economically as possible, a preconceived level of gratification in behavioristic terms. In this respect, the strong affinity of Populism for the rhetorical techniques and imagery of advertising is hardly accidental. Unless one guards against such a convergence, one will confuse the resistant capacity of a critical practice with the demagogic tendencies of Populism.

The case can be made that Critical Regionalism as a cultural strategy is as much a bearer of *world culture* as it is a vehicle of *universal civilization*. And while it is obviously misleading to conceive of our inheriting world culture to the same degree as we are all heirs to universal civilization, it is nonetheless evident that since we are, in principle, subject to the impact of both, we have no choice but to take cognizance today of their interaction. In this regard the practice of Critical Regionalism is contingent upon a process of double mediation. In the first place, it has to "deconstruct" the overall spectrum of world culture which it inevitably inherits; in the second place, it has to achieve, through synthetic contradiction, a manifest critique of universal civilization. To deconstruct world culture is to remove oneself from that eclecticism of the *fin de siècle* which appropriated alien, exotic forms in order to revitalize the expressivity of an enervated society. (One thinks of the "form-force" aesthetics of Henri van de Velde or the "whiplash-Arabesques" of Victor Horta.) On the other hand, the mediation of universal technique involves imposing limits on the optimization of industrial and postindustrial technology. The future necessity for resynthesizing principles and elements drawn from diverse origins and quite different ideological sets seems to be alluded to by Ricoeur when he writes:

No one can say what will become of our civilization when it has really met different civilizations by means other than the shock of conquest and domination. But we have to admit that this encounter has not yet taken place at the level of an authentic dialogue. That is why we are in a kind of lull or interregnum in which we can no longer practice the dogmatism of a single truth and in which we are not yet capable of conquering the skepticism into which we have stepped.

A parallel and complementary sentiment was expressed by the Dutch architect Aldo Van Eyck who, quite coincidentally, wrote at the same time: "Western civilization habitually identifies itself with civilization as such on the pontificial assumption that what is not like it is a deviation, less advanced, primitive, or, at best, exotically interesting at a safe distance."

That Critical Regionalism cannot be simply based on the autochthonous forms of a specific region alone was well put by the Californian architect Hamilton Harwell Harris when he wrote, now nearly thirty years ago:

Opposed to the Regionalism of Restriction is another type of regionalism, the Regionalism of Liberation. This is the manifestation of a region that is especially in tune with the emerging thought of the time. We call such a manifestation "regional" only because it has not yet emerged elsewhere ... A region may develop ideas. A region may accept ideas. Imagination and intelligence are necessary for both. In California in the late Twenties and Thirties modern European ideas met a still-developing regionalism. In New England, on the other hand, European Modernism met a rigid and restrictive regionalism that at first resisted and then surrendered. New England accepted European Modernism whole because its own regionalism had been reduced to a collection of restrictions.

The scope for achieving a self-conscious synthesis between universal civilization and world culture may be specifically illustrated by Jørn Utzon's Bagsvaerd Church, built near Copenhagen in 1976, a work whose complex meaning stems directly from a revealed conjunction between, on the one hand, the *rationality* of normative technique and, on the other, the *arationality* of idiosyncratic form. Inasmuch as this building is organized around a regular grid and is comprised of repetitive, in-fill modules – concrete blocks in the first instance and precast concrete wall units in the second – we may justly regard it as the outcome of universal civilization. Such a building system, comprising an *in situ* concrete frame with prefabricated concrete in-fill elements, has indeed been applied countless times all over the developed world. However, the universality of this productive method – which includes, in this instance, patent glazing on the roof – is abruptly mediated when one passes from the optimal modular skin of the exterior to the far less optimal reinforced concrete shell vault spanning the nave. This last is obviously a relatively uneconomic mode of construction, selected and manipulated first for its direct associative capacity – that is to say, the vault signifies sacred space – and second for its multiple cross-cultural references. While the reinforced concrete shell vault has long since held an established place within the received tectonic canon of Western modern architecture, the highly configurated section adopted in this instance is hardly familiar, and the only precedent for such a form, in a sacred context, is Eastern rather than Western – namely, the Chinese pagoda roof, cited by Utzon in his seminal essay of 1963, "Platforms and Plateaus." Although the main Bagsvaerd vault spontaneously signifies its religious nature, it does so in such a way as to preclude an exclusively Occidental or Oriental reading of the code by which the public and sacred space is constituted. The intent of this expression is, of course, to secularize the sacred form by precluding the usual set of semantic religious references and thereby the corresponding range of automatic responses that usually accompany them. This is arguably a more appropriate way of rendering a church in a highly secular age, where any symbolic allusion to the ecclesiastic usually degenerates immediately into the vagaries of kitsch. And yet paradoxically, this desacralization at Bagsvaerd subtly reconstitutes a renewed basis for the spiritual, one founded, I would argue, in a regional reaffirmation – grounds, at least, for some form of collective spirituality.

31 ALDO ROSSI

'Typological Questions' and 'The Collective Memory'

from *The Architecture of the City* (1982)

TYPOLOGICAL QUESTIONS

The city as above all else a human thing is constituted and of all those works that constitute the true means of transforming nature. Bronze Age men adapted the landscape to social needs by constructing artificial islands of brick, by digging wells, drainage canals, and watercourses. The first houses sheltered their inhabitants from the external environment and furnished a climate that man could begin to control; the development of an urban nucleus expanded this type of control to the creation and extension of a microclimate. Neolithic villages already offered the first transformations of the world according to man's needs. The "artificial homeland" is as old as man.

In precisely this sense of transformation the first forms and types of habitation as well as temples and more complex buildings, were constituted. The *type* developed according to both needs and aspirations to beauty; a particular type was associated with a form and a way of life, although its specific shape varied widely from society to society. The concept of type thus became the basis of architecture, a fact attested to both by practice and by the treatises.

It therefore seems clear that typological questions are important. They have always entered into the history of architecture, and arise naturally whenever urban problems are confronted. Theoreticians such as Francesco Milizia never defined type as such, but statements like the following seem to be anticipatory: "The comfort of any building consists of three principal items: its site, its form, and the organization of its parts." I would define the concept of type as something that is permanent and complex, a logical principle that is prior to form and that constitutes it.

One of the major theoreticians of architecture, Quatremère de Quincy, understood the importance of these problems and gave a masterly definition of type and model:

The word 'type' represents not so much the image of a thing to be copied or perfectly imitated as the idea of an element that must itself serve as a rule for the model ... The model, understood in terms of the practical execution of art, is an object that must be repeated such as it is; type, on the contrary, is an object according to which one can conceive works that do not resemble one another at all. Everything is precise and given in the model; everything is more or less vague in the type. Thus we see that the imitation of types involves nothing that feelings or spirit cannot recognize ...

We also see that all inventions, notwithstanding subsequent changes, always retain their elementary principle in a way that is clear and manifest to the senses and to reason. It is similar to a kind of nucleus around which the developments and variations of forms to which the object was susceptible gather and mesh. Therefore a thousand things of every kind have come down to us, and one of the principal tasks of science and philosophy is to seek their origins and primary causes so as to grasp their purposes. Here is what must be called 'type' in architecture, as in every other branch of human inventions and institutions ... We have engaged in this discussion in order to render the value of the word *type* – taken metaphorically in a great number of works – clearly comprehensible, and to show the error of those who either disregard it because it is not a model, or misrepresent it by imposing on it the rigor of a model that would imply the conditions of an identical copy.

In the first part of this passage, the author rejects the possibility of type as something to be imitated or copied because in this case there would be, as he asserts in the second part, no "creation of the model" – that is, there would be no making of architecture. The second part states that in architecture (whether model or form) there is an element that plays its own role, not something to which the architectonic object conforms but something that is nevertheless present in the model. This is the *rule*, the structuring principle of architecture.

In fact, it can be said that this principle is a constant. Such an argument presupposes that the architectural artifact is conceived as a structure and that this structure is revealed and can be recognized in the artifact itself. As a constant, this principle, which we can call the typical element, or simply the type, is to be found in all architectural artifacts. It is also then a cultural element and as such can be investigated in different architectural artifacts; typology becomes in this way the analytical moment of architecture, and it becomes readily identifiable at the level of urban artifacts.

Thus typology presents itself as the study of types of elements that cannot be further reduced, elements of a city as well as of an architecture. The question of monocentric cities or of buildings that are or are not centralized, for example, is specifically typological; no type can be identified with only one form, even if all architectural forms are reducible to types. The process of reduction is a necessary, logical operation, and it is impossible to talk about problems of form without this presupposition. In this sense all architectural theories are also theories of typology, and in an actual design it is difficult to distinguish the two moments.

Type is thus a constant and manifests itself with a character of necessity; but even though it is predetermined, it reacts dialectically with technique, function, and style, as well as with both the collective character and the individual moment of the architectural artifact. It is clear, for example, that the central plan is a fixed and constant type in religious architecture; but even so, each time a central plan is chosen, dialectical themes are put into play with the architecture of the church, with its function, with its constructional technique, and with the collective that participates in the life of that church. I tend to believe that housing types have not changed from antiquity up to today, but this is not to say that the actual way of living has not changed, nor that new ways of living are not always possible. The house with a loggia is an old scheme; a corridor that gives access to rooms is necessary in plan and present in any number of urban houses. But there are a great many variations on this theme among individual houses at different times.

Ultimately, we can say that type is the very idea of architecture, that which is closest to its essence. In spite of changes, it has always imposed itself on the "feelings and reason" as the principle of architecture and of the city.

While the problem of typology has never been treated in a systematic way and with the necessary breadth, today its study is beginning to emerge in architecture schools and seems quite promising. I am convinced that architects themselves, if they wish to enlarge and establish their own work, must again be concerned with arguments of this nature. Typology is an element that plays its own role in constituting form; it is a constant. The problem is to discern the modalities within which it operates and, moreover, its effective value.

[. . .]

THE COLLECTIVE MEMORY

With these considerations we approach the deepest structure of urban artifacts and thus their form – the architecture of the city. "The soul of the city" becomes the city's history, the sign on the walls of the municipium, the city's distinctive and definitive character, its memory. As Halbwachs writes in *La Mémoire Collective*, "When a group is introduced into a part of space, it transforms it to its image, but at the same time, it yields and adapts itself to certain material things which resist it. It encloses itself in the framework that it has constructed. The image of the exterior environment and the stable relationships that it maintains with it pass into the realm of the idea that it has of itself."

One can say that the city itself is the collective memory of its people, and like memory it is associated with objects and places. The city is the *locus* of the collective memory. This relationship between the *locus* and the citizenry then becomes the city's predominant image, both of architecture and of landscape, and as certain artifacts become part of its memory, new ones emerge. In this entirely positive sense great ideas flow through the history of the city and give shape to it.

Thus we consider *locus* the characteristic principle of urban artifacts; the concepts of *locus*, architecture, permanences, and history together help us to understand the complexity of urban artifacts. The collective memory participates in the actual transformation of space in the works of the collective, a transformation that is always conditioned by whatever material realities oppose it. Understood in this sense, memory becomes the guiding thread of the entire complex urban structure and in this respect the architecture of urban artifacts is distinguished from art, inasmuch as the latter is an element that exists for itself alone, while the greatest monuments of architecture are of necessity linked intimately to the city. "The question arises: in what way does history speak through art? It does so primarily through architectural monuments, which are the willed expression of power, whether in the name of the State or of religion. A people can be satisfied with a Stonehenge only until they feel the need to express themselves in form ... Thus the character of whole nations, cultures, and epochs speaks through the totality of architecture, which is the outward shell of their being."

Ultimately, the proof that the city has primarily itself as an end emerges in the artifacts themselves, in the slow unfolding of a certain idea of the city, intentionally. Within this idea exist the actions of individuals, and in this sense not everything in urban artifacts is collective; yet the collective and the individual nature of urban artifacts in the end constitutes the same urban structure. Memory, within this structure, is the consciousness of the city; it is a rational operation whose development demonstrates with maximum clarity, economy, and harmony that which has already come to be accepted.

With respect to the workings of memory, it is primarily the two modes of actualization and interpretation that interest us; we know that these depend on time, culture, and circumstances, and since these factors together determine the modes themselves, it is within them that we can discover the maximum of reality. There are many places, both large and small, whose different urban artifacts cannot otherwise be explained; their shapes and aspirations respond to an almost predestined individuality. I think, for example, of the cities of Tuscany, Andalusia, and elsewhere; how can common general factors account for the very distinct differences of these places?

The value of history seen as collective memory, as the relationship of the collective to its place, is that it helps us to grasp the significance of the urban structure, its individuality, and its architecture which is the form of this individuality. This individuality ultimately is connected to an original artifact – in the sense of Cattaneo's principle, *it is an event and a form*. Thus the union between the past and the future exists in the very idea of the city that it flows through in the same way that memory flows through the life of a person; and always, in order to be realized, this idea must not only shape but be shaped by reality. This shaping is a permanent aspect of a city's unique artifacts, monuments, and the idea we have of it. It also explains why in antiquity the founding of a city became part of the city's mythology.

32 PATRICK WRIGHT

'Abysmal Heights'

from *A Journey Through Ruins: A Keyhole Portrait of British Postwar Life and Culture* (1993)

> It is possible to gauge the residents' reactions to a housing development to some extent by a variety of small signs such as facial expressions.
> *The Architects' Journal*, 21 July 1965

A few hundred yards from Dalston Lane, due south of the hoarding on which the Araldite poster appeared, stands a sawn-off Victorian pub. The Grange Tavern bears Charrington's name, along with the meaningless inaugural date of 1757, and a sign that struggles to lend a Palladian air to the neighbourhood; it shows a large and classical-looking 'grange' with horses, carriages, and the predictable collection of gentlefolk gathered in the open space before it. Designed like so many pubs of its time, to command the corner at the end of a terraced street, the Grange Tavern is now an amputated stump, a relic stranded in the midst of the famously awful Holly Street Estate, built by Hackney Council in the early Seventies. The old street runs back beside this sad hostelry for a few yards, but is then cancelled out by a monolithic block of flats. To the east there are four huge tower blocks, plonked down on a dingy stretch of 'open space', which can hardly compare with the leafy vistas to which the pub sign alludes. Hackney was among the last councils to recognize the limitations of 'comprehensive redevelopment' as it was practised in the Fifties and Sixties, but the Grange Tavern stands as a monument to the fitful premonitions that came as the final day of awakening approached: a residue of 'community' transferred from one slum to the next; a tiny concession to the old map in the Town Guide Cabinet at Dalston Junction.

I've been looking at the Holly Street Estate for years. Its four, nineteen-storey tower blocks are widely deplored on this side of Queensbridge Road as monstrous eyesores that should be blown up. We survey the burned-out flats with their shattered windows and we shudder. We read of the drug-dealing, the murders, and the sporadic defenestrations, maintaining our own informal body count with the help of the *Hackney Gazette*. When we get burgled we have a way of glancing up accusingly through long-suffering sash windows.

In Dalston, as elsewhere, the council tower block serves as a generator of infernal meanings for people who only look at it from outside. Those who live inside have a different experience, and I recently decided to walk along the length of this banal and heavily mythologized view so that I could turn and consider what it looked like from the other end. One Sunday I set off to meet Reece Auguiste of the Black Audio Film Collective on the twelfth floor of Rowan Court, the last of the four tower blocks to be built. I walked past Grace Jones Close – a small and generally excellent council estate dating from the early Eighties, that short period when the borough showed unmistakable signs of having finally got it right, just as the whole council housing programme was being brought to a shuddering halt. I walked through Mapledene, an attractive area of mid-Victorian terraces, where residents can sit at their kitchen tables (only a mile or so north of the City) and look out into a richly differentiated green world of gardens and wooded 'backlands'. Suddenly this closely textured realm of Victorian detail gives way to the looming cliff-like austerity of the system-built monolith and the dismal uniformity of planned grass and statutory trees. In the early Fifties Nikolaus Pevsner praised

Queensbridge Road for its gracious semi-detached villas, but now it is just a wasted strip of no man's land between visually opposed worlds.

The approach to Rowan Court leads past a stretch of fenced oblivion where the multi-storey car park once stood (tenants quickly learnt better than to leave their cars there, and it was demolished not long after being built). The litter drifts in the air, and one or two dispirited-looking residents prop themselves up on benches in dishevelled simulation of those tiny model citizens who still tend to grace the open spaces around architectural maquettes. Rowan Court sits on its stretch of green void, with a stinking puddle of brown liquid oozing out of the 'textured' (ie marked by the coarse imprint of shuttering boards) concrete lattice-work at its base. The external doors are missing, and there's a cavity where the intercom was ripped out within a few days of being installed. As for the foyer, it is certainly no fashionable 'atrium'. The ceiling is in ruins, with pipes and wire left exposed ever since the council came along to strip the asbestos. The walls, which have been tiled over in a forlorn attempt to improve the environment, are covered with posters; there's more of the unreconstructed Turkish Leninism that ornaments the derelict shop fronts of Dalston Lane, and the paper of the Socialist Workers Party has been pasted up page by page so that tenants can hardly avoid reading it as they wait for the lifts. The lift, which actually arrives quickly enough, is a foul metal box covered over with racist graffiti and, as I quickly find out, the atmosphere inside it is tense. Perhaps it is only on account of my presence as a stranger that everyone seems to be reciting a version of 'Tinker, Tailor, Soldier, Sailor', in their minds: a paranoid inner-city variation that goes 'Racist, Mugger, Rapist, Victim'. Nevertheless, it is already evident that things have slipped since the early Eighties, when Paul Harrison came to the Holly Street Estate in order to draw up a detailed inventory of 'the ecology of the inferno', and found that the real problems were confined to the low-rise buildings beyond the four towers.

Reece Auguiste welcomes me to his mother's flat and shows me a sweeping panorama over the towers of East London. Had the experiment not gone so shockingly wrong, a new kind of connoisseurship would by now have sprung up around views like this: enthusiasts would stand here differentiating point blocks from scissor blocks and cluster blocks with the same eagerness that classical revivalists now show as they identify the figures, both overt and subtle, of analogy, alignment aposiopesis, abruptio, and epistrophe in the facade of a church by Palladio. There are enough tower blocks in view to revive an entire forgotten taxonomy of system-building: Laing's Jesperson system, Taylor Woodrow's Larson-Nielson system (rarer than it was since all the trouble at Ronan Point over there in Newham), Concrete Ltd's Bison Wall Frame, Crudens Ltd's 'Skarn' system. Devotees of post-war planning could also stand here and trace the fading outline of the ideal geometry that post-war public authorities dreamt of imposing on London: the zoning, the distant green belt, the clearances and huge new estates, the roads – from the GLC's frustrated plan for an Inner London Motorway Box, through to the new, or dramatically widened, roads suggested in the East London Assessment Study, which only faded away in 1989. One line goes out immediately to the closely clustered towers of the Nightingale Estate, another reaches out to Clapton Park where five 'Camus' system blocks line up with the Trowbridge Estate to form a triangle around Sutton House, which is hidden among trees between the tower of St John's and a high-rise school building. I turn to see the still respectable towers to the Barbican (originally designed as low-cost public housing, and now home to millionaires like the financial analyst Bob Beckman, who has a three-storey penthouse suite in Lauderdale Tower), but they are out of sight, and I settle for the trusting metal tower of Canary Wharf a couple of miles away on the Isle of Dogs. Looking down, I see the reverse of the view from my own window – the Victorian terraced houses, the street plan, the trees, the secluded gardens. Those, as Reece remarks with gentle irony, are surely the houses to live in; indeed, at the time of our meeting he was about to make his own descent into the more desirable world of the human scale.

During the property boom of the late Eighties, the anarchist squatters of Class War tried to reduce this view to a perspective of pure resentment and envy: 'we', the real people of Hackney, look down from our pokey flats while 'they' (or 'you' as it was by the time the threatening notices went through the letter-boxes), the yuppie gentrifiers and 'rich scum', move into the leafy terraces down there and make a packet as the value of these former 'slums' soars. But resentment is by no means the predominant feeling up here. Auguiste's neighbour on the twelfth floor is Mrs Jan Cooper, a retired West End draper who moved into Rowan Court with her husband when the block opened in 1971. She looks down on the Victorian terraced streets of Mapledene and says, in a voice that still holds a hint of pity, that she could never go back to living in houses like that: cramped, dark, damp, insecure, and claustrophobic. Up here at least the vistas are airy and clear. She looks over to Chingford and the open country of Essex. Her eye then sweeps down over uncounted London boroughs to settle on the rising land of Blackheath to the south. These two allegorical stretches of high ground are the rims of her world, and she takes her pick of the city that lies between them: ignoring the 118 tower blocks of Newham and singling out instead the green dome of the Hackney Empire; a large gasometer that, as she's heard but can hardly believe, has a preservation order on it; the street on which a policeman had recently been murdered. She appreciates the green stretches, too, regretting the school extension that has diminished her view of London Fields, and pointing with unprompted satisfaction to the trees in the gardens and 'backlands' behind the terraced houses across the Queensbridge Road. Even Victorian slums have their attractions from a height.

Mrs Cooper remembers how fortunate she and her husband felt to be joining Hackney's fabled 'skyscraper families' when they first moved in. At that time, as she recalls, 'Everyone wanted to get in here'. People settled down behind that commanding new view with a great sense of confidence. Indeed, Mrs Cooper is adamant that even the tenants who, like herself and her husband, had lost their earlier home to a council 'clearance' scheme were happy to bid farewell to the years they had spent cowering in the slums. As Mrs Cooper points out, her flat, which she keeps in pristine condition, is well designed and generously proportioned. The rooms are a generous size, and the ceilings, which stand at a good height, also offer unrecognized tribute to the minimum standards

recommended by the Parker-Morris report in 1961 and abandoned more recently. It is, without doubt, a good enough machine for living in. The balcony even allows her to grow plants, including some myrtle-like shrubs she has raised from cuttings brought back from her childhood home in Co. Mayo. As Mrs Cooper says, 'I have no fault with the flats'. In fact, she's compared her home with the pokey little boxes, real 'rabbit hutches', the council has built since high rise went so abruptly out of fashion, and she's happy to have the one she's got. She only has to look out of Rowan Court in the other direction to see the later phases of the Holly Street Estate: those squalid, fire-scorched low-rise blocks built around 'courts' that would, as the Housing Committee promised in 1966, be 'landscaped to produce pleasant tranquil areas'. In Mrs Cooper's considered judgement it is the architects who designed those far-inferior blocks who should be shot.

There comes a point, however, at which Mrs Cooper's account joins that of Reece Auguiste, who talks of administrative collapse in the council's management of the estate and the characteristically alienated culture of Rowan Court. As Auguiste describes it, the tower block is a machine for inducing paranoia. The building imposes an 'atomized philosophy' on its tenants, and all its supposedly communal spaces are filled with anxiety, suspicion, and fear. The lift, he confirms, can be dead edgy: keep your eyes down and don't say a word. It has become almost routine for people to die alone in their flats, and to lie undiscovered for days and even weeks. As for the vandalism, Auguiste wonders whether the external doors and entry-phone system, recently fitted by the council in an endeavour to improve conditions in the block, weren't destroyed so quickly because they seemed to reinforce the isolation of the residents. Conceding that he had found the quiet of his mother's flat conducive to work during his student days, he cites sociological classics by the likes of Willmot and Young and Hannah Gavron to describe how the culture of the tower block cuts across any sense of community and confines women all the more viciously to the domestic sphere.

Auguiste suggests that the residents of Rowan Court are increasingly susceptible to a myth of the Golden Age and Mrs Cooper, for all her attachment to those bright modernist days at the outset, certainly shares his view that something has gone terribly wrong in the Eighties. The early years were fine and, as she stresses, they included the first black families. But then a terrifying degeneration set in, and its humiliations are inescapable. Asked to give an example, Mrs Cooper takes a deep breath and starts by pointing to the brown staining that disfigures the ceilings of her immaculate flat. Somewhere upstairs a person turned on the bath and then went out, leaving the water to trickle down through all the flats below. And since, as Mrs Cooper adds in an attempt to illustrate the accumulator effect that tends to take over whenever the smallest thing goes wrong on this estate, the offending person was a squatter, neither the council nor any of the affected tenants' private insurance companies would accept liability for the damage.

Mrs Cooper does her bit on the tenants' association – she's the representative for the top half of the block – but at the same time she takes every opportunity she can to be away from Rowan Court. As she says, 'I'd go mad if I was in'. 'Hail, rain, or snow', she's packed her picnic and her newspaper by eleven o'clock, and she tries not to be back before nine: 'You name any park within reach of London Transport and I've been there.' Many tenants tell of the embarrassment of inviting visitors or bringing friends home to Rowan Court, and Mrs Cooper has a story that seems to speak for them all. She was recently visited by her brother, who had left Co. Mayo for California rather than Hackney, and now lives in far superior conditions near San Francisco. He and his wife came over to visit and since Hackney has no real hotels to speak of, there was no choice but to put him up in Rowan Court. Mrs Cooper will not forget the humiliation of getting up before dawn in order to sneak out and clean the lifts (people have been known to do 'everything' in there) before taking her visitors out in the morning.

She leads me on a tour of the upper floors to prove her point about the degradation of the block. The evidence is everywhere: the graffiti, the filth around the rubbish chutes, the stairways littered with the debris of adolescent drug abuse, the open access to the roof where the pirate radio stations bring their transmitters. On one particularly squalid floor, she indicates the elaborate, solid metal barricades tenants have erected in their doorways and suggests that we might just as well be in the Bronx, Beirut, or Alcatraz. The generally accepted communal decencies of the Seventies have gone, and in their place comes an endless succession of horror stories: there's the woman who has given up on civilized waste disposal, and simply heaves all her rubbish out of the kitchen window; there's the psychotic tenant who recently took against his neighbours' children and poured a lake of scarlet paint on the floor outside their door – a confused and agitated trail of footprints testifies to the success of his mad ploy.

Trying to find a pattern in this picture of mounting chaos, I ask for some benchmarks, some dates, some clearly defined milestones on this road to ruin. Reece Auguiste mentions the recent hardening of the racist graffiti into a coherent and deliberately formulated fascist form. Both he and Mrs Cooper mention the time when the respectable families disappeared. There used to be 'loads of kiddies' here, as Mrs Cooper remarks before moving on to suggest the moment when the council gave up enforcing its rules forbidding pets, and the dogs started to arrive. Back issues of the *Hackney Gazette* reveal that the council was allowing, or at least turning a blind eye to, cats on its estates from as early as 1971, when Rowan Court opened: it was hoped that they would keep the warfarin-resistant mice under control. But it was later, as the break-ins, burglaries, and assaults spiralled, that the dogs started to arrive. Not just the usual inner-city mongrels, but the whole pedigree succession of inner-city survivalism: the traditional Alsatians were looking hopelessly soppy by the end of the Seventies, giving way first to Dobermanns and Rottweilers and finally, as the Eighties peaked, to that ideal high-rise beast, the pit bull terrier. These symbolic creatures pace around in their modernist flats, howling and snarling, stinking the block up, shitting in the hallways, and occasionally coming out to strut about on the stretch of green void outside.

The decline of that 'open space' is another symbolic marker of degeneration. It may have been planned as an amenity for the flat dwellers, but now it's a truly indefensible space. The drifting airborne rubbish would outsmart even the most diligent cleaner, and the land has become public in the worst possible sense. Just as the people from the Victorian houses opposite use Rowan Court as a throughfare, walking straight through the foyer on their way to the shops (as Mrs Cooper asks, 'how would they like it if we started walking though their houses from front door to back?'), the 'open space' outside has become a dog patch for half the borough. The resident curs are bad enough, but over the last few years Mrs Cooper reckons she must have seen half the population of Hackney drive up in battered cars and let out their dogs so that they, too, can shit all over her doorstep.

Finally, and really only in the last couple of years, there are the squatters – a 'floating population' drifts through the towers of Holly Street like a band of outlaws camping out on the high plains. They move around within and between the blocks, getting evicted by the council and then settling into another empty flat and claiming their rights for a year. How do they get through the solid metal of the anti-squatter doors? On some occasions the council workers have simply failed to turn the key, and on others, at least so many tenants suspect, they seem to have joined the well-established black-market trade in illicitly copied keys. I was shown flats that have been systematically wrecked by squatters, and others that are just gaping burned-out voids. Some of the squatters spend all their time 'drugging' – including, as Mrs Cooper suspects, the Vietnamese student who recently died after jumping off the balcony of his blazing flat. Others are 'perfectly decent' people who 'have to live somewhere' (Mrs Cooper has a delightful man with a cello on her floor), but their arrival still causes extensive breakdown in the system (to say nothing of the council's policy of housing the homeless families on its waiting-list). For a start they enjoy free heating and hot water, provided by a central boiler for which the official tenants have to pay. And just as the squatters have the indirect effect of cancelling the tenants' insurance, their refusal to open the door even to sanitation officials means that it is nigh on impossible to rid the block of the cockroaches and other pests that infest it. In the end, however, the squatters are not at the root of the problem. Mrs Cooper reads the *Sun*, but she doesn't need that rag to tell her where the accusing finger should be pointed: 'The council are not doing their job. That's the all of it.'

There are some exceptional moments of brightness on the Holly Street Estate – a fine adventure playground, and some brave murals that manage to lift the place a bit. These flimsy attempts to stem the decline are joined by scrawled and anonymous notices urging residents to pull back from the abyss. On one floor of Rowan Court after another there are anonymous messages by the rubbish chute. Scribbled on cardboard, they plead that 'neighbours' should not just dump their filth on the floor or bung up the chute unnecessarily. A similar plea stands by the demolished car park, part of which has recently been turned into a community play centre and garden. It points out that this modest but also unimaginably heroic attempt to counter the degeneration is actually the voluntary work of 'neighbours', and pleads that it should not be vandalized, filled with dogshit, or otherwise ruined. On the fifth floor of Rowan Court, the floor tiles in the 'communal' hallway have been repaired on a DIY basis by the couple who accepted the government's offer and bought their flat in this hell-house (only to find themselves the owners of an unsaleable flat and lumbered with a maintenance charge that, while not necessarily being unreasonable, threatens to exceed the old rent). These brave holding measures may be hopeless, but people persist in them knowing that while it is easy to slip down the slope it is almost impossible to climb back up. There's no turning round on the road to ruin. As Mrs Cooper said repeatedly, 'you can never get it back'.

When the time came for me to leave, Mrs Cooper insisted, with old-fashioned courtesy, on seeing me to the front door. There were six people already in the lift, but this time it didn't seem to matter whether they were black or white, men or women, squatters, tenants, or owner-occupiers. Somebody made a remark about the interior condition of the lift, and everyone came together in the helpless humour of the modern chasm dweller. These people of the new abyss are the victims not just of a council that can't manage its own resources, but of a national government that has enjoyed making symbolic capital out of the disorders of Labour authorities like Hackney, while at the same time forcing through policy changes designed to reduce council housing to a residual welfare net. In 1971, when Rowan Court was opened to its first tenants, council housing was still accepted as a respectable form of housing for the skilled working class: one didn't have to be disaffiliated, marginal, incompetent, or just plain desperate to end up there. But no longer. Ten years ago, Hackney Council planned to spend £16 million redeveloping the Holly Street Estate and landscaping its green voids, but the government wasn't having any of it. In the age of 'opting out', Mrs Cooper's situation has become representative in a new way. Ageing and unable to move out of the estate that is collapsing around her, she looks back over a life of respectable endeavour and feels both betrayed and abandoned. There are elderly people like this all over the inner city: decent and hard-working in their active lives, as they feel obliged to repeat over and over again, they spend their retirement sinking down as, one after another, the promises of the early Welfare State give way beneath them. The London in which they expected to be living in their old age has become as remote as any other homeland on the block – from Ireland to Vietnam or, for that matter, the Lincolnshire of the *Market Rasen Mail*. Tempted by tabloid stereotypes, and the dubious consolations offered by such figures as the Revd Donald Pateman, vicar of St Mark's, these are the people who turned out to welcome the Queen Mother when she made her nostalgic return to Victoria Park on her ninetieth birthday, and to watch the Spitfire and Lancaster Bomber fly over on the way to Buckingham Palace on the fiftieth anniversary of the Battle of Britain. In the early Seventies they might have conformed to sociologist Ruth Glass's conception of 'the people of the public sector': feisty, aware of their rights, not easily conned, articulate, and demanding. By 1990 they had been reduced to a very different stereotype as 'zombies of the welfare state'. So much for the Queen Mother's famous demand, made during the blitz, that something must be done to improve the lot of East Londoners.

33 FRANK MORT

'Boulevards for the Fashionable and Famous'

from *Cultures of Consumption: Commerce, Masculinities and Social Space in Late Twentieth-Century Britain* (1996)

'The Soho sex shops . . . of the last two decades will not be mourned,' insisted *The Times* in 1986, 'but what new Soho will emerge?' Throughout the mid-1980s competing visions of the area's future vied for ascendancy. Many local residents rallied behind the arguments advanced by the Soho Society. The organisation maintained that the premises of former sex establishments should be reserved for small retail outlets, thus providing links with the original local craft and service industries. But such aspirations looked increasingly utopian in the light of a different type of commercial development which was already making its mark. In all the contemporary debates over Soho's future one group more than any other was cast as the driving force for change. It was media professionals – dubbed by *The Times* as the affluent media men – who were seen to be in the vanguard of Soho's renaissance. Eccentrics and bohemians were being edged out, noted *The Daily Telegraph*'s columnist John Wyman in 1986; sleaze had given way to respectability. The *Financial Times* flattered its own readers the following year when it told how 'people power' had saved Soho, tilting the balance back towards legitimate commerce. And the paper noted approvingly that media folk were encouraging further economic growth with their own specialist demands. Theirs was a milieu of expense-account restaurants, personalised shopping and the thousand and one luxury commodities which were mushrooming in Soho. The discerning customers for these goods and services were identified as a familiar band – taste leaders in the world of fashion. *Midweek*, the weekly free paper for London's office workers, sang the praises of the new arrivals in Soho's square mile:

Cosmopolitan, bohemian, wildly trendy – Soho is London's very own rive gauche. This square mile of style is a kingdom unto itself: the land of the brasserie lunch and the after-hours watering hole, the land of accessories and attitude, where fashion relentlessly struggles to become style and image is simply everything; the glittering heart of media land where the worlds of art, journalism, film, advertising and theatre blend into one glamorous heady cocktail . . . How does one become part of this bohemian world? How can you . . . cut a dash as a get-ahead young turk effortlessly oozing that quintessential Soho style – style that you have only glimpsed at in the pages of *The Face*? . . . There is a whole unspoken language to be learnt, a code of behaviour and dress, a long list of do's and don'ts.

Here was a universe driven by dynamic but feverish consumption. Presented as a popular version of bohemia, it was peopled by a number of key characters. Young entrepreneurs and professionals were the personalities who moved so effortlessly across these landscapes. Like the disciples of *The Face*, their key to success was style. As we have seen, such coverage became the stock-in-trade language of consumer journalism in the late 1980s. Whether in the magazine sections of the Sunday newspapers, or through the gossip of television chat shows, media discourses began to represent a specific type of urban experience which focused on Soho.

This rhetoric of Soho as the centre of media style needs to be decoded with caution. A powerful metaphor, it appealed to metropolitan writers precisely because it testified to the energy of their own culture and the centrality of their terrain. Yet all this was not simply a journalistic fantasy. Under the generalised references to the media were subsumed a whole range of more concrete developments, for Soho was fast becoming the site of an expanding commercial infrastructure. It was the growth of advertising and public relations companies, as well as the publishing and film industries, which were most influential. Together they pointed to a professional network which laid claim to the area. They also hinted at a particular lifestyle shaped by the world of goods. Many of these changes have already featured in our history of *commerçants* in the 1980s. But their crystall-isation in the social geography of London was distinctive. This is the key to unlocking Soho's significance at the time.

Such transformations had their own economic rationale. Market-based philosophies, already in the ascendant at Westminster Council by the middle of the decade, favoured an aggressively business-oriented approach to Soho's future. Centring on speculative property development, this involved the council encouraging ventures which promised higher rateable values and rents. In the summer of 1986 Conservative councillor Peter Hartley saw the choices facing his administration as increasingly stark:

> If it's a choice between upgrading the area with a possibility that it would become a bit too upmarket or trendy, or leaving the place to become an absolute junk-heap, the council has taken the view that the first was better.

This emphasis on the positive aspects of com-mercial regeneration, as opposed to predictions of inevitable decline if the area were simply left to local residents, was underlined three years later by David Weeks, the council's deputy leader. Responding to accusations that Soho's growing popularity was producing an affluent but anodyne culture, Weeks affirmed his faith in a market-led approach to redevelopment. Sounding a note of mock resignation, he confided to *The Evening Standard Magazine*: 'One hopes a mixed character will remain, but all cities change and evolve ... We can only respond to market forces.' The market forces identified by Weeks were generally media-related in character.

Viewed from the vantage point of the newly arrived professionals, Soho was an attractive proposition. A survey by local estate agents Allsop and Co., in 1987, noted that, while rents were rising, they were still miniscule when compared to the charges in neighbouring West End areas. Many offices and studios migrated from adjacent locations in response to mounting fixed costs. Encouraged by growing media deregulation, a clutch of independent film companies and related services (such as dubbing, mixing and cutting) set up in Soho, on account of the area's established links with this sector. Members of the style press also staked-out their interest in the same space. *The Face* was produced from basement offices in Broadwick Street, west Soho, in the early 1980s, before moving slightly further north. In 1987 Neville Brody set up his studios in Tottenham Court Road, on Soho's eastern limit. Most significantly, a number of the most prominent advertising agencies laid claim to prestige office develop-ments in the district. Since the 1960s the industry's nerve centre had been sited on and around Charlotte Street, to the north of Oxford Street, together with sites in Mayfair and later in Covent Garden. But the continuing attraction of lower property costs drew advertisers and their clients towards the Soho net. Bartle Bogle Hegarty took up residence in Great Pultney Street in 1986, Ogilvy and Mather nestled in Soho Square, as did the design consultants Fitch and Co. Of the twenty top creative agencies named the following year, six had offices in or adjacent to Soho. The glamorous reputations of these companies were embodied in the archi-tectural facades and design styles of their offices. David Ogilvy had long preached the importance of the outward appearance of an advertising agency for attracting potential clients. *Campaign* whispered that it had been granted a private view behind the exterior of Bartle Bogle Hegarty. Titallating their readers, journalists uncovered a characteristic mix of fashionable consumerism. Beautiful girls glided through contemporary but understated interiors, which were personally designed by Hegarty himself. Clients were offered tea from plain white Wedgwood in uncluttered, minimalist surroundings. *Campaign* concluded that this was the perfect place to entertain German executives for a Audi car launch!

Prestige developments of this kind were skil-fully managed by a number of recently formed property companies and estate agents, eager for further clients. Rumours of an impending rise in property values and rents intensified the atmosphere of feverish speculation. Paul Raymond, the acceptable face of the reformed sex industry (who continued to own a Soho portfolio larger than any other private landlord), celebrated this activity as a testament to the power of the open market. Once established, the new businesses demanded a further range of services, especially leisure and entertainment facilities. A plethora of restaurants, cafés and bars opened in quick succession. All of these ventures were strategically marketed to the immigrants, while old established venues were remodelled. There was a brisk flow of commercial energy within a compressed urban space.

Alastair Little's celebrated restaurant in Frith Street appeared in 1985. Little had previously worked round the corner, as chef at L'Escargot in Greek Street. Specialising in so-called 'new British cuisine', with a menu which was changed twice daily, his new venue quickly became a favourite with diners from the record and advertising industries. Less intimate, but more imposing, was Braganza's, sited further along Frith Street, which had started the previous year. For this venture architects and interior designers had been specially commissioned to give each of the three floors a particular atmosphere. Opening its doors a the same time, the Soho Brasserie, in Old Compton Street, was one of Soho's first deliberate emulations of a Parisian café-bar. Furnished with the ubiquitous chrome and stainless steel interior, marble-topped tables and authentic foreign waiters, the café offered a 'continental' ambience. Ordering a Kir, or perhaps a *salade de fruits de mer*, customers could seat themselves at window tables opening directly onto Soho's main thoroughfare. From this vantage point they could watch the boulevard like latter-day *flâneurs*.

Much of the new Soho raided Parisian models of urban behaviour. But the overall effect was not simply a continental pastiche. It was a distinctly English reading of Gallic culture. Metropolitan London was kaleidoscoped together with signifiers of French taste. Similar hybrid forms were produced in relation to Italy. Run by the Polledri family since the 1950s, the Bar Italia, in Frith Street, was hailed as the 'most authentic Italian café in Soho'. During the soccer season, satellite broadcasts of Italian football matches beamed down from a huge television screen, while the walls of the bar were covered with posters of post-war Italian sporting heroes. Once again this was not simply the recreation of

a Milanese or a Neapolitan environment. It translated an 'Italianate' experience into a London setting. These mimetic renderings of other European cultures proliferated in Soho in the 1980s, as they did in London's other fashionable quarters, such as Camden or Hampstead in the north of the city. But Soho's sedimented history of 'continentalism' gave its foreign atmosphere a more piquant and authentic flavour.

In all the flurry of leisure activity two particular venues symbolised the new Soho. These landlords were The Groucho Club, opened in 1985, and the Limelight, which was launched a year later. The Groucho, in Dean Street, cemented an alliance between literary culture and the media industries. Carmen Callil and Liz Calder, of Virago feminist publishers, approached the leisure consultant and member of the chocolate manufacturing family, Tony Mackintosh, with a brief to create a new type of professional club. It was to be a venue which welcomed women and avoided the stuffiness of the traditional male bastion, the gentlemen's club. Mackintosh's choice of a Soho location was justified on the grounds of its rising status. As he put it: 'We chose Soho because it is enjoying a new lease of life.' The Groucho serviced journalists, writers and their agents, as well as the media world. It aimed to provide a relaxed space, where commercial alliances and more informal social relations could be cemented. Opening membership was £200 a year, with a reduced fee for applicants under 28, reflecting the relatively youthful profile of many of the newcomers. In addition to its professional customers, the club rapidly acquired an influential position among members of the style community. *The Face* reported club gossip and the *bons mots* of celebrities in its lounges and dining rooms, especially after Madonna chose The Groucho for her birthday celebrations in 1987. Burchill also testified to the club's importance as a haunt of freelance writers and media intriguers. Such venues lay at the heart of her metropolitan fantasy, places where 'CVs and calling cards filled the . . . sky'.

While The Groucho Club provided an intimate setting for the *habitués* of Soho society, the Limelight, on Shaftesbury Avenue, was a much more public arena in which taste and fashionability could be asserted. We have already encountered the nightclub as the venue chosen by *The Face* to host its fashionable 'party of the year' in 1986. Launched by Peter Gatien, a 32-year-old Canadian multi-millionaire, its opening involved the dramatic conversion of a former Welsh Presbyterian chapel. The Limelight was designed to appeal precisely to the group christened by Gatien the new 'Sohoemians'. This was an untapped market of affluent 25- to 40-year-olds who, he observed, were missed by the contemporary London scene. Gatien's reading of his customer profile was based on experience gained in New York. There, he argued, a broad spectrum of older professionals took advantage of such nightspots, both to relax and to conduct business. Decorated with murals and with a strong general emphasis on art and architecture, the Limelight enshrined its own caste-like hierarchy within the spatial environment. The 'VIP suite' restricted access to 'celebrities', or to those able to buy into celebrity status. From an elevated gallery the élite could either look out across London, or gaze down from a position of superiority on the dance-floor below.

The tropes of modishness and innovation were endlessly reproduced in surroundings of this type. But to what extent did Soho's earlier bohemian legacy inform the new urban vision? There was sharp disagreement between the older population and the newcomers about the value of the recent changes, which also condensed differences of age and status. Jeffrey Bernard was bitter in his complaint about the environmental and cultural pollution caused by the immigrants and their caravan of goods. Soho was now dead, he proclaimed, killed by a massive overdose of advertising executives with pocket bleepers and a taste for cheap wine! Bernard's polemic was unashamedly élitist. He was resolute in his insistence on the superiority of his own circle, as opposed to the superficiality of those he dubbed the new plebeians. Some of Bernard's contemporaries were less hostile in their appraisal of the changes. Melly, along with journalist Daniel Farson, was generally supportive of Soho's redevelopment. But both of them also worried about 'a cleaned up, hygienic Soho . . . a fraud, a tourist attraction'. Yet to the newcomers the situation appeared quite different. From their perspective, it was precisely the informal mixing of the area's established culture with more

contemporary inputs which provided the allure. Estate agents and developers carefully wrote this repertoire into their advertising copy. 'Sohoemians like the idea of an area that is not too manicured', local estate agent Laurence Glynne confided to *The Guardian* in 1989. Favourable comparisons were frequently made with neighbouring Covent Garden, where a sanitised environment had obliterated any traces of the original vegetable and flower market. Christopher New, whose first clothes shop opened in Dean Street in 1985, pointed out that many retailers moved to Soho not simply on account of lower rents, but because the district was 'considerably more interesting'. New felt that the area had a genuine village feel. Fred Taylor, club manager of 'Freds' in Carlisle Street, believed that the secret of Soho's creative dynamism lay with its bohemian atmosphere: 'From the word go' he had found it full of 'incredible diversity' and an 'intangible ambience'. Talent thrived and, as a result, coming into Soho was always a voyage of discovery. Soho was 'like a club'; anyone with style could apply for membership.

Let us briefly trace the circuit of these new bohemians as they wove their way through the network of Soho's spaces. Emerging from the design studio or the media conference with colleagues or prospective clients, professional life might continue in the more informal setting of one of the area's media pubs. For a special occasion, or a more prestigious client, the choice might be the Champagne Bar of Kettners restaurant, in Romilly Street. If there was no after-hours business then shopping, for personal accessories or for gifts, could be the alternative. There were the eye-catching items displayed in American Retro in Old Compton Street, where owner Sue Tahran brought together an abundance of classic design items: from Braun alarm clocks and Zippo lighters to more esoteric commodities, such as a 'matt black hand-held photocopier'. Or for personal organisers there might be a visit to Just Facts, in Broadwick Street, reputedly the only shop in the world to stock the entire Filofax range. Distinctive 'eye-wear' could be bought at Eye Tech in Brewer Street, designer clothes picked up at Workers For Freedom or the Academy, both showcases for young fashion talent. Later in the evening the party might wind up with an authentic *espresso* at the Bar Italia, or take in a jazz set at nearby Ronnie Scott's.

Such vignettes of contemporary life were among the standard narratives circulated about Soho's renaissance during the 1980s. According to their register, they were either stories offered up to Soho initiates, with the accompanying pleasure of self-recognition, or accounts produced for a more distant tourist gaze. Like nineteenth-century *feuilleton* literature or the early guides to Paris, Soho journalism taught readers how to find their bearings – to locate a position within these representations of social space. Forms of reportage might stimulate the desire for actual metropolitan experience, but they could also supply the mental furniture for a more imaginary participation in the city. If Soho could not actually be concretely apprehended, it could be visited in fantasy, via the structures of journalism and literary leisure.

Yet a historical reading of these urban myths reveals the extent to which they were selective and partial. They were in fact carefully edited tales of Soho's redevelopment during these years. Written into their formula were a number of important silences and concealments. Selecting a series of emblematic artefacts and personality types, these accounts evoked a landscape of consumer desire which was instantly recognisable and glamorous, and at the same time seamless and coherent. They were discreetly regulatory in two related senses. First, they worked with a system of topographical closure. Under their rubric the urban map was homogenised and unified. What were marginalised, or deliberately ignored, were a number of more heterotopic forms of city life; a series of other worlds and spaces which jostled for attention on the boulevards. Further, these narratives remained silent about gender. Following standard patterns of urban story-telling, the inhabitants of London's West End were generally assumed to be men, but there was little exploration of these masculine identities. Yet Soho in the 1980s did not display such an unproblematic syntax. Rather it presented the geography of not one but several different taste communities. They were linked by their gendered hold on public space.

'Re-imagining the Global: Relocation and Local Identities in Cairo'

from A. Öncü and P. Weyland (eds), *Space, Culture and Power*

(1997)

'Praise the Prophet. Once upon a time, there was an old woman who used to live in an apartment that was as small as that tiny table [pointing to the small table in their living room]. Each time the old woman swept the floor, she found either one pound or 50 piastrs [an Egyptian pound is worth around 34 cents] that she kept hidden in a place in her window. The old woman was saving to buy a larger apartment. But one day, a thief stole all the money that she saved. She was very sad. An *afriit* [demon or ghost] appeared and asked the old woman what she would like to have. She asked for a larger apartment. The *afriit* asked her, "Would you like the apartment to have a balcony?" and she answered yes. He asked her, "Would you like a television set, a fan and a bottle of water?" [describing some of the things that were in front of us in the living room]. The old woman said yes. Then he asked her, "And would you like some pictures of Samira Said and Latifa?" [two popular female Moroccan and Tunisian singers whose posters were decorating the wall of the living room]. The woman again answered yes. The *afriit* brought all these things for the old woman. She was very happy and cried out of joy. In the same day, however, she smelled the *birshaam* that was hidden behind the television set which caused her heart to collapse [*gham ala albaha*] and the old woman died' (a story told to me by the five-year-old Amal in Cairo, 1994).

Amal's narrative was contextualized by her family's attempts to find a larger housing unit to move into from the one-bedroom apartment that she, her four sisters and their parents have been occupying since 1980, when the family was displaced from their home in the centre of the city and relocated in al-Zawiya al-Hamra in northern Cairo. Amal's images of the desired home are constructed, as is the case with many other children, from global images transmitted to them through television programmes, school textbooks and visits to different parts of the city. Her dreams, as well as those of her sisters, of the future apartment are informed by the movies and soap operas that they like to watch: a big apartment with a balcony, a spacious kitchen, modern furniture and organized spatial arrangements inside and outside the unit. These images contradict the objective realities of Amal's life and create desires that cannot be satisfied even through some magical means. Like the dreams of many other low-income people, Amal's discourse 'proceeds in a jagged line, the leaps into day-dream being followed by relapses into a present that withers all fantasies'. Death and destruction is the ultimate answer.

Amal, her family and the rest of their neighbours are not fax-users, e-mail receivers, jumbo jet travellers or satellite-owners. They are part of Cairo's working class whose experience of 'the global' is structured by their economic resources and position in social space. In addition to the many consumer goods, especially television sets, that are desired by people and are becoming signs of distinction, Amal's family and many other families experienced the force of the global in their displacement from their 'locality' in the centre of the city. Their houses were demolished to be replaced by buildings and facilities that cater to upper-class Egyptians, international tourists and the transnational community. In this chapter, I focus on relocation, utilized as part of the state efforts to 'modernize' Cairo and its people, to show how global discourses and forces are articulated in contradictory ways at the national and local levels. In the first section, I present a brief review of the history of the relocation of roughly 5000 working-class families from 1979 to 1981 and the state public discourse utilized to justify the project. This discourse strategically appealed to the global in the state's attempts to implement its different economic policies and to construct a modern national identity. In the second part, I draw on my recent ethnographic research in Cairo, or *Umm al-Dunya* (the mother of the world) as Egyptians like to refer to it, to map some of the identities that are attached to and formed by Amal's group to show how the displacement of the local by global processes and national policies brought new changes that paved the way to redefine local communal feelings in ways that help people live in the modern world. I argue that religious identity, as a hegemonic identity in the formation, was consolidated by the changes brought by the global as experienced by the people and as filtered in national policies.

MODERNITY AND THE STRUGGLE OVER URBAN SPACE

In *Search for Identity* (1978), Anwar el-Sadat presents a strong critique of Nasser's policies that kept the country isolated from its neighbours and the rest of the world and destroyed Egypt's economy. To remedy the country's chronic economic and financial problems, Sadat reversed Nasser's policies by suspending relationships with the Soviet Union and reorienting Egypt towards the West. He turned to the United States in particular for aid in resolving Egypt's conflict with Israel as well as the economic and technological development of the country. After his victory in the 1973 war (at least it was a victory for him), Sadat crystallized his new visions and ideas in declaring to 'the open-door policy' or *infitah* in 1974. This policy aimed to 'open the universe . . . open the door for fresh air and remove all the barriers (*hawajiz*) and walls that we built around us to suffocate ourselves by our own hands'. As he explained to a group of young Egyptian men, Sadat's *infitah* was motivated by his belief that each one of them would like to 'get married, own a villa, drive a car, possess a television set and a stove, and eat three meals a day'.

Sadat's policy strived to modernize the country through speeding planned economic growth, promoting private investment, attracting foreign and Arab capital, and enhancing social development. Private local and international investments were expected to secure the capital needed to construct modern Egypt. Egyptians were encouraged to work in oil-producing countries and invest their remittances in the building of the country. At the same time, laws were enacted to secure the protection needed to encourage foreign investors and to facilitate the operation of private capital. Investments in tourism were especially important because they were expected to 'yield high economic returns and provide substantial foreign exchange and well-paid employment'.

This orientation to the global, the outside, or the 'universe' as Sadat describes it, required a 'distinctive bundle of time and space practices and concepts'. Many changes were needed to facilitate the operation of capital and meet the new demands that were created. For example, the growing demand for luxury and middle-class housing for the transnational community and Egyptians who work in oil-producing countries inflated the price of land, especially in the centre of the city, and increased the cost of construction materials. The promotion of private and foreign investment also increased the demand for offices and work-oriented spaces. High-rises proliferated around Cairo, using Western design principles and Los Angeles and Houston, Sadat's favourite American cities, became the models that were to be duplicated.

Two tendencies were expressed in the discourses and policies of urban planning that aimed to promote the *infitah* policies and to rebuild modern Cairo. The first tried to integrate into the modern city areas of significance to Egypt's glorious past (for example, the pyramids and Islamic monuments), which Sadat loved to emphasize and which were visited by tourists. The second tendency, which is the subject of this chapter, attempted to reconstruct the 'less desirable' parts, especially popular quarters, that did not represent the 'modern' image of Egypt and were not fit to be gazed at by upper-class Egyptians and foreign visitors.

The state, the global and the creation of the 'modern' city

As part of Sadat's larger plan to restructure the local landscape and build 'modern' Cairo, around 5000 Egyptian families were moved during the period from 1979 to 1981 from Central Cairo (Bulaq) to housing projects built by the state in two different neighbourhoods: 'Ain Shams and al-Zawiya al-Hamra. Bulaq, once the site of the winter houses of the rich, then a major commercial port and later an industrial centre, had become unfit for the modern image that Sadat was trying to construct. This area, which over the years had housed thousands of Egyptian low-income families, is adjacent to the Ramsis Hilton, next to the television station, around the corner from the World Trade Center, across the river from Zamalek (an upper-class neighbourhood), over-looks the Nile, and is very close to many of the facilities that are oriented to foreign tourists. The area then occupied by low-income families became very valuable because Sadat's policies, as he proudly announced, increased the price of the land which was needed to facilitate the operation of capital. The old crowded houses were to be replaced by modern buildings, luxury housing, five-star hotels, offices, multi-storey parking lots, movie theatres, conference rooms, and centres of culture (*al-Ahram* 27 December 1979). Officials thus emphasized the urgent need to remove the residents of this old quarter because many international companies were ready to initiate economic and tourist investment in the area. Expected profits from these investments would contribute to national income and assist the state in securing money to build new houses for the displaced groups (*al-Ahram*, 27 December 1979). The residents' efforts to stop their forced relocation did not materialize and their calls upon the government to include them in the reconstruction of their area were denied. Voices that protested the relocation were quickly silenced and objections raised by the displaced population were considered 'selfish'. Officials emphasized that the benefit of the 'entire nation' should prevail over everything else (*al-Ahram*, 9 July 1979).

The 'local' was also displaced to protect the state orientation to the global. The relocation project took place two years after the famous 1977 riots that protested the increase in the prices of basic daily goods, especially bread. Protesters targeted *infitah*-related facilities such as five-star hotels and nightclubs, and chanted slogans against Sadat's policies. The neighbourhood, with its narrow lanes and crowded streets, made it impossible for the police to chase those who participated in the riots. The relocated group was seen by the state and the state-controlled press as part of 'a conspiracy organized by communists' that aimed to distort the achievements of the *infitah* and their housing became an obstacle to the promotion of Sadat's policies and to police attempts to crush protest against these policies.

The rhetorical strategies employed by the state were largely based in the appeal to the global. This appeal was manifested by the emphasis on modernity and its objectification in material forms, rational planning, the importance of visual aspects and the tourists' gaze in representing Cairo, the separation of the home from the workplace, international investment, science, health, hygiene, green areas and clean environment, consumer goods and the importance of the productive agent in the construction of a modern national identity. The global was strategically used to offer the people a 'Faustian bargain' which forced the relocated group to pay a high price for Egypt's opportunity to be 'modernized'. Using force (police) and seduction (by appealing to the global and offering alternative housing), the project removed them from the centre of the city and deprived them of the benefits associated with the modern facilities and the new changes that promised prosperity for everyone. The group lost a major part of its economic and 'symbolic capital', to use Bourdieu's term, which was their central geographical location. Relocation destroyed most of the group's informal economy, altered their access to many cheap goods and services, and destroyed their social relationships and reordered their personal lives.

As previously mentioned, the group was divided into two parts, each relocated to a different neighbourhood away from the gaze of tourists and upper-class Egyptians. One part, the focus of this study, was moved to public housing (*masaakin*) units constructed for them in al-Zawiya al-Hamra in northern Cairo. The move into these units, which were labelled as 'modern', promised to improve the lives of the people and turn them into 'healthy modern productive citizens who will contribute in the construction of their mother country' (*al-Ahram*, 27 December 1979). The state's project assumed a transparent relationship between space and identity and totally ignored the role of social actors in transforming and resisting its policies and ideologies. Rather than creating a unified modern city, I argue that these policies created a more fragmented urban fabric and paved the ground for other competing collective identities. Religious identity in particular has successfully presented itself as a powerful alternative that can articulate the various antagonistic identities that are constructed in the relocation site.

[. . .]

OLD PLACES, NEW IDENTITIES

To understand the local identities that are in the process of formation in al-Zawiya al-Hamra, it is important to remember that state practices and discourses were based on what Foucault calls 'dividing practices'. The project started by separating and stigmatizing the targeted population as an expedient rationalization of policies that aim to modernize, normalize and reintegrate them within the larger community. Not only were the housing conditions attacked by state officials, but the people themselves were stigmatized and criticized. A 'scientific' social study conducted to determine the needs of the relocated group revealed, as stated by the Minister of Construction and New Communities, that the area of Bulaq in general and one of its neighbourhoods (al-Torgman) in particular have been shelters for *qiradatia* (street entertainers who perform with a baboon or monkey), female dancers, pedlars and drug dealers (*al-Ahram*, 27 December 1979). The 'locals' were also represented as passive, unhealthy and isolated people who did not contribute to the construction of the mother-country and who had many social ills. After resettlement, these publicized stereotypes fostered a general feeling of antagonism towards the newcomers. In addition to repeating the same words that were circulated in the media,

residents of al-Zaiwya added other stereotypes to describe this group such as *labat* (trouble-makers) and *shalaq* (insolent). Women, in particular, were singled out (as they were also singled out by the Minister who described them as dancers or *Ghawazi*); they were described as rude and vulgar, and were used in daily conversation as an analogy for bad manners.

These negative constructions of the relocated group are supported and perpetuated by the physical segregation of their housing (*masaakin*) from the rest of the community. Their public housing is clearly defined and separated from other projects and private houses (*ahali*). Public housing is characterized by a unified architectural design (the shape, the size of the buildings, as well as the colours of walls and windows), whereas private housing has more diversified patterns. This unity in design and shape sharply defines and differentiates public housing from private houses and makes it easier to maintain boundaries that separate the relocated from other groups. In short, neither the discourse of the state nor the shape and location of the housing project enhance the dialogical relationship between the relocated group and other groups in al-Zaiwya. After fourteen or fifteen years of resettlement, the relocated group continues to be stigmatized and its interaction with the rest of the neighbourhood is restricted.

The identity of Bulaq

With their stigmatization in the state discourse and by the residents of al-Zawiya, and with the hostility that faced them, the relocated population rediscovered their common history and identification with the same geographical area. While people used to live in Bulaq and identify strongly with their villages of origin, after relocation Bulaq became an anchor for the group's sense of belonging and took precedence over other identifications. The attachment to the old place is not single or one-dimensional and Bulaq is remembered and related to differently by gender and age groups. These differences are beyond the scope of this chapter but it is sufficient here to say that Bulaq is of great significance for most of the group in reimagining their communal feelings. Currently, their public housing is called after one of Bulaq's neighbourhoods (*masaakin al-Torgman*) and people express their strong attachment to their old place in songs and daily conversations. Despite the fact that relocation reordered relationships within the group and destroyed a major part of their support system, the old neighbourhood still structures parts of the people's current interaction. They still refer to the people who used to live in Bulaq as '*min 'andina*' (from our place) which not only creates a common ground for identification but also indicates certain expectations and mutual obligations between the people in the current area of residence. At the same time, Bulaq is the point of reference for their identification with those who still live in parts of Bulaq and those who moved to 'Ain Shams.

Through relocation, the group lost, among other things, a major part of its 'symbolic capital'. This is mainly manifested in two important aspects related to group members' identification with the old location. First, they used to live next to an upper-class neighbourhood, Zamalek. Young men and women, as emphasized by the people themselves and documented in a famous old movie (*A Bride from Bulaq*), could even claim that they were from Zamalek because only 'a bridge' separated (or connected) the two neighbourhoods. People also lost the pleasure and satisfaction associated with looking at the beautiful buildings and knowing that people of Zamalek – and much to their shame, as described by one informant – used to see Bulaq with its old and shabby houses.

Second, the group used to live in an 'authentic popular' or *baladi* area and perceives its relocation in al-Zawiya as moving down the social ladder. In Bulaq, the 'authentic popular' quarter, people used to live next to each other, separated only by narrow lanes that allowed close interaction and strong relationships. They remember the old place in the way people used to cooperate and 'eat together'. Their rootedness in the same place over a long period of time provided people with a strong support system, open social relationships, and a sense of security and trust. In contrast, al-Zawiya is a relatively new neighbourhood. It was mainly agricultural fields until the 1960s, when the area started to expand rapidly with the state construction of the first public housing project. This project housed families from different parts of Cairo who could not afford to live in more central locations.

Immigrants (mostly Muslims) also came to al-Zawiya from different parts of the countryside and many live in private housing. The heterogeneity of its population is used by its residents, especially members of the relocated group, and people around them to indicate that al-Zawiya is not 'an authentic popular quarter'. Its people are 'selfish', 'sneaky' and 'untrustworthy'. It is seen as located between *baladi* and *raqqi* (upper-class areas) which places it, as described by a male informant, in a tedious or annoying (*baaykh*) position. Al-Zawiya, thus, is geographically and socially marginal compared to Bulaq.

A key word in understanding the differences between what is seen as an 'authentic' neighbourhood such as Bulaq and 'less authentic' newer neighbourhoods such as al-Zawiya is *lama*. This word refers to the growing mixture and gathering of people from different backgrounds who live in the same locality. People from various quarters, villages and religions are coming to live in the same neighbourhood, hang out at the same coffee shop, visit the same market, and ride the same bus. These spaces are defined as *lamin* as compared to a more homogeneous or less *lama* places such as the village and the 'authentic popular' quarter based on long established relationships. Being rooted in a certain area, that is, localized in a particular place, allows the development of strong relationships between people. *Lama* is used to classify different localities and points to the difference between a neighbourhood where people know each other by name and face as opposed to more heterogeneous areas where people are strangers and not to be trusted. *Masaakin* is *lama* as opposed to *ahali* housing. Al-Zawiya is *lama* compared to Bulaq and Cairo is *lama* compared to the villages where the inhabitants originally came from.

RELOCATION AND RELIGIOUS IDENTITY

Despite the significance of Bulaq in how people reimagine their communal feelings, this identity does not facilitate the group's interaction with the rest of the people who live in al-Zawiya al-Hamra. Relocation rearranged local identities and added to the old identifications: people are now identified with a village (the place of origin), as locals of Bulaq (where they resided for generations), as occupiers of *masaakin* (which is stigmatized by dwellers in private housing) and as inhabitants of al-Zawiya al-Hamra (not known for its good reputation in Cairo). But above all, they are mainly Muslims. Religion, rather than nationalism, neighbourhood and the village of origin, became a powerful discourse in articulating and socially grounding the various identities of the different groups residing in al-Zawiya al-Hamra. Only the religious identity promises to articulate these identifications without destroying them. Displaced families, *ahali* and *masaakin* inhabitants, people of Bulaq and al-Zawiya, rural immigrants, *Fallahin* (peasants who come from villages in Lower Egypt) and *Sa-'idis* (immigrants from Upper Egypt), who are largely pushed from their villages to Cairo in their search for work and a better life, as well as residents who moved from other areas of Cairo can all find commonality in religion that is expressed in practices such as a dress code and the decoration of houses and shops.

Islam brings people together on the basis of a common religion. Despite the fact that Muslims do not know each other on a personal basis, religion creates a 'safe' space (the mosque), a common ground where they are connected to each other, and a sense of trust and rootedness. This is clearly manifested in how the mosque, of all public spaces, is gaining importance in facilitating the interaction of various groups and the formation of a collective identity. To start with, the mosque's growing centrality in daily life is manifested in the many modern services that are provided to the people in it. Through charitable organizations (*jam'iyyat khayriyya*), the mosque provides socially required services such as affordable education, health care and financial support to the poor. It is also the place where discourses circulate that prescribe and/or forbid daily practices. Above all, it is the most acceptable and safest social space where various groups can meet and interact.

To understand the importance of the mosque, we need to go back to the world *lama*. As previously mentioned, people tend to distrust areas and public spaces that are labelled as *lama* such as the market, the coffee shop and the bus. These spaces are seen as 'dangerous' and people are very careful when visiting or utilizing them. Compared to such spaces, the mosque, which is

a historical space that is legitimated through its naturalized relationship with religion, is currently being actively articulated to frame the interaction between members of different groups as well as to empower emerging meanings, identities and relationships. Those who are labelled as trouble-makers and rude (people who come from Bulaq and live in *masaakin*) as well as the untrustworthy and selfish (people of al-Zawiya as described by people of Bulaq) can all meet in the mosque and collectively identify themselves as Muslims.

Thus, the power of the mosque is being currently reinforced through its promise of an equal and unified community out of a heterogeneous urban population. It is accessible to all Muslims and brings them in on equal terms. The unity of prayers and the importance of communal feelings is manifested in the unifying discourse and the similar movements that are performed simultaneously. The Imam leads the prayer and coordinates the movement of all the attendees through his pronounced signals that indicate when one should bend forward on the knees or stand up straight, and so on. Emphasis is placed upon standing in straight lines, very close to other attendees, in a way that leaves no empty spaces through which the devil could enter among the devout and divide their collectivity.

The feelings that are associated with being part of a collectivity were cited by many, especially by women, as one of the main reasons for going to the mosque. As is the case with most of her neighbours, relocation shattered most of the support system that connected the fifty-five-year-old Umm Ahmed with friends and neighbours who were relocated to 'Ain Shams or to different parts of the new housing project in al-Zawiya al-Hamra. Although she used to perform her religious duties on a regular basis in Bulaq, Umm Ahmed's religiosity gained a different meaning in al-Zawiya al-Hamra. In addition to her adoption of the *khimar* (a head garment that covers the hair and the shoulders), which is seen as the 'real Islamic dress', Umm Ahmed began attending local mosques on a daily basis. She explained that she goes to mosques because the presence of other people strengthens her will and provides her with more energy than when praying alone. Currently, Umm Ahmed frequents five local mosques to perform four out of the five daily prayers. For Friday prayer, she usually selects a large mosque, located within the boundaries of the *masaakin* but that is also attended by some worshippers from the *ahali*. She also visits two small mosques that are identified with an Islamic group active in al-Zawiya al-Hamra. She attends these two mosques, which are located in the *ahali* area, to listen to weekly lessons and participate in Qur'ān recital sessions. Another mosque, which is located next to the vegetable market in the *ahali* area, is a convenient site for the midday prayer when Umm Ahmed is shopping for the family's daily food. For the evening prayer, she chooses a smaller mosque on the edge of the housing project that is attended by a mixture of worshippers from *ahali* and *masaakin* areas. She prefers this mosque, as she explains, because she meets 'wise' women who like to talk to her. Over the last five years, Umm Ahmed has formed strong relationships with other women from different parts of the neighbourhood, especially from the *ahali* area, who attend the same mosque. If one of them does not come to the evening prayer, she goes with other women to ask about their absent friend. At the same time, the mosque not only brings people together from the same neighbourhood but also encourages people to move from one part of the city to the other. Young men and women, for example, use the city bus to tour the city in their search for the 'truth'. They cross the boundaries of their localities to go to other neighbourhoods to attend certain mosques where popular sheikhs preach.

The mosque is also becoming more open to women in al-Zawiya al-Hamra. This is perceived by some Islamic activists as essential to counter other spaces that are open to women, such as universities, the workplace, cinemas and nightclubs. Women are identified by men as more vulnerable to the influence of global (defined here as American) discourses and practices. Women's actions, dress and access to public life are seen as threatening the harmony of the Islamic community and as the source of many social ills. Women have internalized these ideas and hold themselves, and not men, responsible for the safety of the morals of the community. As women repeatedly emphasize, men are weak creatures and cannot resist the seduction imposed on them by women who do not adopt Islamic dress. At the same time, women can be very active in the construction of the Islamic community. More voices have emphasized the positive aspects associated with opening the mosque to women who, as mothers, sisters and wives, can be active agents capable of altering their own practices as well as shaping the actions and values of other family members. Thus, to contain the destructive potential of women and promote their constructive power in the formation of the Islamic community, more attempts are made by Islamic activists to incorporate women within the mosque. Currently, women, especially those without jobs and small children, go to the mosque on a regular basis for prayer and to attend weekly lessons, while working women usually attend the Friday prayer. Women are also becoming more active in the mosque through their roles as teachers, students, workers and seekers of social, educational and medical services. In addition, more women help in taking care of the mosque and participate in mosque-related activities such as preparing food and distributing it to the needy.

[. . .]

CONCLUSION

I have tried to show in this chapter how the articulation of global discourses and processes is producing contradictory identities at the national and local levels. By destroying old neighbourhood relationships, stigmatizing and physically segregating the relocated population, the project that aimed to construct modern subjects has paradoxically produced antagonistic local identities that empowered the basis of a collective identity which is based on religion. I have also aimed to show the important role of active social agents in mediating the different global practices and selectively articulating certain global discourses in the formation of their local identities. Social agents face the global in collectivities rather than as individuals and the struggle between the local and the global is not simply taking place in 'human minds'. In general, although 'new regimes of accumulation' are appealing to the individual, alienation, racism and uprootedness are being faced collectively. In fact, being part of a collectivity is necessary to feel at home in the modern world with its rapid global changes. Thus, the local is not passive and local cultural identities are not waiting to be wiped out by globalization as some authors suggest.

Amal's dreams should continue to remind us that people experience the global in structured ways. It should also draw our attention to the fact that many of the writings on globalization are conducted by people who feel at home in the global and tend to celebrate the growing efficiency of transportation, electronic communication and the growing connectedness of the globe. The freedom of travel, however, while experienced by the privileged, is denied for millions of people who find borders of the global (especially, the USA and Europe) closed to them. The relationship between the local and the global cannot be brushed aside by assuming that they are 'articulated as one'. Such statements reduce the complexity of the interaction between the global and the local and ignore the asymmetrical relationship that is still central to this interaction. The analysis presented in this chapter points to the need for more attention and sensitivity to the structured nature of globalization processes. When people experience the global as a violent attack on their cultural identities and self-images, it is not strange that they do not embrace global discourses and its representatives (such as international tourists). In short, more attention should be devoted to those who live on the margin of the marginal: those who are displaced in their own 'culture' and the millions who cannot find solutions to the growing number of desires that are brought by the global except through magical means, death and destruction or religion that at least promises them a better life and the glories of eternal existence in Paradise.

REPRESENTATIONS OF THE CITY

INTRODUCTION

Any representation of the city is, necessarily, always going to be a reductive entity; the very size, complexity and ever-changing nature of the city mean that any attempt made to capture its essence is going to have to leave something out. As Michel de Certeau points out in *Critique of Everyday Life* (1991, Verso), maps, for example, may show a comprehensive overview of the city, but, by using an abstraction based on vison and facticity, also erase qualities of urban experience and street life. Any representation is a partial exercise.

This does not mean that representations are wrong. Far from it. Representations of all kinds are means by which we come to know the city, by which we come to understand and control it, and, above all, the means by which we come to revel in its possibilities and adventures. Nor are representations things which are somehow 'separate' to the city. Rather, representations can be carried within us continuously, part of how we negotiate the city on an everyday basis. Representations, therefore, are absolutely embedded in the culture of cities.

Henri Lefebvre, the French Marxist philosopher and perhaps one of the most influential writers on the city this century, considers in *The Production of Space* (1991, Blackwell) one of the most fundamental aspects of the city: space. Lefebvre sees space not as a natural or god-given entity but as a historical and social product. Space, he proposes, is made up of spatial practices (buildings and actions), representations of space (conscious theories and figures) and representational spaces (imaginations, experiences). Lefebvre's representation of the city is then an intellectual one, a metatheoretical schema that accounts for all kinds of space-production in all cities at all times. The second part of the extract is more specific, showing how thirteenth century Tuscany involved a dialectic between those who lived there and the space in which they lived, and which together led to the production of a new kind of space, one that was at once urban and rural. This, in turn, led to new forms of cultural production (including perspectival painting) – representations of the city are therefore at once contingent and necessary to the production of space as a whole.

If all forms of representation are related to the development of the city, as Lefebvre argues, then this would also be true of all forms of intellectual formations, including seemingly rarefied disciplines such as philosophy. This is the reasoning maintained in Heinz Paetzold, 'The Philosophical Notion of the City' (1997). Paetzold shows how the city has been used metaphorically by philosophers like Wittgenstein, Descartes and Nietzsche, as well as the more obvious relation where city can be the object or place of study for philosophy. As a result, philosophers have, for example, frequently rejected the metropolis and praised the small town.

In two short pieces from his book *Empire of Signs* (1970 [1982], Hill and Wang), the literary theorist and semiologist Roland Barthes treats the spaces of the city in a rather different and more direct way. In 'Center-City, Empty Center', Barthes considers that Tokyo's city plan offers a way of reading the city from above, noting that the imperial centre of that city is present, but is seemingly absent or empty – an interpretation made intelligible when one considers that all traffic and urban life circumnavigate it. In 'No Address', Barthes notes that Tokyo is largely bereft of street names and building numbers; reading and communicating the city must, therefore, be accomplished by other means: sketches, nearby landmarks, continuous instructions. The representation of the city, he concludes, is less visual and more active, and, hence, to locate a place is to begin to write it.

Representations of the city do not have to be undertaken in words, however. In the extract from 'Alternative Space' (1991), art historian Rosalyn Deutsche considers how the artist Martha Rosler in a project entitled 'If You Lived Here . . . ?' sought to represent a city's 'buried life' – such as tenants' lives and homelessness – through video, film, photographs, installations, painting and so forth. Rather than show-piece examples of 'public art' commissioned for corporate buildings and plazas, such representations use direct involvement, image-crystallization and documentaries to disclose the contested, changing and often forgotten aspects of the city. The extract ends by noting the continuous tension between the object of representation and those who seek to represent. What is the mediation between the two? What is their distance? A comparison with 'The Power of Place' by Dolores Hayden, reprinted in Section III, The Culture Industries, is also instructive in this respect.

Perhaps the most pervasive of all representations of the city is the architecture that makes up that city. Most buildings, after all, have been implicitly designed with urban context in mind. But how can architecture be a conscious representation of a city? One answer is to consider the notions of memory and typology suggested by Aldo Rossi in the previous section of this book. Another is offered in 'Brief Encounters', by the British architect Nigel Coates. Here Coates suggests that spaces like theatres and gardens offer clues to new forms of architecture, one that uses metaphor, theatricality and conceptual clues above simple function and the mirroring of reality. Referring both to his older Narrative Architecture Today (NATO) and to more recent work, Coates sees cities and architecture as creators of chaos, discord, difference, movement, identity, etc. – conditions very different to a city culture of stability, heritage and certainty.

35 ROLAND BARTHES

'Center-City, Empty Center' and 'No Address'

from *Empire of Signs* (1970)

CENTER-CITY, EMPTY CENTER

Quadrangular, reticulated cities (Los Angeles, for instance) are said to produce a profound uneasiness: they offend our synesthetic sentiment of the City, which requires that any urban space have a center to go to, to return from, a complete site to dream of and in relation to which to advance or retreat; in a word, to invent oneself. For many reasons (historical, economic, religious, military), the West has understood this law only too well: all its cities are concentric; but also, in accord with the very movement of Western metaphysics, for which every center is the site of truth, the center of our cities is always *full*: a marked site, it is here that the values of civilization are gathered and condensed: spirituality (churches), power (offices), money (banks), merchandise (department stores), language (agoras: cafés and promenades): to go downtown or to the center-city is to encounter the social "truth," to participate in the proud plentitude of "reality."

The city I am talking about (Tokyo) offers this precious paradox: it does possess a center, but this center is empty. The entire city turns around a site both forbidden and indifferent, a residence concealed beneath foliage, protected by moats, inhabited by an emperor who is never seen, which is to say, literally, by no one knows who. Daily, in their rapid, energetic, bullet-like trajectories, the taxis avoid this circle, whose low crest, the visible form of invisibility, hides the sacred "nothing." One of the two most powerful cities of modernity is thereby built around an opaque ring of walls, streams, roofs, and trees whose own center is no more than an evaporated notion, subsisting here, not in order to irradiate power, but to give to the entire urban movement the support of its central emptiness, forcing the traffic to make a perpetual detour. In this manner, we are told, the system of the imaginary is spread circularly, by detours and returns the length of an empty subject.

NO ADDRESS

The streets of this city have no names. There is of course a written address, but it has only a postal value, it refers to a plan (by districts and by blocks, in no way geometric), knowledge of which is accessible to the postman, not to the visitor: the largest city in the world is practically unclassified, the spaces which compose it in detail are unnamed. This domiciliary obliteration seems inconvenient to those (like us) who have been used to asserting that the most practical is always the most rational (a principle by virtue of which the best urban toponymy would be that of numbered streets, as in the United States or in Kyoto, a Chinese city). Tokyo meanwhile reminds us that the rational is merely one system among others. For there to be a mastery of the real (in this case, the reality of addresses), it suffices that there be a system, even if this system is apparently illogical, uselessly complicated, curiously disparate: a good *bricolage* can not only *work* for a very long time, as we know; it can also satisfy millions of inhabitants inured, furthermore, to all the perfections of technological civilization.

Anonymity is compensated for by a certain number of expedients (at least this is how they look to us), whose combination forms a system. One can figure out the address by a (written or printed) schema of orientation, a kind of geographical summary which situates the domicile starting from a known landmark; a train station, for instance. (The inhabitants excel in these impromptu drawings, where we see being sketched, right on the scrap of paper, a street, an apartment house, a canal, a railroad line, a shop sign, making the exchange of addresses into a delicate communication in which a life of the body, an art of the graphic gesture recurs: it is always enjoyable to watch someone write, all the more so to watch someone draw: from each occasion when someone has given me an address in this way, I retain the gesture of my interlocutor reversing his pencil to rub out, with the eraser at its other end, the excessive curve of an avenue, the intersection of a viaduct; though the eraser is an object contrary to the graphic tradition of Japan, this gesture still produced something peaceful, something caressing and certain, as if, even in this trivial action, the body "labored with more reserve than the mind," according to the precept of the actor Zeami; the fabrication of the address greatly prevailed over the address itself, and, fascinated, I could have hoped it would take hours to give me that address.) You can also, provided you already know where you are going, direct your taxi yourself, from street to street. And finally, you can request the driver to let himself be guided by the remote visitor to whose house you are going, by means of one of those huge red telephones installed in front of almost every shop in the street. All this makes the visual experience a decisive element of your orientation: a banal enough proposition with regard to the jungle or the bush, but one much less so with regard to a major modern city, knowledge of which is usually managed by map, guide, telephone book; in a word, by printed culture and not gestural practice. Here, on the contrary, domiciliation is sustained by no abstraction; except for the land survey, it is only a pure contingency: much more factual than legal, it ceases to assert the conjunction of an identity and a property. This city can be known only by an activity of an ethnographic kind: you must orient yourself in it not by book, by address, but by walking, by sight, by habit, by experience; here every discovery is intense and fragile, it can be repeated or recovered only by memory of the trace it has left in you: to visit a place for the first time is thereby to begin to write it: the address not being written, it must establish its own writing.

'Plan of the Present Work' and 'Social Space'

from *The Production of Space* (1974)

PLAN OF THE PRESENT WORK

[...]

The third implication of our initial hypothesis will take an even greater effort to elaborate on. If space is a product, our knowledge of it must be expected to reproduce and expound the process of production. The 'object' of interest must be expected to shift from *things in space* to the actual *production of space*, but this formulation itself calls for much additional explanation. Both partial products located *in space* – that is, things – and discourse *on space* can henceforth do no more than supply clues to, and testimony about, this productive process – a process which subsumes signifying processes without being reducible to them. It is no longer a matter of the space of this or the space of that: rather, it is space in its totality or global aspect that needs not only to be subjected to analytic scrutiny (a procedure which is liable to furnish merely an infinite series of fragments and cross-sections subordinate to the analytic project), but also to be *engendered* by and within theoretical understanding. Theory *reproduces* the generative process – by means of a concatenation of concepts, to be sure, but in a very strong sense of the word: from within, not just from without (descriptively), and globally – that is, moving continually back and forth between past and present. The historical and its consequences, the 'diachronic', the 'etymology' of locations in the sense of what happened at a particular spot or place and thereby changed it – all of this becomes inscribed in space. The past leaves its traces; time has its own script. Yet this space is always, now and formerly, a *present* space, given as an immediate whole, complete with its associations and connections in their actuality. Thus production process and product present themselves as two inseparable aspects, not as two separable ideas

It might be objected that at such and such a period, in such and such a society (ancient/slave, medieval/feudal, etc.), the active groups did not 'produce' space in the sense in which a vase, a piece of furniture, a house, or a fruit tree is 'produced'. So how exactly did those groups contrive to produce their space? The question is a highly pertinent one and covers all 'fields' under consideration. Even neocapitalism or 'organized' capitalism, even technocratic planners and programmers, cannot produce a space with a perfectly clear understanding of cause and effect, motive and implication.

Specialists in a number of 'disciplines' might answer or try to answer the question. Ecologists, for example, would very likely take natural ecosystems as a point of departure. They would show how the actions of human groups upset the balance of these systems, and how in most cases, where 'pre-technological' or 'archaeo-technological' societies are concerned, the balance is subsequently restored. They would then examine the development of the relationship between town and country, the perturbing effects of the town, and the possibility or impossibility of a new balance being established. Then, from their point of view, they would adequately have clarified and even explained the genesis of modern social space. Historians, for their part, would doubtless take a different approach, or rather a number of different approaches according to the individual's method or orientation. Those who concern themselves chiefly with events might be inclined to establish a chronology of decisions affecting the relations between cities and their territorial dependencies, or to study the construction of monumental buildings. Others might seek to reconstitute the rise and fall of the institutions which underwrote those monuments. Still others would lean toward an economic study of exchange between city and territory, town and town, state and town, and so on.

To follow this up further, let us return to the three concepts introduced earlier.

1 *Spatial practice* The spatial practice of a society secretes that society's space: it propounds and presupposes it, in a dialectical interaction; it produces it slowly and surely as it masters and appropriates it. From the analytic standpoint, the spatial practice of a society is revealed through the deciphering of its space.

What is spatial practice under neocapitalism? It embodies a close association, within perceived space, between daily reality (daily routine) and urban reality (the routes and networks which link up the places set aside for work, 'private' life and leisure). This association is a paradoxical one, because it includes the most extreme separation between the places it links together. The specific spatial competence and performance of every society member can only be evaluated empirically. 'Modern' spatial practice might thus be defined – to take an extreme but significant case – by the daily life of a tenant in a government-subsidized high-rise housing project. Which should not be taken to mean that motorways or the politics of air transport can be left out of the picture. A spatial practice must have a certain cohesiveness, but this does not imply that it is coherent (in the sense of intellectually worked out or logically conceived).

2 *Representations of space*: conceptualized space, the space of scientists, planners, urbanists, technocratic subdividers and social engineers, as of a certain type of artist with a scientific bent – all of whom identify what is lived and what is perceived with what is conceived. (Arcane speculation about Numbers, with its talk of the golden number, moduli and 'canons', tends to perpetuate this view of matters.) This is the dominant space in any society (or mode of production). Conceptions of space tend, with certain exceptions to which I shall return, towards a system of verbal (and therefore intellectually worked out) signs.

3 *Representational spaces*: space as directly *lived* through its associated images and symbols, and hence the space of 'inhabitants' and 'users', but also of some artists and perhaps of those, such as a few writers and philosophers, who *describe* and aspire to do no more than describe. This is the dominated – and hence passively experienced – space which the imagination seeks to change and appropriate. It overlays physical space, making symbolic use of its objects. Thus representational spaces may be said, though again with certain exceptions, to tend towards more or less coherent systems of non-verbal symbols and signs.

SOCIAL SPACE

[. . .]

From about the thirteenth century, the Tuscan urban oligarchy of merchants and burghers began transforming lordly domains or latifundia that they had inherited or acquired by establishing the *métayage* system (or *colonat partiaire*) on these lands: serfs gave way to *métayers*. A *métayer* was supposed to receive a share of what he produced and hence, unlike a slave or a serf, he had a vested interest in production. The trend thus set in train, which gave rise to a new social reality, was based neither on the towns alone, nor on the country alone, but rather on their (dialectical) relationship in space, a space which had its own basis in their history. The urban bourgeoisie needed at once to feed the town-dwellers, invest in agriculture, and draw upon the territory as a whole as it supplied the markets that it controlled with cereals, wool, leather, and so on. Confronted by these requirements, the bourgeoisie transformed the country, and the countryside, according to a preconceived plan, according to a model. The houses of the *métayers*, known as *poderi*, were arranged in a circle around the mansion where the proprietor would come to stay from time to time, and where his stewards lived on a permanent basis. Between *poderi* and mansion ran alleys of cypresses. Symbol of property, immortality and perpetuity, the cypress thus inscribed itself upon the countryside, imbuing it with depth and meaning. These trees, the criss-crossing of these alleys, sectioned and organized the land. Their arrangement was evocative of the laws of perspective, whose fullest realization was simultaneously appearing in the shape of the urban piazza in its architectural setting. Town and country – and the relationship between them – had given birth to a space which it would fall to the painters, and first among them in Italy to the Siena school, to identify, formulate and develop.

In Tuscany, as elsewhere during the same period (including France, which we shall have occasion to discuss later in connection with the 'history of space'), it was not simply a matter of material production and the consequent appearance of social forms, or even of a social production of material realities. The new social forms were not 'inscribed' in a pre-existing space. Rather, a space was produced that was neither rural nor urban, but the result of a newly engendered spatial relationship between the two.

The cause of, and reason for, this transformation was the growth of productive forces – of crafts, of early industry, and of agriculture. But growth could only occur via the town–country relationship, and hence via those groups which were the motor of development: the urban oligarchy and a portion of the peasantry. The result was an increase in wealth, hence also an increase in surplus production, and this in turn had a retroactive effect on the initial conditions. Luxurious spending on the construction of palaces and monuments gave artists, and primarily painters, a chance to express, after their own fashion, what was happening, to display what they perceived. These artists 'discovered' perspective and developed the theory of it because a space in perspective lay before them, because such a space had already been produced. Work and product are only distinguishable here with the benefit of analytic hindsight. To separate them completely, to posit a radical fissure between them, would be tantamount to destroying the movement that brought both into being – or, rather, since it is all that remains to us, to destroy the concept of that movement. The growth I have been describing, and the development that went hand in hand with it, did not take place without many conflicts, without class struggle between the aristocracy and the rising bourgeoisie, between *populo minuto* and *populo grosso* in the towns, between townspeople and country people, and so on. The sequence of events corresponds in large measure to the *révolution communale* that took place in a part of France and elsewhere in Europe, but the links between the various aspects of the overall process are better known for Tuscany than for other regions, and indeed they are more marked there, and their effects more striking.

Out of this process emerged, then, a new representation of space: the visual perspective shown in the works of painters and given form first by architects and later by geometers. Knowledge emerged from a practice, and elaborated upon it by means of formalization and the application of a logical order.

This is not to say that during this period in Italy, even in Tuscany around Florence and Siena, townspeople and villagers did not continue to experience space in the traditional emotional and religious manner – that is to say, by means of the representation of an interplay between good and evil forces at war throughout the world, and especially in and around those places which were of special significance for each individual: his body, his house, his land, as also his church and the graveyard which received his dead. Indeed this *representational space* continued to figure in many works of painters and architects. The point is merely that *some* artists and men of learning arrived at a very different *representation of space*: a homogeneous, clearly demarcated space complete with horizon and vanishing-point.

37 ROSALYN DEUTSCHE

'Alternative Space'

from B. Wallis (ed.), *If You Lived Here* (1991)

[. . .]

ARTISTS IN THE CITIES

What variety of means *is* available in the effort to persuade and convince? How can one represent a city's "buried" life, the lives in fact of most city residents? How can one show the conditions of tenants' struggles, homelessness, alternatives to city planning as currently practiced – the subjects of "If You Lived Here . . ."? These have been the central issues shaping this project. Its forums, of course, provided an opportunity for direct speech. The three shows, however, also featured varieties of "direct evidence" and argumentation about the grounding of urban life. Artists, community groups, and activists made their points through an array of materials, from video-tapes, films, and photographic works to pamphlets and posters to paintings, montages, and installations.

Certainly the conventionalized picture of the postmodern city, with its fortresses and deeply impoverished ghettos, with its epidemics of drugs and AIDS, reinforces the imagery of the urban frontier and discourages even partial approaches to poverty and homelessness. For artists, the image of the city's mean streets may feed a certain romantic Bohemianism. Yet, because artists often share city spaces with the underhoused, they have been positioned as both perpetrators and victims in the processes of displacement and urban planning. They have come to be seen as a pivotal group, easing the return of the middle class to center cities. Ironically, however, artists themselves are often displaced by the same wealthy professionals – their clientele – who have followed them into now-chic neighborhoods.

The "percent for art" programs put in place in a number of U.S. cities have also brought artists into the urban-planning blueprint, at a time when even the idea of public art – like the notion of public space – is being severely attacked. This isn't the place for a broad consideration of public art, but what is worth mentioning is the current high-profile version of "beautification," an ambition to improve the "quality of life" often invoked by anxious city administrations in canceling both taxes and unsightly urban elements for the benefit of powerful corporations. This sort of public art project is exemplified by Battery Park City, a megaproject on New York's Lower West Side. Financed by international capital (in this case, Olympia & York, the corporate entity of Montreal's Reichmann Brothers), Battery Park City imagines itself to be a fantasy enclave of residences, offices, parks, and gardens – something like the ruling-class rooftops in Fritz Lang's film, *Metropolis*. What is of interest here is the regularized incorporation of art by the authority running it – precisely as though this exclusive preserve reinvented the public, on privatized but publicly subsidized turf. Although the art program has been touted as showing risky "socially conscious" art, such work seems severely compromised by its context.

Irrespective of such public or corporate commissions, artists have always been capable of organizing and mobilizing around elements of social life; the city is art's habitat. But how do artists address directly the issues of city life and homelessness in which they are implicated? Most directly, of course, many artists engage in activism, including working with homeless people in shelters and hotels, as do Nancy Linn and Rachael Romero; producing posters and street works on urban issues, as do Robbie Conal, Ed Eisenberg, Janet Koenig, and Greg Sholette; or engaging in other forms of political activism, as do Marilyn Nance, Mel Rosenthal, and Juan Sanchez. Krzysztof Wodiczko and the Mad Housers work with homeless people in projects whose stop-gap solutions to homelessness show up the absurdity of official responses. But there are many other approaches as well.

Postmodern life is characterized by the erasure of history and the loss of social memory. Social life includes multiple streams of contesting momentary images, which, detach from particular locales, join the company of other images. Images, in appearing to capture history, become the great levelers, the informational counterpart of money, replacing material distinctions with their own "depthless" (that is, ahistorical) logic. One of the social functions of art is to crystallize an image or a response to a blurred social picture, bringing its outlines into focus. Many artists and critics engage with these dislocating politics of the image through critiques of signification. Such critical practices temporarily check the flow of (what passes for) public discourse. But such critiques-in-general, crucial as they are to a reorientation of social understanding, don't exhaust the avenues to urban meaning.

Consider the city once again. It is more than a set of relationships and a congeries of buildings, it is even more than a geopolitical locale – it is a set of unfolding historical processes. In short, a city embodies and enacts a history. In representing the city, in producing counter-representations, the specificity of a locale and its histories becomes critical. Documentary, rethought and redeployed, provides an essential tool, though certainly not the only one.

The arguments for documentary apparently need to be made anew. Image politics and still-contested notions of difference have prompted serious philosophical critiques of the claims to transparency and univocality of news, documentary, and photography in general – critiques made in the context of the growing distance between imagery and social meaning in the culture at large. Even past documentary works, which have taken on new meanings in textbooks, art history books, and gallery sales, are a matter of perpetual reinterpretation.

The "problem" of documentary is com-pounded by the art-world distrust of populist forms (for various reasons, some of which are valid and others simply manifestations of professional snobbery). Who could possibly deride a healthy skepticism in regard to the propaganda of the obvious that characterizes the myths of documentary transparency? On the other hand, the agitational intentions of activist social documentary aren't sufficient in themselves to secure a conviction except in the court of formalist aestheticism.

It would be ironic if those of us seeking a more complex account of experience and meaning were enjoined by our own theoretical strictures from presenting evidence in support of social meaning and social justice. Documentary practices are social practices, producing meanings within specific contexts. Rejecting various entrenched documentary practices hardly amounts to a negation of documentary in toto. The critical minefield surrounding practices rebuked for empiricism calls for careful negotiation. Social activists, certainly, continue to recognize the importance of documentary evidence in arguing for social change. It is the necessity to acknowledge the place – and time – from which one speaks that is an absolute requirement for meaningful social documentary. This requirement allows for an unspecifiable range of inventive forms but doesn't dispose of the historically derived ones. Naturally, this shifts the terrain of argument from the art object – the photograph, the film, the videotape, the picture book or magazine – to the context, to the processes of signification, and to social process. An underlying strategy of the project "If You Lived Here . . ." (of which this book is a part) has therefore been to use and extend documentary strategies.

A documentary photograph of a member of a social group composed of undifferentiated stereotypes – the "homeless," say – today serves the same purposes as did similar images at the inception of social documentary as a public photographic practice: it "humanizes" by particularizing. It suggests the character of a person's existence, in which material circumstances contradict human worth, and the more dire the conditions, the more the photo may have to tell us. Sometimes the "condition" is invisible, a conceptual understanding laid over the image by the viewer. But the problem is that of projection, of imagining that the characteristics we "see" in the person or scene are those that are "there." For that reason, the more patent the image, the more it accords with "common-sensical" presuppositions, the less it may have to tell us. This is not a condition that should make us vacate the territory of image making, for it is precisely the role of the con-text – especially the verbal text (written or otherwise supplied) linked with *this* image-text – that establishes a meaning beyond a simple ground for projection.

Documentarians – unlike "street photo-graphers," another sort of practitioner entirely – have hardly relied on images alone to tell the right story. The development of high-profile, commercial, professional photojournalism, and the art-world appropriation of all kinds of photography into its own procrustean canon, paved the way for a photographic practice passing for social documentary to shorten its circuit from the street to the gallery wall. Lost along the way were more than symbolic claims for agitational intentions. The dead hand of "universalism" has lain heavily on docu-mentary's shoulder, for a documentary work alibied as revealing an underlying human sameness becomes simply an excuse for spectacle. That is the basis of one of the most telling critiques of documentary, particularly of the subgenre exotica – a form of anthropology that masquerades as humanism when the subject is the down-and-outer in advanced Western society or in its familiar margins (Mexico or Bensonhurst). One of the problems of repre-sentations of the city is to make an argument without betraying people.

In one of the exhibitions for "If You Lived Here . . .," a pair of texts placed side by side on the wall argued for and against photographing the homeless. The first text, an excerpt from an essay of mine on documentary photography, criticized "victim photography" for rarely ser-ving the purpose which (presumably) its makers intended – namely, to gather public support, to generate outrage, and to mobilize people for change. Rather, I argued, documentary photography may inadvertently support the viewers' sense of superiority or social paranoia. Especially in the case of homelessness, the viewers and the people pictured are never the same people. The images merely reproduce the situation of "us looking at them."

In the other text, "On Photographing the Homeless," photographer Mel Rosenthal argued for photographing the homeless. Although, he wrote, he was troubled by photographing people in desperate straits – people who, even when they gave their consent, may not have had much idea of how their photos would be used – on balance he felt that images of real individuals can dispel the numbness many people feel. Context, however, still remains crucial, and Rosenthal acknowledges this. (I've remarked elsewhere that political photography is repressed in our culture by being hung in a gallery.) Rosenthal's projects are never geared toward the gallery-museum circuit. His South Bronx photographs, for example – made during a period when he worked at a health clinic in the Bathgate area where he grew up – were published in activist and grass-roots magazines. Rosenthal gave prints to the people photo-graphed, who often had no other photos of themselves. In exhibition form, these photos of resiliency in a war zone are accompanied by an array of quoted remarks (some of which are reproduced in this book) providing the necessary – damning – information.

It would be reductive to insist that no levels of mediation can exist between those who experience a situation and those who view it. In a fragment of an interview with Alexander Kluge reprinted in this book, Kluge takes up precisely this question of participatory versus supportive mediations – by chance, in relation to the eviction of squatters in Germany. There has to be room for an interested art practice that does not simply merge itself into its object. Interestingly, though, Bienvenida Matias, in Loisaida (the Hispanic Lower East Side of New York), and Nettie Wild, in Vancouver, B.C., were each invited to live in the housing communities whose struggles they were documenting on film and videotape (Matias in *El Corazon de Loisaida*, or *The Heart of Loisaida*, and Wild in *The Right to Fight*). Both accepted.

Ultimately, there's no denying that no matter how the works in "If You Lived Here . . ." originally were woven into the social fabric, the venue of the exhibitions was an art gallery, even if partly "transformed." The idea of these shows wasn't simply to thicken the context for the reception of "photographs of the Other." It was, first, to allow for a consideration of an underreported, underdescribed, multidetermined set of conditions producing simple results: homelessness and sadly inadequate housing. Perhaps no less importantly, the project intended to suggest how art communities (might) take on such questions. Since the problem of homelessness, like all social problems, exists in a stream of conflicting representations, it is not possible to change social reality without challenging its simplifying overlaid images. That was a main task of "If You Lived Here . . ."

38 HEINZ PAETZOLD

'The Philosophical Notion of the City'

from Heinz Paetzold (ed.), *City Life* (1997)

The main purpose of my essay is to contribute to the question of the city, urban life and urban culture from a philosophical perspective. There is enough evidence for the assumption that the question of the city has played an important role within the different formative stages of Western philosophy. Giambattista Vico once stated that the notions of Western metaphysics originated in the marketplace of ancient Athens, which means that they are of urban origin. A consequence of this claim would include the methodological device of deciphering such key notions of Western metaphysics as the dichotomies of essence and appearance, of principle and dependency, of unity and manifold, of public and private, as a set of notions signifying the reality of urban life in the ancient Greek polis. This does not mean, however, that philosophy should be equated with cultural history.

In books such as *The Fall of Public Man* (1976), *The Conscience of the Eye* (1990) and *Flesh and Stone* (1994), Richard Sennett revealed the intrinsic relationship between the various ideas concerning the city and urban design in Western civilisation and the main stations and steps of Western cultural history. My discourse here will not be concerned with this relationship. Rather, what I want to do is to relate the idea of the city to the core of philosophy, which means philosophy's systematicity. As an indication of the main drift of my essay, I will start with a few metaphorical references to the very idea of the city at various moments of Western philosophy. My examples from Wittgenstein, Descartes and Cassirer are nothing more than indications of the fact that philosophy and urbanity have been and still are, or at least should be, closely interrelated with each other. Then I will deal with more specific questions of a philosophy of urbanity.

I will distinguish between three basic philosophical attitudes towards the question of the city within the modernist movement. In the final part of my essay I will provide some points of reference concerning the question of urbanity within postmodern philosophies. I would argue that philosophy not only originates from urban life, but should also find means to influence city life.

I

When he published his *Philosophical Investigations* in 1953, Ludwig Wittgenstein was on the road leading to a philosophy of language centred on the diversity of pragmatic usages. Instead of conceiving language as being built out of basic semantic elements, as the *Tractatus* from 1921 tells us, Wittgenstein had discovered the multiverse of 'language games'. In a rather famous passage of his *Philosophical Investigations* Wittgenstein refers metaphorically to the image of the city. He reflects upon the question of how, and to what degree, a language can be judged as being complete. While our knowledge is growing, new languages, sort of suburbs, are added to the existing old one, which Wittgenstein parallels with an 'ancient city'. There is no such thing as one singular language, as the Tractatus had presumed, but only divergent languages. Wittgenstein says literally: 'Do not be troubled by the fact that [some, H.P.] languages . . . consist only of orders. If you want to say this shows them to be incomplete, ask yourself whether our language is complete – whether it was so before the symbolism of chemistry and the notion of the infinitesimal calculus were incorporated in it, for these are, so

Grant Street, New York City

to speak, suburbs of our language. (And how many houses or streets does it take before a town begins to be a town?)

Our language can be seen as an ancient city: a maze of little streets and squares, of old and new houses, and of houses with additions from various periods, and this is surrounded by a multitude of new boroughs with straight regular streets and uniform houses.

Let me just add that Wittgenstein occupies on the chessboard of 20th century philosophy quite a prominent field. He introduced what Richard Rorty has labelled the linguistic turn of philosophy. Philosophical problems are to be spelled out in terms of language, since it is through language that we may or may not understand the world. Linguistic structures mirror urban structures, their growth and their relatedness. Wittgenstein goes on to say, in another context, that 'Language is a labyrinth of paths. You approach from one side and know your way about; you approach the same place from another side and no longer know your way about.' The moving force of language, language's dynamics, is comparable with orientating and disorientating walks through the city.

It was René Descartes who gave philosophy in the dawn of modern times a decisive push. Together with Galileo Galilei and with Gottfried Wilhelm Leibniz, Descartes introduced a powerful paradigm of philosophy, namely philosophy as 'mathesis universalis'. The result of this is that philosophy's methods and contents came under the sway of mathematical physics. In his *Discours de la Méthode*, which was first published in 1637 in Leiden, Descartes praises an urban design which has been brought about by only one engineer or architect.

The slow growth of a city, Descartes argues, hardly leads to coherence and perfection. The beauty of singular houses remains without effect. Just as science is more convincing the more it constitutes a logical homogeneous space, an analogous rule is valid with regards to the construction of a whole city. The American urbanist Lewis Mumford touched on this analogy when he said that Descartes reinterpreted 'the world of science in terms of the unified order of the Baroque city'.

Descartes wrote: 'So it is that one sees that buildings undertaken and completed by a single architect are usually more beautiful and better ordered than those that several architects have tried to put into shape, making use of old walls which were built for other purposes. So it is that these old cities which, originally only villages, have become, through the passage of time, great towns, are usually so badly proportioned in comparison with those orderly towns which an engineer designs at will on some plain so it is that, although the buildings, taken separately, often display as much art as those of the planned towns or even more, nevertheless, seeing how they are placed, with a big one here, a small one there, and how they cause the streets to bend and to be at different levels, one has the impression that they are more the product of chance than that of a human will operating according to reason. And if one considers that there have nevertheless always been officials responsible for the supervision of private building and for making it serve as an ornament for the public, one will see how difficult it is, by adding only the constructions of others, to arrive at any great degree of perfection.'

As already mentioned, Descartes' considerations have to be weighed against the background of Renaissance or Baroque urbanism, which, according to Lewis Mumford, are indistinguishable. During the time of Descartes, who lived from 1596 till 1650, it was usually a prince, a lord or a king who inspired or ordered the erecting of a new city in a uniform design. We have to think of the foundation of La Valletta on Malta (1593), Palma Nuova in Italy (1593) or Coevorden (1600), Charleville (1606) or Hanau at the end of the 16th century. The city of Naarden received after 1579 a totally new urban design by Thomas Thomasz and Adriaen Anthonisz.

For my argument here it is sufficient to state that a clear relationship can be seen between the idea of philosophy as 'mathesis universalis' and the emerging urbanism of the Renaissance and Baroque period. A city should be constructed by the will of one engineer only. And he has to follow the rules of reason.

I could continue listing further metaphorical references in philosophy to the idea of the city. If one objects that metaphorical references cannot count on an equal footing with literal ones, then we have to realise that clear-cut differences between a literal and a metaphorical meaning are not so easily established. With Nietzsche and Rorty one could say that even truth is nothing other than a 'moving army of metaphors'. However, in order to come closer to my own discourse here, let me give you just one last example. It was Ernst Cassirer who gave philosophy in the first half of our century a new shape by arguing that philosophy should be transformed into a philosophy of culture. This would imply that the paradigmatic role of mathematical physics as a model for philosophy as a whole has to be given up. Even the linguistic turn does not reach far enough. We have to build up philosophy from a new angle, the symbolic, and we have to realise that science and language

Shop window, New York City

are 'symbolic forms', but without any privileged value.

They are 'symbolic forms' among others, like art, myth, morals, politics, technique, economics. All these forms together constitute what we call culture. In Cassirer's philosophy of culture two cultural periods are of special value for modernity and its dynamics. In the period of the Renaissance, an explicit modern culture took shape, that is, one which substituted a democratic attitude concerned with all facets of secular and worldly life for the legacy of mediaeval scholasticism with its hierarchies and pedantries. The Enlightenment period introduced a different positioning of culture within social life. Culture lost its former separation from everyday life and became a decisive part of it. While discussing the social and political theories of the philosophies of Enlightenment, Cassirer stresses urbanity as the main characteristic, especially of the French thinkers. They were concerned with politics, ethics and aesthetics as well. But the demands for philosophy, namely clarity and distinctness, were to be related to the new ideal of urbanity.

'Not only political, but also theoretical, ethical and aesthetic ideals are formed by and for the salons. Urbanity becomes a criterion of real insight in science. Only that which can be expressed in the language of such urbanity has stood the test of clarity and distinctness.'

The worldly realm of the various symbolic forms finds its final expression in the enlightened urban public space. This move of philosophy was already indicated by Kant, when he demanded that the 'scholastic conception' of philosophy, it's 'Schulbegriff' should be replaced by a 'conceptus cosmicus', a conception as related to the world, philosophy's 'Weltbegriff'. This move can only be exercised to its full extent within the urban space.

I would argue that what these three examples

show is the fact that philosophy, at decisive stages and turning points of its own development, had a clear relationship with the idea of the city and urban life and culture. The introduction of the paradigm of philosophy in the sense of a 'mathesis universalis' was accompanied by a metaphorical reference to urbanity. According to this paradigm of philosophy, urbanity should be based on and should depend on an idea of rationality which has its core in methodicity. Even the linguistic turn of philosophy in our century was related to the very idea of the city. Our ideas concerning the growth of language and language's completeness can only be worked out if we look at the realm of a city. Finally, the shift from philosophy understood as transcendental criticism to the philosophy of culture implied in one way or another a reference to the idea of urbanity. Urbanity was a practical goal of enlightened philosophy.

II

Although, as I have said, it is difficult to make a strict distinction between the literal and the metaphorical meaning of a notion, I would like to concentrate in this part of my essay on the more literal or more explicit meaning of the concept of the city within modern philosophy. The internal contradictions of the social and cultural processes of modernisation resulted in the inherited strong tie between philosophy and urban life, which can be traced back to Plato and Aristotle, being called into question. With regards to a more explicit notion of modernity or modernism, I would like to differentiate three lines within modern philosophy and its relationship to the urban.

III

Vis-à-vis the already established industrial society it was Friedrich Nietzsche who broke decisively with the metaphysical legacy of philosophy. This starting-point of Nietzsche should be recognised if we want to understand his critique of the big city. In his *Thus spoke Zarathustra* from the 1880s, Nietzsche's hero bypasses the city, refusing to enter it. The fool, Zarathustra's ape, tells the story of the moral and cultural decline of urban life. The merchant's spirit has infected the big city. Morals have been perverted into the mentality of narrow-minded shopkeepers. The culture of the city has turned out to be subservient to the shopkeeper's gold.

Zarathustra, on his part, after having listened

Side street, Tokyo

to the fool's warning of the big city's decline is disgusted. Zarathustra, Nietzsche's counter figure to Plato's Socrates, does not enter the gates of the city. Whereas Plato's Socrates had confessed that he behaves in a strange manner outside the city walls, and that he does not learn from the trees and a lovely countryside, but only from people in the city, Zarathustra on the contrary rejects the city. But he is not so much impressed by the fool's litany. He asks the fool: 'Why did you not go into the forest? Or plough the earth? Is the sea not full of green islands?'

Although Zarathustra is disgusted by the fool's talk he nevertheless agrees with the fool's condemnation of the city. He says: 'Woe to this city! And I wish I could see already the pillar of fire in which it will be consumed!'

At first sight it would seem as though Nietzsche only wants to dismantle the metaphysical tradition in which Christianity had merged with Platonism. The Platonic dualism of 'idea' and 'semblance', of 'philosophical thought' and 'opinion', of 'essence' and 'appearance', had been mixed up with Christian distinctions between the 'eternal divine' and the 'earthbound flesh' or 'heavenly transcendence' and 'transitory immanence'. But in that Zarathustra passes by the fool as well the city; Nietzsche not only wants to loosen the alliance between traditional philosophy and urban life, but to go a step further. His Zarathustra removes himself from the city and its culture. He seeks to create the state of Super-Man in the loneliness of the mountains. He has such desires after, as Nietzsche literally says, having made his way through diverse towns. The post-metaphysical culture of modernity cannot find its origin in the realm of the modern metropolis. We have to leave the city if we are to rediscover the integrity of our bodies, since the city remains a metaphysical concept and can by no means get rid of the restraints implied. This philosophical line of rejecting the city of modernity was followed by authors of our century, such as Oswald Sprengler and Martin Heidegger. For Heidegger, the event of being is experienced only in the forests and fields of the countryside far away from the big cities and their unauthentic ways of living.

In the figure of Zarathustra we can grasp one strong tendency in modern culture. This tendency has given up all hope for a reform of the industrial and commercialised modern city.

Urban culture is rejected as being superficial and as being under the sway of a narrow-minded mentality of shopkeepers.

IV

We have to distinguish this line of philosophy from a second one. Here the rejection of the existing big city includes the praise of a small town. The small town shows a high degree of social transparency and it is beloved for that reason. This position can be traced back to Jean-Jacques Rousseau in the 18th century.

The importance of Rousseau as a philosopher lies in the fact that, in giving the theory of contract a new foundation, he renewed political philosophy. Moreover, it was Rousseau who introduced cultural critique as a new philosophical genre. Some years after having published his two Discourses – the *Discourse on the Sciences and the Arts* (1750) and the *Discourse on the Origin of Inequality* (1755) – and only a few years before his *The Social Contract* (1762), Rousseau published a book in which he developed his critique of metropolitan life. This book of more than 150 pages was sold out only a few weeks after its appearance in 1758. Titled *Letter to M. D'Alembert on the Theatre*, it criticises metropolitan life on the grounds that it destroys the immediacy of moral feelings.

People in big cities pretend to be more or other than they really are. They live in a world of semblance. Theatre, one of the main institutions of urban culture since ancient times, produces only an indirect, reflective attitude with regard to moral life.

The actor is not that person which he performs in his role and is for that reason alienated from an authentic existence. On the grounds of such considerations, Rousseau criticises d'Alembert, the famous editor of the *Encyclopaedia*, for arguing in an Encyclopaedia article on 'Geneva' that Geneva should have a theatre as a cultural tool for refining both the moral sentiments of its citizens and those of the professional actors.

Rousseau, on the contrary, praises the wisdom of Geneva, precisely because it lacks a theatre. He argues that the republican state of Geneva would be undermined if the institution of the theatre were be introduced. The real geniuses, Rousseau continues, are to be found in small towns like Geneva, because here their lives are transparent to others and for that reason morality is in accord with everyday life. It is not necessary for the individual to appear differently than what is in reality the kernel of his being. In big cities, however, everyone hides his or her internal sentiments. The beautiful morality, 'die schöne Sittlichkeit', to quote Hegel, of Geneva, would be destroyed if the theatre were introduced. In the metropolis, theatre does its work, because the performance of a play interrupts for a short while the idleness of the roaming people and the unending crimes. Metropolitan man leads a life of idleness and wickedness.

In ancient Athens, in contrast, theatre fulfilled a fruitful role for the urban community. The tragedies and comedies in Athens exposed narratives which were related to the city and for that reason theatre had rather a political and religious function aimed at uniting the citizenry rather than being pure entertainment. Rousseau praises the wisdom of ancient Rome where actresses were regarded as prostitutes. In Athens the male actors were rather priests performing an urban ritual.

In order to understand the full extent of Rousseau's position it is necessary to touch on his philosophy of culture. Rousseau notices a decline of culture in modern times, partly because culture had fallen prey to the power of money. Man has lost the state of nature and can never return to it. The only possibility which can be reached consists in an equilibrium between the lost state of nature on the one hand and the state of culture on the other. To use the technical terms of Rousseau: the 'amour de soi' has not been totally perverted into the 'amour propre'. Love of oneself has not yet become egocentricism.

Such a state of cultural equilibrium, however, can only be lived, maintained and embodied in a small town which has a republican constitution. The general will, the 'volonté général', governs such a community and not the numerical majority, the 'volonté des tous'.

The scope of this essay does not allow me to analyse Rousseau's position to any greater extent. But let me give you some indications as to how this line of treating urbanism developed further later on. In a vein informed by Rousseau,

we can place what Françoise Choay has called the 'culturalist model' within 19th century urbanism. The 'culturalist model' was opposed to the Haussmann pattern of regularisation and the socialist utopian models (Fourier, Owen, Cabet, Soria y Mata, Tony Garnier). We have to think of William Morris (1834–1896) who proposed a limitation to the growth of the city. Morris was in favour of a city which is aware of its site within a countryside. In particular, the ecological damages and devastations of the industrial cities should be overcome, craftsmanship renewed, etc. Whether Camillo Sitte's *City Planning According to Artistic Principles* (1899) fits into this line is questionable, because the social aspects of urban life are not Sitte's main concern. In any case, Ebenezer Howard's *Garden City* (1850–1928) is more in accordance with Rousseau. Limitation of the number of inhabitants (to around 32,000 persons), exchange between city and countryside, centring of cultural institutions, are the key concepts. In our times it was Lewis Mumford who gave the Rousseau line a strong voice again. Mumford died 5 years ago. Contrary to Rousseau, however, Mumford does not repudiate any urban institution. He favours theatre, library, school, museum, etc. Mumford discusses contemporary urban culture in relation to what Patrick Geddes calls Megalopolis with all its various aspects, encompassing economics, hygiene, politics, planning procedures, social composition of the inhabitants and aesthetics. According to Mumford, we live in a period of obvious decline of gigantic super cities. He pleads for limitations of growth.

Urbanism has to be placed within a framework of questions concerning regionalism in the cultural and political sense. But Mumford favours necessary balances. In order to exclude any regressive version of regionalism, the regional site of a city has to be balanced with cosmopolitanism. The cosmopolitan mentality is an outcome of modernity and can by no means be rolled back. Furthermore, the city should be in a cultural, political and economic balance with the surrounding countryside, in terms of exchange, rather than of exploitation. Finally, a city should realise a balance between culture and nature.

Mumford's concept was intended to combat nationalism with its menaces of war, totalitarianism and uncontrolled power. The city is for him 'a collective work of art', in which 'the universalising forces and the differentiating forces' are at work. The following definition of the city by Lewis Mumford could be read as a critical comment on Rousseau, while still transforming central concerns of Rousseau into a position of discussible value today: 'The city in its complete sense, then, is a geographic plexus, an economic organisation, an institutional process, a theatre of social action, and an aesthetic symbol of collective unity.' Mumford continues: 'On one hand it is a physical frame for the common place domestic and economic activities: on the other, it is a consciously dramatic setting for the more significant actions and the more sublimated urges of a human culture. The city fosters art and is art, the city creates the theatre and is the theatre. It is in the city, the city as theatre, that man's more purposive activities are formulated and worked out, through conflicting and co-operating personalities, events, groups, into more significant culminations.' Although I share many of Mumford's convictions – the stress on a concrete urban community, the stress on democratic participation, the stress on the relevance of the symbolic-aesthetic dimension of our environment, the stress on an interrelationship between culturally determined regions as an alternative to national and international acts of unification from above – I find his approach in the end too idealistic. It should be clear, of course, that any definition of urban culture has a normative basis.

V

These remarks bring me to the third line of urbanism within modern philosophy. This third line takes as its starting point the conviction that modern cities not only offer the articulation of utopian desires, but also provide possibilities for catastrophes.

The utopian aspect of a modern city could be traced back to Max Weber and Karl Marx. For Weber, the cities in the span of the Middle Ages up to Renaissance and Baroque periods were on their way to constituting an autonomous social and economic space of freedom in a mediary space between the household economics of a prince or a church and the national economics. These cities were autonomous political bodies opposed to legal-bureaucratic, traditional and charismatic modes of domination. The occidental city was, as Weber put it, 'a place where the ascent from bondage to freedom was possible by means of monetary acquisition'. ['Die okzidentale Stadt war...ein Ort des Aufstiegs aus der Unfreiheit in die Freiheit durch das Mittel geldwirtschaftlichen Erwerbs']. The cities

Central Park, New YorkCity

articulated the sense of a community which involved processes of political participation in power on the part of the citizenry.

But the point is that the cities which had fought against the lord, king or church and had gained political, economic and cultural freedoms in the span of the late Middle Ages and during the dawn of modernity in the Renaissance period lost these freedoms and their autonomy through the emergence of the modern national bourgeois state. In her book *The Nature of the City*, Heide Berndt drew our attention to this point nearly 20 years ago. Rousseau's question had been how to continue and maintain the freedoms of republican urban citizens under the conditions of the modern absolute and bourgeois state and under the conditions of an emerging economics of exchange? Rousseau failed partly in giving a concrete answer. He could only helplessly try to reanimate the republican freedoms in Geneva. For that reason we have to introduce some insights of Marx and the tradition of critical theory. Within this tradition, the modern city and the processes of modernisation have a double-sided design. On the one hand, untempered and uncontrolled processes of capitalisation lead to ecological devastations, social crises and a strengthening of the city's command over the countryside. On the other hand, the huge amounts of commodities which are concentrated within the urban space can only be reappropriated within the cities themselves.

As Heide Berndt has put it, the development of a 'total individual' (Marx) is something which takes place in the city. To this formula of Marx (and Engels) lots of new ideas and concepts have been added, ranging from Georg Simmel, Walter Benjamin, Henri Lefebvre, to the contemporary writings of Mike Davis. Georg Simmel conceived his *Philosophy of Money* (1900) as a complement to 'Historical Materialism', in that it deals with the psychological aspects of human social existence. 'The essence of modernity as such is psychologism', Simmel argues. This quotation already indicates Simmel's concerns. He explores modernity in its various cultural aspects. Modernity, however, is deeply rooted in and related to metropolitan life. Subjects of Simmel's interest, such as fashion, lifestyles, everyday aesthetics like 'Door and Bridge', feminine culture, the destiny of socialist movements and the adventure in modern life bear the impact and imprint of modern urban life.

For Simmel, the modern metropolis is the site of money economy. The predominance of this economy causes the loss of the substantiality of things. Their substantiality is replaced by functionality and fluid relationships. Cassirer will describe ten years after Simmel a parallel in modern epistemology in that it moves from

Park, Tokyo

'substantial forms' to 'functional concepts'. Money economy, Simmel continues, leads towards a predominance of formalised rationality. Rationality becomes a kind of psychological 'protecting organ' against what Simmel calls the 'threatening currents and discrepancies of (the) ... external environment'. Simmel's central question in his essay 'The Metropolis and Mental Life' (Die Grosstädte und das Geistesleben, 1903) is the following: How can an individual maintain his freedom in the presence of a jostling crowd?

Simmel argues that two attitudes are internalised by metropolitan man. The blasé attitude is the individual's answer to the rapidity and the contradictoriness of the stimulation of the nerves. Metropolitan man is exposed to such an environment as the modern urban one. The blasé attitude is in the first place rooted in the predominance of the money economy. In his *Philosophy of Money* Simmel says: 'Whoever has become possessed by the fact that the same amount of money can procure all the possibilities that life has to offer must ... become blasé.

Since all things become available they appear in the end in a grey tone. Money levels out all things and values. But the blasé attitude also produces the craving for new excitements, extreme impressions, greater rates of change. Metropolitan environments produce, Simmel says, also the emotional reserve. Whereas the blasé attitude is seated in the subjectivity's relationship towards objects, the reserve has its site within the intersubjective relationship. Reserve, Simmel says, is necessary for metropolitan man to bear the anonymous crowds around him. The reserve against the others is paradoxically the basis for the individual's freedom. Reserve does not mean that metropolitan man does not experience all the nuances of emotional life – sympathy, empathy, antipathy, etc. – but he has to hide behind a protective screen of reserve in order to survive.

Reserve as a metropolitan mentality is explained by Simmel through the dissociation of our senses in urban life, the visual being separated from the audible, and through the predominance of the visual in modern urban life. The blasé attitude and reserve are, for Simmel, the two main characteristics of the metropolitan psychogram. The reserve creates personal

Tokyo

freedom and independence, but it is also responsible for the 'specifically metropolitan extravagances of mannerism, caprice, and precariousness'. Being different in a striking manner is a prerequisite of urban life. Here we find the roots of fashion.

Simmel practised a post-metaphysical mode of philosophy, caused by his concentration upon metropolitan urban life. He was convinced that only a shift in the form of writing would be adequate to cope with urban culture. He did not write commentaries, as the scholastic philosophers in the Middle Ages did, nor did he employ the form of an 'encyclopaedia', as the French philosophers of the Enlightenment or Hegel did. His was the form of the 'essay'. The essay is not meant to attain metaphysical principles or first grounds, but rather is meant, as Adorno said, to move within a given cultural setting, an urban setting, which should be influenced by the essay's main drift.

This trait of urban philosophy is shared by both the lines that inform Simmel's work very deeply. The one leads to urban sociology, and especially to the Chicago School. The founder of this school, Robert Park, attended Simmel's lectures in Berlin. Park and his follower Louis Wirth concentrated on the question of freedom in urban modern life. Park asked how the city as a geographical space can be transformed into a 'moral order'. For Park as well as for Louis Wirth, the civilisatory function of big cities lies in the fact that they allow possibilities for deviant people and outsiders. As Sennett noted, for Baudelaire as well as for the sociologists of the Chicago

W. 50th Street, New York City

School, 'urban differences' are understood as 'provocation of otherness, surprise, and stimulation'. The measure of the freedom of a big city consists in its capacity to enable outsiders and deviant forms of behaviour to exist.

The second line starting from Simmel concentrates on him as an author of Kulturphilosophie, philosophy of culture. This is the starting point of Benjamin, who attended both Simmel's as well as Cassirer's lectures at Berlin University. But let me mention one other central point of Simmel's philosophy of the city, which leads immediately to Walter Benjamin. For Simmel the metropolis is the place where the crisis of modern culture, 'culture's tragedy', is occurring. Simmel observes an internal contradiction within modern culture. On one hand we experience an unbelievable increase in the culture of objects, 'the wonder and comforts of space-conquering technology', the many 'buildings', and museums, where 'crystallised and impersonalised spirit' is invested. This culture of objects makes life comfortable and easy. On the other hand, however, the culture of subjectivity cannot keep pace with this. Subjectivity becomes impoverished. An alienation or splitting between the objective culture of things and the subjective culture of individuals takes place. The metropolis is the place where this 'tragedy' of modern culture, as Simmel says, takes place. It goes without saying that many of Walter Benjamin's concerns can be traced back to Simmel. We have to think of Benjamin's concentration upon the metropolitan city as the site where modernism's preoccupation with

presence is exercised. We have to think of Benjamin's general approach to the experience of modernity with its two edges, the dream-world and the catastrophe in the urban space.

But instead of scrutinising Benjamin's dependence upon Simmel, which has been demonstrated particularly by David Frisby, I will concentrate on two central issues in Benjamin, in order to come closer to my own discourse. Drawing on Charles Baudelaire's 'Les Fleurs du Mal' Benjamin argues that modernity is accessible only in terms of allegorical images. The predominance of the production of commodities causes an erosion of meanings. Things lose their substantiality and become meaningful within a multitude of pragmatic situations. Metropolitan man has to come to grips with these frequent shifts and dramatic changes of meaning. Speaking in terms of

Times Square, New York City.

philosophy, the predominance of allegorical meanings is related to a relativity of things, images, even persons. But in order to avoid nominalistic and relativistic consequences Benjamin reintroduces, as Susan Buck-Morss has shown, a reference to Jewish Kabbalah on the one hand and to Historical Materialism on the other. The arbitrariness of the allegorical can only be overcome if it can be related to the 'corpus symbolicum' of the Kabbalah. According to the Kabbalah, mankind lost adequate symbols after the fall. What has to be done is to create a plurality of allegorical signs which have to be related to the Messianic light of redemption in order to attain their final symbolic meaning. This symbolic meaning is intuitively perceived in a mystical 'now', in the present.

It is essential, however, to realise that, rather than falling back into metaphysics, Benjamin employs a post-metaphysical philosophic stance. Benjamin's reference to Historical Materialism comes into play here. It is by political actions exercised by a group or a collective that a commonly shared perspective and a commonly shared world of meaning are constituted. Benjamin believed that the arbitrariness of allegories is only surmountable if social actions, in the sense of Historical Materialism, could be linked with the Messianic idea and light of redemption. Benjamin was not so much thinking

of a unification of man with God, but rather a reconciliation of the human with nature. And the means of reaching such a state are through the production of symbols.

This is the right place to introduce Benjamin's second central notion. The world of allegorical meanings of the modern big city is experienced by the flâneur who moves through its streets, squares, arcades, etc. It is by way of a politics of strolling that a modernist subjectivity of the city comes into existence.

The Benjaminean flâneur is clearly distinguished from Jean-Jacques Rousseau's solitary promenader and from Plato's Socrates. Rousseau's promenader, walking at the border between the urban space and the surrounding countryside, longs for solitude and is confused by the urban crowds. The flâneur's attitude towards the crowd is ambivalent. He longs for an identification with the urban crowds on the street and for being one with the collective around him, but suffers isolation.

The flâneur is not a person of dialogues as was Plato's Socrates. The flâneur has the idleness of Socrates, but does not have the addressees with whom Socrates was able to relate. The guarantees of the latter's idleness, slave labour, have gone in modernity. The modern flâneur is not provided the luxury of the 'vita contemplativa' on solitary promenades. Nor is

Temple, Tokyo

there a modern public in the urban space to whom the flâneur can offer his ideas.

The flâneur strolling around in the streets of the metropolitan cities exercises a post-metaphysical idea of philosophy outside the school and the academic space. Here we grasp a decisive transformation. Philosophy in the period of Enlightenment received its 'urbanity' – as Cassirer said – within the salons of Paris. A modern, and I would add, a postmodern idea of philosophy is exercised by the flâneur on the streets of the metropolitan city.

As a cultural worker, the flâneur communicates his experiences in the streets within the cultural apparatuses of journals, magazines and radio stations. This was already the case with the results of flânerie as seen by Benjamin, Siegfried Kracauer, Robert Park and Ernst Bloch.

In the final part of my essay I will add some additional features. What is important for me is the following conclusion, drawn from Benjamin: The split image of modernity, modernity's promises for social and individual emancipations, as well as modernity's failures, are coded in a special mode of the symbolic, the allegory. Allegories offer tentative and provisional meanings. They are to be brought in accord with the Messianic on the one hand and with social actions on the other hand. The sites where the allegories of modernity emerge are the spaces of metropolitan cities. The subject who experiences these allegories is the flâneur. The flâneur's experiences are part of a politics of strolling which relate these experiences with the cultural apparatuses.

Before reaching the final part of my essay I would like to mention some crucial issues in Henri Lefebvre's philosophy of the urban. Among the social philosophers of our days Lefebvre is one who relates the question of philosophy explicitly to the question of the city. Starting from a point of departure informed by Nietzsche as well as by Marx, Lefebvre sees all elements of urban spatiality as double-faced. The streets can be read – in the vein of Guy Debord and the Situationists as well as Constant's 'New Babylon' – as a place of encounter, as a meaningful spectacle, of which I am performer and spectator at the same moment. 'Use value' would then dominate 'exchange value'. The street as the performance of a ballet, as Jane Jacobs has put it. On the other hand, streets have

to be conceived in the vein of Herbert Marcuse as 'display' of 'commodities'. The urban spaces of today are colonised by the 'bureaucratically controlled consumer society' – Lefebvre coined this notion parallel to, but independent of, Marcuse's 'One-Dimensional Man' – in that 'images' and 'commercials' are part of an 'aesthetics' and 'ethics of consumer organisation.' The same concrete contradiction is valid for the monument. On the one hand a monument is repressive in that it expresses the praise of a ruler or state institution. On the other hand, however, Lefebvre agrees with Nietzsche that 'beauty and monumentality' belong together. Which means a monument, let's say a cathedral, always embodies a 'transcendence', by which it surmounts its socially repressive function.

Such considerations do not add anything totally new to reflections found already with Benjamin. But Lefebvre introduces a methodological clarity within a perspective informed by Marx and critical theory. With regard to an explicit concept of urban philosophy, first of all Lefebvre argues that the 'Marxist project of a revolution within the industrial revolution has to be completed by a revolution of the city.' Such an urban revolution would transform the universal process of urbanisation which has been brought into existence by the industrial society. But two methodological steps are necessary. First of all, philosophy has to stick to everyday life. Lefebvre calls this step the move from philosophy to 'meta-philosophy'. The dwelling becomes important strategically due to Hölderlin's saying that only as a poet does man reside on earth. Lefebvre reminds us of the Japanese custom of 'tokonoma'. The 'tokonoma' is a precious or a simple object which, according to the seasons, has to be placed within the corner of a house ... From Lefebvre's meta-philosophy it becomes understandable why the squatters' movement in Amsterdam, Berlin, Hamburg and elsewhere had such a convincing social power. The question of dwelling immediately turned out to be related to issues concerning social relationships between groups and within the sphere of material production.

The dwelling space as a private space is always related to the public space of a region or the global space of the state and its power. For a renewal of the sense of urbanity the dwelling space has a political and strategic priority.

Meta-philosophy, that is, philosophy's transformation through sticking to everyday life, is the first step in what Lefebvre calls the sublation of philosophy into social practice. Marx himself spoke of the realisation of philosophy by transforming or sublating philosophy's attitude towards totality into social practices. Lefebvre's philosophy of the urban starts with philosophy as a totalising project. This does not imply a return to the ancient Greek polis as the place of origin of philosophy, but rather it signifies philosophy's desire for a transcendence of fragmentarisation and devastation as brought about by the industrial city. As meta-philosophy, however, philosophy goes beyond its own limits in that it realises its own inspiration within the urban praxis (in the sense of Marx).

Lefebvre speaks of a 'right to the city', by which the exclusion of the proletariat from the centre of the city, codified by Haussmann's urbanism and renewed by Le Corbusier's functionalism, should be undone. Within Lefebvre's project of the realisation of philosophy through its transformation into the realm of urban space, the reference to works of art and philosophy are decisive. Art works can function as paradigms for the processes of appropriating time and space. Music for instance, makes use of counting, order and numerical measurements so as to express its lyricism. Parks, gardens and landscapes have been part of urban culture everywhere. 'Art can become praxis and poiesis on a social scale: the art of living in the city as work of art.' Lefebvre speaks of an 'ephemeral city', 'the perpetual oeuvre of the inhabitants, themselves mobile and mobilised for and by the oeuvre.' In a perspective deeply informed by Nietzsche, Lefebvre says that 'festival' (fête) signifies a realm beyond the reach of capital. And it is the urban space wherein 'La fête' is realised and performed:'The eminent use of the city, that is, of the streets and squares, edifices and monuments is la fête.'

Without doubt, this perspective is informed by Nietzsche. But we should not forget the Marxian perspective. Urban practices are based on the city as a productive power. First of all, the city is the place where all kinds of products are gathered and centralised in markets. Mumford used in this respect the notion of a container, the city as a container. There is the market of agrarian and industrial products, there is the labour market, there is the market of capital, there is the market of signs and symbols. There is the market of real estate.

The city is above all, as Lefebvre puts it, the power of synthesis and multiplication of the divergent and the disparate. But the most important thing is the appropriation of all products and their use. It is by use that the contract as the juridical coding of social use is surmounted.

Lefebvre says literally: 'In the urban the use comprises the customs and gives to the custom a priority with regards to the contract. The utilisation of urban objects (this sidewalk, this street, this passage, this illumination, etc.) is a usage and it is not contractually fixed and determined by the state.'

If the scope of this essay would allow it, one could illustrate these sentences by Mike Davis' book on Los Angeles. Even in the desert of this metropolis there exists a possible coalition, as Davis sees it, between Rap and the L.A. School of Fredric Jameson and Edward Soja, which Davis has called 'an emergent twenty-first century urbanism'.

VI

In this final part of my essay I would like to conclude with a few additions to the third attitude of modern urban philosophy, the line of argument starting with Simmel through the Chicago School, Benjamin and Lefebvre, and clarify some of the presuppositions implied here. It is a line of thought to which I myself subscribe.

First, following the Benjamin and Lefebvre line, we should structure the urban experience and the experience of urbanity along schemes of symbolic activities. Symbolic activities are beyond metaphysical hierarchies. There is no essential centre which would govern the periphery as being something purely accidental. That is to say, acts of symbolisation are by virtue of their own structure beyond metaphysics.

What Roland Barthes, in his book *Empire of Signs* (1982), found fascinating in Tokyo, a city similar to Los Angeles in that it has an empty centre, could be seen as a characteristic of a post-metaphysical experience of urbanity anywhere. Benjamin's stress on allegory as the main trait of urban experience would then be universalised with a cross-cultural end. Jinnai Hidenobu's book on Tokyo offers inside support to Barthes, for, according to him, Tokyo was indeed designed not as directed towards a centre but rather as a collage with independent areas on an equal footing yet in one way or another related to each other. Jinnai furthermore enriches the understanding of Benjamin's flânerie. His aim is to provide a historically informed reading of Tokyo from the Edo period on. He calls his approach a 'spatial anthropology' which means a reading of urban geography in its multi-layeredness. There are of course some constant structures, namely the relevance of an urban site as something following geographical givens such as the waterside and the mountains on the one hand, and the relevance of landmarks outside the city, like Mount Fuji-San on the other. Jinnai furthermore stresses the sites in front of a bridge as a locality providing a meeting place for youth and outsiders. Here the 'vulgar' is enabled to embody itself. But these urban sites have to be read and understood in terms of their meaning. According to Jinnai, this reading involves the reading of city maps, especially of former times, and the linkage of urban sites to the people's memories laid down in poetical narratives or in paintings.

But Benjamin's and Lefebvre's second notion also receives in Jinnai a new significance. Jinnai practices city walks in order to explore different areas and in order to grasp their specific geographical peculiarities, which in turn have to be read in the light of poetical narratives and the memories of the inhabitants there. This is clearly a return to Benjamin's flânerie, although in a totally different cultural context. There are different, even divergent politics of strolling but they all presuppose a body which walks and strolls around in search of the symbolic meaning of an urban space or site.

I should like to argue, then, that the search for a contemporary philosophical notion of the city has to imply a new idea of philosophy. After the linguistic turn in 20th century philosophy, a new philosophy is required, one which has its core in the symbolic and in symbolic activities. But there is a specific subjectivity, too, that needs to be involved here. We can understand urban subjectivity along with Benjamin, Lefebvre, Jinnai, but also in the line of Michel de Certeau

as an activity of walking, of passing by. De Certeau speaks of a rhetorics of walking and characterises it as centred in a style as a way of being in the world, as well as in a use of the symbolic.

In conclusion let me stress some aspects, which are crucial for a philosophy of urbanity and of urban culture. Such a philosophy would have to take as its core the notion of symbolic activity. This notion seems to be attractive because it involves at least two presuppositions. Symbolic activities are characterised by moves in different and even divergent or contradictory dimensions. The mythical is often merged with the poetical, the social and the political. In a word, the symbolic is always of a highly complex structure with allusions and multiple references. We have to think of how Benjamin and de Certeau deal with street names and the names of neighbourhoods, the poetics of geographies. Furthermore, symbolic activities as related to urban life consist in sudden and abrupt changes in perspectives. The processes of constituting a meaning do not run through just one per-spective, but shift to others which might contradict the first one. The perspectives taken are never totally exhausted, but rather are abandoned and replaced by the next ones.

Experiencing the urban implies movements of one's own body. We cannot neglect the rhythms of the moving body involved here. What Benjamin described as the intoxication of flânerie – leading to a kind of secular 'illumination' – and Lefebvre tried to cope with in studies concerning the experience of rhythms, are still highly significant for urban experiences today.

In rounding off my essay I should like to sum up my own position by stressing three points. 1) A contemporary notion of city life presupposes a post meta-physical, fragmented subjectivity. City life makes us aware of the multiple facets of human culture, including high and mass culture on an equal footing. Only a fragmented subjectivity would be able to cope with the complexities of the postmodern, indicated in the preeminence of fashion, the over-stylisations of street objects, the return of arcades, the aesthetisation of shopping malls and the rise of fast food culture. 2) City life implies a radical perspectivism of all world views and actual visual perceptions. The rapidly changing urban scenes require a flexibility and openness with

Shop, Tokyo

regard to the readings and rereadings of the symbolic meanings embodied in the urban spaces. 3) Symbolic meanings are related on the one hand to the phenomenology of bodily-centred experiences as well as to cultural inscriptions, the syntactic and semantic orderings of culture, on the other. The agent, however, of unending symbolic readings and rereadings is the moving and walking body which is culturally articulated within the frame of a politics of strolling, a flânerie. A politics of strolling does not consist just in walking around, but leads to ever more new images and new readings of the contemporary culture inscribed in ethnographies, reportages, essays, paintings and other cultural works and designs.

Benjamin already drew our attention to the fact that flânerie is not a given, in the sense that it could be equated with tourism or shopping, but that it is an eminent 'art' and 'learning' which emerged along with high modernism during the 19th century in urban metropolitan life. But a flânerie which is dismantled from its former male genderedness, as well as from its Eurocentricism, is still a crucial ingredient of today's urbanity.

39 NIGEL COATES

'Brief Encounters'

from Iain Borden, J. Kerr and J. Rendell with A. Pivaro (eds), *The Unknown City: Contesting Architecture and Social Space* (2000)

[. . .]

A city ought to be a place which encourages the acknowledgement of differences, not necessarily addressing particular bits to particular people, but at least asking questions, being playful enough to permit the other. How to reconcile the mass and substance of architecture with this kind of event is difficult, but any space will have components that can be used to amplify a sense of unexpected connection – and that can happen on lots of different scales. Our own office, for example, is an old industrial space. The rigid organisation of desks emphasise this, but we introduced blue sails, pushing the inside out and allowing the outside to come in. They are a parenthesis of the space that indicates a dynamic of observation, signs of movement, so as you walk through a building you feel a part of something else.

There is a way of generating a process of design that is to do with moving through space, and being a figure in a space. This can bring about quite different ideas that begin to modulate and therefore design that space. We don't work from the plan but from situations that are set up; we then start to interpret the situations from a pragmatic point while searching for what they allude to. In a small theatre project in Poland we wanted to add a vital sense of the present. By repeating the proscenium arch on the outside of the building – so that you enter through it – a component of the inside is imported to set up a tension in space and meaning. For me a narrative component is derived from the thing itself. In other projects the tension is created by the character of spaces rubbing up against another, where each space may be affiliated to very different things. In the

National Centre for Popular Music in Sheffield, the main museum is housed in four drums, and between them is a cross. The building comes together like a piece of machinery, the intention being that the cross is the connection to the streets and passages, a continuation of the genetic coding that forms the street pattern of the city. While the drums are more industrial in character, more static – you enter each of them as into a cave. It is a very simple idea, but it sets up a complex dynamic. While designing I didn't really understand this complexity – I was working intuitively. My hunch was that the enclosed qualities of the domes would be very appropriate for the museum, when off-set and emphasised by the open elevated, free-flowing cross shape.

This sort of duplicity in form, through what it feels like, makes the experience at once familiar and not familiar, somehow commonly unfamiliar. This leads to what *narrative* architecture is really about. It is not an architecture that tells stories, so much as an architecture that has additional fragments of choreography and insinuation that contradict the first order vocabulary. The museum and other current projects are more abstract with their narrative. Today, our work is getting cleaner. It doesn't have this sense of collage that our work may have had before, but the transformational mechanisms and illusions are there – just simplified. To strip down is a process of amplification, and the way things work on the mind and body are far more important than what they look like in a photo. This is more of a cinematic sensibility than one of the frozen image.

But in some of our earlier work there was an over-riding desire to assault the senses. For example, the Katharine Hamnett shop was very piled on – a plethora that doesn't seem to work anymore. Similarly, Jigsaw in Knightsbridge used the Italian palace as a narrative, which seemed to be the right concept for that part of town. In some way you had a feeling that the place was attracting you and involving you, distracting you, and turning around what you are actually there for – the clothes. Many people go into a shop, dive at the rails and won't have a clue where they are but they take the environment in peripherally. These shops were formed around the idea of places in the city that are stimulating but at the same time familiar. In some odd way you belong, which contrasted with straight chic or minimalism. This attitude comes from never seeing architecture as a cultural end in itself – it is always contingent. Not to see it as such is out of tune with the way people use the city – which is all sorts of signs threaded together from the car dashboard to the views from the top of a bus, hoardings, closed down shops, people . . .

REFERENCE, IDENTITY, CHANGE, MEANING

In Japan we did not set out for people to really understand our references. As we were just pillaging our own tools box, never intending our work to be read as one meaning, we were creating environments rich and stimulating enough to be interpreted differently by whoever went to them. In a curious way this sort of multiplicity was exactly right for Japan. People didn't really know what the bits meant, but they knew that there was something they liked about how they added up. We were one of few practices building in Japan who had looked around at the sense of what made a Japanese town – and took the mish-mash in. When many Japanese architects were going on about purity and Zen, creating little islands of contemplation, effectively turning their backs on the city – we enjoyed the fact that there were electrical cables hung festively along every street, massive zebra crossings, traffic then people switching, choreographed chaos.

As in London, the people and the traffic are the blood pulsing through the city. But many historic cities have had their centres cleaned up and "heritagised" – and so have had that very vital sense taken away, suppressing the chance

danger and unpredictable experience that makes the essence of the urban experience.

Duplicitous by nature, the city is something you can never know or understand completely, can never want to predict. Like us, the city constantly wrestles for control and the loss of it, always wanting something new to happen while wanting security to preside. This I see as an existential parallel to the practice of architecture. The way we are now is so very interesting in that the environment is becoming more and more contrived and controlled, but we are far more capable now of thinking laterally, of interacting in between our thoughts, expression and desires, of being attuned to a much more intangible environment, through information space, markets, clubs and music – even relationships are much less prescribed. I just think that all this can influence the culture that you work with in architecture and if it doesn't – if you try to tidy up – it falls back into the same old form.

Whilst in buildings as well as cities there must be components for orientation, there must also be a sense of getting lost – traditionally an anathema to architects who always strive to make things clear. I put my own identity in my work to a huge degree – there is a deliberate ambivalence in finding sources in things I like. If you are completely detached from what you do, and don't use your own experiences as a laboratory, then that touch isn't there. But at the same time I always undo that part, so that I can let go. Our work sends out different signals for different people. My sexual orientation is not fundamentally important in our work – except that maybe an understanding of duplicity provides something extra. There are components of the masculine and feminine in what I do – never meant specifically for men or women – that indicate a sense of evolution, towards confounding interpretation. This may not always be obvious, it may be that I'm as chauvinistic as Le Corbusier was, but that is not what I'm trying to do. I try to include elements of self-criticism and retreat.

Architecture is a public art, a setting up of frameworks which are never absolute in use or interpretation. Each project is different. I don't mind that some of the interiors in Japan no longer exist. Some things come together at a time, cause a stir, then conditions change. I think that is all part of the way cities evolve. What is important is to do with what the original project intended. Like Rachel Whiteread's *House* – we knew it was going to be demolished, and its passing reinforced what was important about it: it is a memory.

I never want to build monuments. I have an excitement for the way that places are, and therefore try to extrude what is there and then pile in narrative metaphors for what it is, building up a condition which isn't just read once but also has a sense of the way it is used and added to. There is a need for architects to bring together conflicting layers of signs, layer components that set off triggers, generate erotic conditions in space. But for me these conditions are always familiar – the hallucinating effect of night clubs, the way Soho has changed, ships and the Thames and its bank (HMS Belfast is the best building in London). They create a frisson in a place, in small scale and large scale with a constant switching of effects. The cultural role of architecture has huge potential, but people will become more interested in architectural expression only when it comes to parallel something intimate in their lives.

SECTION VIII

CULTURES AND ECOLOGY

INTRODUCTION

Ecological and environmental concerns have become central to recent debates about the city. This has stemmed from both the general rise of environmentalism in recent years and the specific recognition of the negative impacts of city design, patterns of urban life and production on the environment. It is commonly accepted that cities absorb an enormous amount of the world's non-renewable resources and contribute greatly to the world's pollution and toxic emissions. Further, daily patterns of urban life necessitated by the decentralized urbanization characteristic of twentieth-century cities in the developed world, particularly daily commuting from suburbs to city centres, is a major threat to their sustainability. Ecology has become a new paradigm, fundamental to urban thought, and one which has been advocated as an underpinning logic to future urban design.

However, it is crucial to recognize that these debates are not *acultural* or separate from the concerns of this book. Such issues are crucial to any understanding of urban culture. This is so for three reasons. First, as one contributor to this section, Victor Papanek, has recognized: 'Ecology and the environmental equilibrium are the basic underpinnings of all human life on earth; there can be neither life nor human culture without it.' Put simply ecological imbalance and degradation is a threat to urban culture. Second, urban cultures, of whatever kinds, must be recognized as having differential ecological impacts. Despite the failure of previous commentary on urban culture to recognize this, the patterns of living characteristic of different cultures have specific impacts on the ecological balance of localities and the environment. No urban culture is ecologically neutral. Finally, ecological thought represents a culture, or rather a series of cultures, in itself. It may seem a strange thing to say, but thinking ecologically, or about the environment in whatever way, is not natural, rather, it is a human, cultural construction. The ways we think about ecology and the environment reflect (and affect) the state of the environment, but they also reflect human perception and more generally wider culture trends. When we think about the environment we cannot step outside of our cultural contexts. Ecological thought is a cultural construction like any other facet of human thought, and one we can interrogate in the same ways as we do other facets of human thought.

There are three key aspects of recent debates on cities, city ecologies and the environmental impacts of urban development. These are reflected in the contributions in this section. First, there is the question of the negative impact of cities on the environment. Since the Industrial Revolution the impact of cities on the environment has become increasingly and disturbingly apparent. Research has demonstrated, for example, the exponential rise in the emission of CFCs and other toxins from industry over the course of the twentieth century. Environmental problems have become more apparent, both within scientific research agendas and in the landscape. The blame for problems such as acid rain and global warming has been traced squarely to cities and to the industrial production systems which have fuelled their development and which dominant concerns argue are necessary for their maintenance. Ongoing political debates about international regulation of pollution and toxic emissions are, implicitly at least, economic and urban as well as ecological. The whole question of sustainability is increasingly central to debates about cities, their forms, economies, transport networks, industries, patterns of life and building design. This is an issue exemplified by Victor Papanek's contribution from *The Green Imperative* (1995, Thames and Hudson). Papanek considers the issue of design in packaging, a major source of non-recyclable waste with harmful effects on city, and wider, ecologies. Much of the extract is a polemical attack on the wasteful quantities and qualities of product packaging and their effects on the environment, which is followed by an exploration of recyclable and organic alternatives. Such advocacy of ecological considerations is typical of much recent writing on the ecological implications of urbanization and urban society. The problem is effectively demonstrated and practical alternatives and appropriate policies outlined.

While polemic, Papanek's contribution can be read as more than this. During this discussion Papanek reveals the ways in which culture is intermeshed with these issues. The packaging of products serves a number of needs. The obvious need it serves is functional, to protect products from damage, decay and vermin. However, Papanek argues packages are also cultural products that serve a number of needs beyond the functional. These needs include product differentiation, advertising and imbuing meaning into essentially meaningless products, elaborate packaging instilling distinction and 'class' into overpriced perfume for example. Much non-renewable resource and non-recyclable material is used in fulfilling the cultural needs addressed by product packaging. As Papanek's extract demonstrates, the problems of ecological sustainability derive from cultural as well as functional issues.

A second aspect of debates on cities, cultures and ecology is the question of how attitudes to the environment, ecology or nature have shaped city form or the spaces of the city. An aspect of recent debates, sparked off by the realization of the damage caused by carbon monoxide emissions by cars, has been what is ideal city form in ecological terms. There have been numerous calls for a return to compact cities characterized by mixed uses, for example. The decentralized urbanization characteristic of twentieth-century, developed-world cities has been fuelled by capital's relentless search for investment opportunities, making suburban development profitable, and facilitated by the development of the car. The effects of this have been predominantly mono-functional zoning, the separation of work from home, and the generation of unsustainable patterns of daily life (especially commuting). By contrast, advocates of the compact city have argued that its form and the mix of uses within zones negate the necessity for such damaging patterns of living. The following two extracts, in different ways, outline and advocate ecological alternatives to the damaging urbanization processes that have dominated in the twentieth century.

The problems of producing urban spaces, for example in the form of housing, are acute in non-affluent countries. Egyptian architect Hassan Fathy, in a text derived from his speech to the Right Livelihood Awards in Stockholm in 1980, argues for a 'no-cost' architecture to replace the import of expensive and ecologically destructive technologies of the industrialized countries. Fathy's work revived traditional skills of building in mud brick, as at the village of New Qurna, which he planned and designed in the 1940s, and which is now a successful habitat. Although Fathy's example has not been widely followed in his own country, it offers much from which the affluent world might learn. William McDonough's contribution 'The Hannover Principles'

(1997) succinctly outlines a set of principles that place ecology firmly as the central concern that should drive urban design and architecture, a long overdue call for sustainability and ecology to guide the production and shaping of urban space.

Finally, different cultural attitudes towards the environment or nature have shaped the ways that nature is presented in the city. Nature is never just 'there', it is always shaped to particular cultural ends. The neatly manicured lawns and carefully tended flower beds of civic parks take the form they do for cultural reasons. This landscape addresses issues of aesthetic appreciation, which are different within and between cultures, and helps project an ideology of civic virtue and value. The shapes and forms of the landscapes of nature in the city are ideological. This is something picked up in the extract by Michael Hough from *Cities and Natural Process* (1995, Routledge). In this extract Hough draws attention to the importance of the neglected 'fortuitous' landscape of the city, a landscape of weeds and wildlife found in the forgotten and ignored spaces of the city – 'cracks and gratings in the pavement, on rooftops, walls, poorly drained industrial sites or wherever a foothold can be gained' – and in the eclectic human landscapes of ethnic neighbourhoods. Hough argues that attention to the manicured nature of civic landscapes represents a neglect of the richest urban landscapes ecologically and culturally and instead celebrates a landscape that is both ecologically poor and ideologically exclusive. From here Hough goes on to criticize planning and urban design for their admission of utopian and functional values above those of nature and ecology.

Christian Norberg-Schulz' (1976) 'The Phenomenon of Place', the first extract in this section, gives the broadest notion of ecology. Norberg-Schulz reflects other humanistic thinkers in arguing that the notion of place, encompassing the relationship between the natural and the manmade, is of fundamental importance to our existence. He argues, reflecting Heidegger, that place is more than location, and escapes functionalist attempts at description. Others have emphasized that place is a repository of feeling or 'a centre of felt value' (Yi-Fu Tuan, 'Humanistic Geography', *Annals of the Association of American Geographers*, 1976). In this article Norberg-Schulz takes us through an approach – phenomenology – for understanding place and uncovering its meanings and significance to everyday existence. Very much characteristic of the humanist approach that swept a significant strand of the social sciences in the 1970s, such approaches have come in for some severe criticisms subsequently for their tendency towards essentialism and their failure to engage critically with social structures. However, Norberg-Schulz' contribution remains of value in demonstrating that failure to attend to ecology as a paradigm of urban thought poses threats existentially perhaps before it does so physically.

40 CHRISTIAN NORBERG-SCHULZ

'Place'

from *AA Quarterly* (1976)

THE PHENOMENON OF PLACE

Our everyday life-world consists of concrete 'phenomena'. It consists of people, of animals, of flowers, trees and forests, of stone, earth, wood and water, of towns, streets and houses, doors, windows and furniture. And it consists of sun, moon and stars, of drifting clouds, of night and day and changing seasons. But it also comprises more intangible phenomena such as feelings. This is what is 'given', this is the 'content' of our existence. Thus Rilke says: 'Are we perhaps *here* to say: house, bridge, fountain, gate, jug, fruit tree, window, – at best: Pillar, tower ...' Everything else, such as atoms and molecules, numbers and all kinds of 'data', are abstractions or tools which are constructed to serve other purposes than those of everyday life. Today it is common to mistake the tools for reality.

The concrete things which constitute our given world are interrelated in complex and perhaps contradictory ways. Some of the phenomena may for instance comprise others. The forest consists of trees, and the town is made up of houses. 'Landscape' is such a comprehensive phenomenon. In general we may say that some phenomena form an 'environment' to others. A concrete term for environment is *place*. It is common usage to say that acts and occurrences *take place*. In fact it is meaningless to imagine any happening without reference to a locality. Place is evidently an integral part of existence. What, then, do we mean with the word 'place'? Obviously we mean something more than abstract location. We mean a totality made up of concrete things having material substance, shape, texture and colour. Together these things determine an 'environmental character', which is the essence of place. In

general a place is given as such a character or 'atmosphere'. A place is therefore a qualitative, 'total' phenomenon, which we cannot reduce to any of its properties, such as spatial relationships, without losing its concrete nature out of sight.

Everyday experience moreover tells us that different actions need different environments to take place in a satisfactory way. As a consequence, towns and houses consist of a multitude of particular places. This fact is of course taken into consideration by current theory of planning and architecture, but so far the problem has been treated in a too abstract way. 'Taking place' is usually understood in a quantitative, 'functional' sense, with implications such as spatial distribution and dimensioning. But are not 'functions' inter-human and similar everywhere? Evidently not. 'Similar' functions, even the most basic ones such as sleeping and eating, take place in very different ways, and demand places with different properties, in accordance with different cultural traditions and different environmental conditions. The functional approach therefore left out the place as a concrete 'here' having its particular identity.

Being qualitative totalities of a complex nature, places cannot be described by means of analytic, 'scientific' concepts. As a matter of principle science 'abstracts' from the given to arrive at neutral, 'objective' knowledge. What is lost, however, is the everyday life-world, which ought to be the real concern of man in general and planners and architects in particular. Fortunately a way out of the impasse exists, that is, the method known as *phenomenology*. Phenomenology was conceived as a 'return to things', as opposed to abstraction and mental constructions. So far phenomenologists have been mainly concerned with ontology, psy-chology, ethics and to some extent aesthetics, and have given relatively little attention to the phenomenology of the daily environment. A few pioneer works however exist but they hardly contain any direct reference to architecture. A phenomenology of architecture is therefore urgently needed.

Some of the philosophers who have approached the problem of our life-world, have used language and literature as sources of 'information'. Poetry in fact is able to concretize those totalities which elude science, and may therefore suggest how we might proceed to obtain the needed understanding. One of the poems used by Heidegger to explain the nature of language, is the splendid *A Winter Evening* by Georg Trakl. The words of Trakl also serve our purpose very well, as they make present a total life-situation where the aspect of place is strongly felt:

A Winter Evening
  Window with falling snow is arrayed,
Long tolls the vesper bell,
The house is provided well,
The table is for many laid.
  Wandering ones, more than a few,
Come to the door on darksome courses,
Golden blooms the tree of graces
Drawing up the earth's cool dew.
  Wanderer quietly steps within;
Pain has turned the threshold to stone.
There lie, in limpid brightness shown,
Upon the table bread and wine.

We shall not repeat Heidegger's profound analysis of the poem, but rather point out a few properties which illuminate our problem. In general, Trakl uses *concrete* images which we all know from our everyday world. He talks about 'snow', 'window', 'house', 'table', 'door', 'tree', 'threshold', 'bread and wine', 'darkness' and 'light', and he characterizes man as a 'wanderer'. These images, however, also imply more general structures. First of all the poem distinguishes between an *outside* and an *inside*. The *outside* is presented in the first two lines of the first stanza, and comprises *natural* as well as *man-made* elements. Natural place is present in the falling snow which implies winter, and by the evening. The very title of the poem 'places' everything in this natural context. A winter evening, however, is something more than a point in the calendar.

As a concrete presence, it is experienced as a set of particular qualities, or in general as a *Stimmung* or 'character', which forms a back-ground to acts and occurrences. In the poem this character is given by the snow falling on the window, cold, soft and soundless, hiding the contours of those objects which are still recognized in the approaching darkness. The word 'falling' moreover creates a sense of *space*, or rather: an implied presence of earth and sky. With a minimum of words, Trakl thus brings a total natural environment to life. But the outside also has man-made properties. This is indicated by the vesper bell, which is heard everywhere, and makes the 'private' inside become part of a comprehensive, 'public' totality. The vesper bell, however, is something more than a practical man-made artifact. It is a symbol, which reminds us of the common values which are at the basis of that totality. In Heidegger's words: 'The tolling of the evening bell brings men, as mortals, before the divine'.

The *inside* is presented in the next two verses. It is described as a house, which offers man shelter and security by being enclosed and 'well provided'. It has, however, a window, an opening which makes us experience the inside as a complement to the outside. As a final focus within the house we find the table, which 'is for many laid'. At the table men come together, it is the *centre* which more than anything else constitutes the inside. The character of the inside is hardly told, but anyhow present. It is luminous and warm, in contrast to the cold darkness outside, and its silence is pregnant with potential sound. In general the inside is a comprehensible world of *things*, where the life of 'many' may take place.

In the next two stanzas the perspective is deepened. Here the *meaning* of places and things comes forth, and man is presented as a wanderer on 'darksome courses'. Rather than being placed safely within the house he has created for himself, he comes from the outside, from the 'path of life', which also represents man's attempt at 'orienting' himself in the given unknown environment. But nature also has another side: it offers the grace of growth and blossom. In the image of the 'golden' tree, earth and sky are unified and become a *world*. Through man's labour this world is brought inside as bread and wine, whereby the inside is 'illuminated', that is, becomes meaningful. Without the 'sacred' fruits of sky and earth, the inside would remain 'empty'. The house and the table receive and gather, and bring the world 'close'. *To dwell in a house therefore means to inhabit the world.* But this dwelling is not easy; it has to be reached on dark paths, and a threshold separates the outside from the inside. Representing the 'rift' between 'otherness' and manifest meaning, it embodies suffering and is 'turned to stone'. In the threshold, thus, the *problem* of dwelling comes to the fore.

Trakl's poem illuminates some essential phenomena of our life-world, and in particular the basic properties of place. First of all it tells us that every situation is local as well as general. The winter evening described is obviously a local, nordic phenomenon, but the implied notions of outside and inside are general, as are the meanings connected with this distinction. The poem hence concretizes basic properties of existence. 'Concretize' here means to make the general 'visible' as a concrete, local situation. In doing this the poem moves in the opposite direction of scientific thought. Whereas science departs from the 'given', poetry brings us back to the concrete things, uncovering the meanings inherent in the life-world.

Furthermore Trakl's poem distinguishes between natural and man-made elements, whereby it suggests a point of departure for an 'environmental phenomenology'. Natural elements are evidently the primary components of the given, and places are in fact usually defined in geographical terms. We must repeat however, that 'place' means something more than location. Various attempts at a description of natural places are offered by current literature on 'landscape', but again we find that the usual approach is too abstract, being based on 'functional' or perhaps 'visual' considerations. Again we must turn to philosophy for help. As a first, fundamental distinction Heidegger introduces the concepts of 'earth' and 'sky', and says: 'Earth is the serving bearer, blossoming and fruiting, spreading out in rock and water, rising up into plant and animal ...' 'The sky is the vaulting path of the sun, the course of the changing moon, the glitter of the stars, the year's seasons, the light and dusk of day, the gloom and glow of night, the clemency and inclemency of the weather, the drifting clouds and blue depth

of the ether ...' Like many fundamental insights, the distinction between earth and sky might seem trivial. Its importance however comes out when we add Heidegger's definition of 'dwelling': 'The way in which you are and I am, the way in which we humans *are* on the earth, is dwelling ...' But 'on the earth' already means 'under the sky'. He also calls what is *between* earth and sky *the world*, and says that 'the world is the house where the mortals dwell'. In other words, when man is capable of dwelling the world becomes an 'inside'.

In general, nature forms an extended comprehensive totality, a 'place', which according to local circumstances has a particular identity. This identity, or 'spirit', may be described by means of the kind of concrete, 'qualitative' terms Heidegger uses to characterize earth and sky, and has to take this fundamental distinction as its point of departure. In this way we might arrive at an existentially relevant understanding of *landscape*, which ought to be preserved as the main designation of natural places. Within the landscape, however, there are subordinate places, as well as natural 'things' such as Trakl's 'tree'. In these things the meaning of the natural environment is 'condensed'.

The man-made parts of the environment are first of all 'settlements' of different scale, from houses and farms to villages and towns, and secondly 'paths' which connect these settlements, as well as various elements which transform nature into a 'cultural landscape'. If the settlements are organically related to their environment, it implies that they serve as *foci* where the environmental character is condensed and 'explained'. Thus Heidegger says: 'The single houses, the villages, the towns are works of building which within and around themselves gather the multifarious in-between. The buildings bring the earth as the inhabited landscape close to man, and at the same time place the closeness of neighbourly dwelling under the expanse of the sky'. The basic property of man-made places is therefore concentration and enclosure. They are 'insides' in a full sense, which means that they 'gather' what is known. To fulfill this function they have openings which relate to the outside. (Only an *inside* can in fact have openings.) Buildings are furthermore related to their environment by resting on the ground and rising towards the sky. Finally the man-made environments comprise artifacts or 'things', which may serve as internal foci, and emphasize the gathering function of the settlement. In Heidegger's words: 'The thing things world', where 'thinging' is used in the original sense of 'gathering', and further: 'Only what conjoins itself out of world becomes a thing'.

Our introductory remarks give several indications about the *structure* of places. Some of these have already been worked out by phenomenologist philosophers, and offer a good point of departure for a more complete phenomenology. A first step is taken with the distinction of natural and man-made phenomena. A second step is represented by the categories of earth–sky (horizontal–vertical) and outside–inside. These categories have spacial implications, and 'space' is hence re-introduced, not primarily as a mathematical concept, but as an existential dimension. A final and particularly important step is taken with the concept of 'character'. Character is determined by *how* things are, and gives our investigation a basis in the concrete phenomena of our everyday life-world. Only in this way we may fully grasp the *genius loci*; the 'spirit of place' which the ancients recognized as that 'opposite' man has to come to terms with, to be able to dwell. The concept of *genius loci* denotes the essence of place.

41 HUSSAN FATHY

'Palaces of Mud'

from *Resurgence* (1984)

We need a new way of knowledge. The enforced academic knowledge of schools has alienated us from nature just as industrialisation by force has taken away the possibilities of our participating in satisfying our needs. We have only ready-made solutions, prefabricated ideas to be carried out. In the fields of life which need a high cash outlay, like housing, we have been cut off from solving our problems by using our own hands and own potential. We have been integrated into the cash economy. By this integration we have imposed on the poor the cash economy without the cash. The annual income per capita in the Third World is between £25 and £30. How can someone with such an income hire an architect and a contractor to build his house for him with industrialised materials which need cash? Imposing the cash economy on these people has created a class I call the economic untouchables, because they cannot be integrated into the cash economy and have been deprived from doing anything themselves. The effect of this is that, according to U.N. statistics, 20 years ago there were 800 million people in the Third World doomed to a premature death because of bad housing alone, not to mention nutrition and other needs. I thought that this figure must have surpassed the billion by now. But they tell me it is 'only' 900 million.

The system prevailing now is the architect/contractor system by which the owner has been completely set aside as the architect designs and the contractor builds. To solve the problem of the 900 million we have to have not low cost housing but no cost housing. We must subject technology and science to the economy of the penniless, the people, instead of the other way round. This is the role of the conscious modern architect, this is our great responsibility. Up to

now many governments and international organisations have tried to solve the problem by trying to find some means of reducing the costs of building and of the industrialised materials. They have produced what they call aided self-help for the Third World, by providing concrete shakers and vibrators to make prefabricated panels for building. But after 20 years of experimenting with this system they have had to confess that it does not work. Because a man with an income of £25 a year cannot afford any industrialised materials like cement and concrete. The problem is not in the shaker and the vibrator, the problem is what to shake and vibrate. We have to rely on the materials we have and can afford to have, on our labour, on our own hands, on what we find under our feet. Nature itself has provided the solution. The cave man noticed after threshing corn that straw mixed with earth makes big lumps which hold together. The earth molecules do not hold together enough so we have to have a stabilising factor. The straw mixed with mud at harvest time showed man how to make mud bricks, Adobe, to build walls. When he came to the problem of roofing primitive man used timber or other materials. But timber was not always at hand.

In Iran, Egypt, Libya and Tunisia they found a solution. If you build a boat, every ring is pulling on the other, and it is working entirely under tension. There is no compression or it would crumble. If you reverse this upwards it will be working entirely under compression. Mud brick can take compression but not tension. They invented a system to build roofs with just bricks end to end, leaning the vertical a little against an end wall so that the brick is on an inclined plane. The sticking power is the weight of the brick multiplied by the cosign of the angle divided by the area of the brick. They found that they had to have very light adobe bricks, 25 cm. by 15 cm. and only 5 cm. thick.

Once all this is recognised, modern science can help by giving us the qualities of mud, the physical, mechanical qualities and so on, and even solve the problem that mud brick does not last long in more humid areas through stabilisation with bituminous emulsions.

Those who want to play with mud brick ought to be a trio. The cellist would be the soil engineer deeply in tune with the vibrations of the soil. The violinist, highly strung, would be the structural engineer. The architect would be the conductor. We can use mud-brick adobe, which costs nothing except the hand labour, with the same security as steel and concrete. We have examples which have lasted from antiquity. Bolting goes back in history to the very earliest period. The first example I know in Egypt is from the Third Dynasty, something like 5,000 years ago. They used the parabolic bolt as a centering and built an arch. We have another example in the oasis in Upper Egypt. In the 4th century A.D. the Christians were persecuted by the Romans and a group fled into the desert. They had nothing, only what was under their bare feet, but they built something like 250 structures, all vaulted with domes, using mud brick from under their feet. These models are still standing. The 'experts' say that mud brick would not last and the maintenance costs would be astronomical, due to the fragility of the mud. But these buildings have been studied by architects and engineers.

To my mind the value of any project, any idea, lies in the answer to the question: is it for people or for politics, economics, etc? When we think about housing people, we have to think about the quality of life of the people we are serving. For example, when you have 20 people sleeping in one room the airing requirements are different. We have to consider the aesthetic factor. When they were working with their own hands men used to beautify everything they made. Even if it was a war ship it was carved with the most fantastic designs because man was interacting with the wood. But machinery does not care for beauty.

It takes time for certain changes to show their effects. If we could jump from the sixth floor and our legs would not break until six months later, we would have many people with broken legs because they would not associate cause and effect! Some of the mistakes we commit need time to reveal themselves. If a family of five can farm five acres, and somebody gives them a tractor and a mechanical plough so they can farm twenty times five acres, it is seen as progress. But we have dispensed with 19 other families. What are they going to do for a living? We have dispensed with the plough carpenter, the village weaver and all the crafts that were being satisfied in the village. The tractor does not eat from the ground like the cow, or give milk, nor does it give any manure, only poisonous gases. It needs fuel and spare parts and changes the economy of the countryside. God created man in Nature, surrounded by plant and animal life. In our cities we have only asphalt, steel, aluminium and concrete. The best material you can surround yourself with when considering cosmic radiation is wood. The worst is concrete which stops the beneficial radiation. Water is affected by the cosmic rays coming from the moon and as our bodies are almost all liquid, all water, they are affected too. But we never think about these things. Modern man has lost this cosmic consciousness. The cathedrals of France were built on the geographical area reflecting the sign of the Virgin in the sky reflected on earth. Why? We are part of a system. If I integrate myself into the system all the elements in the system will come to help me. If I cut my finger all the elements of my body will come to heal it. But if my finger were isolated it would never heal.

How do we go from the architect/contractor system to the architect-owner/builder system? This needs quite a change in the relationships between the people concerned. In the communities with £25 per head income, nothing can explain their remaining alive, unless they live outside the cash economy, depending mainly on co-operation. One man cannot build a house, but ten men can build ten houses very easily, even a hundred houses. We need a system that allows the traditional way of co-operation to work in our society. I cannot co-operate in a city if the moment I get out of the door I am launched into the anonymity of millions. We must create new neighbourhoods where I build for you and you build for me (i.e. I will have the same help from you when I come to build my house).

What a waste of energy not to use our muscles properly for building, for culture, for modelling, beauty! When I think of the energy wasted on football: take a ball and run after it and have a goal and finito ... If we could only make the millions have the same interest in construction as in football!

Instead of hatred and destruction we would have love and construction, because construction itself has that impact. Every act, anything we do has an impact on our basic nature. In the old societies they used to have the temple architecture reflecting the sky. So when the sun changed signs they would dismantle the temple and rebuild it according to the new measurements and directions! What is our standard of reference when we build? The findings of modern science, of physics? But we do not even take those into account. We put huge windows in the 'modern' houses we build in the desert nowadays, each one letting in thousands of kilo calories of heat an hour. They need a lot of air conditioning, a lot of cash, and when this cash runs out what will the people do with their houses and with themselves?

We must subject technology and science to the economy of the poor and penniless. We must add the aesthetic factor because the cheaper we build, the more beauty we should add to respect man. When man built on his own he used to beautify everything with his own hands. When architects build for the poor what do we give them from the aesthetic point of view? There is a book called *Architecture without Architects*. When I see the present architecture, the regular architecture, I don't know which is which. Which is the architecture with or without the architect? Because we have over-simplified and over-reduced our efforts. The modern house is the paid portrait of the owner. When we are designing for the rich we take care of the aesthetic factor, the functional and the demographic. But when we design for the 900 million, we design one house and have it multiplied by the million in Europe as it is in Africa, as it is in India and everywhere, because we are using concrete and concrete does not allow any manipulation of space or articulation of the material.

I would like to introduce in our villages and our cities musicality and harmonics. The eye physiologically does not see more than one point at a time and sends these to the brain, one point after the other. We hear music, one note after the other, and have the melody in our brain. We have the image in our brain. It happens very quickly so we think it is instantaneous but it is not. When I look around the room my eyes go round the lines. If they are harmonic I feel happy. But if they are hectic I feel nervous, but do not know why. I wish that the eye would suffer like the ear and when it sees ugliness become red and have tears! Unconsciously we feel the dissonance.

The material is amorphous, neutral. With half a cubic metre of clay, Rodin made the Thinker. The palaces of the Pharaohs were all in mud brick. In New Mexico we have a style of architecture all in mud brick from the time of the Indians. In Iran they have used a most interesting technique. I have seen a village school built in adobe covered by three vaults, one next to the other, to catch the breeze. The span was 6 metres! By combining the modern science of soil mechanics and structures with the skills of master masons we can have such vaulting in millions of houses. Instead, in hot humid zones, we get corrugated iron roofs that have to be paid for in cash and are not insulated from heat. We once invited all the architects and engineers in Egypt to present ideas for rural housing. Model buildings were put up in the grounds of the building research centre in Cairo. There was one entirely prefabricated, ultra-modern, and one in mud brick. Air temperatures in the prefabricated house were 7 degrees centigrade higher than in the mud-brick one in April. The temperature in the mud-brick model didn't fluctuate more than 2 degrees in 24 hours and never came out of the acceptable temperature zone. In the ultra-modern concrete model the temperature didn't enter into that temperature zone except during one hour in the morning and one hour in the evening. It was at times even higher than the outside temperature! So this 'modern' house ignored the findings of modern physics, aerodynamics, sociology, social psychology, physiology and so on. If you want to be modern, you have to consider all these sciences. In architecture the human sciences are the most important.

This is what we mean by Right Livelihood. God has not changed the design of the face of man, having the nose above the mouth or in the back of the neck, just to be modern. When God created man out of mud brick he asked the angels to bow down to Adam. They all bowed down except Satan who wanted God to make man out of concrete! Because we have reached the moon we think we can discard our physique, our values, our traditions, our nature. But even the astronauts have egg and bacon for breakfast. They don't have something from space to feed on.

Speech given at the Right Livelihood Award Ceremony, 1980, Stockholm.

42 WILLIAM McDONOUGH

'The Hannover Principles'

from C. Jencks and K. Kropf (eds) *Theories and Manifestos of Contemporary Architecture* (1997)

1 Insist on rights of humanity and nature to co-exist in a healthy, supportive, diverse and sustainable condition.

2 Recognize interdependence. The elements of human design interact with and depend upon the natural world, with broad and diverse implications at every scale. Expand design considerations to recognize even distant effects.

3 Respect relationships between spirit and matter. Consider all aspects of human settlement including community, dwelling, industry, and trade in terms of existing and evolving connections between spiritual and material consciousness.

4 Accept responsibility for the consequences of design decisions upon human well-being, the viability of natural systems, and their right to co-exist.

5 Create safe objects of long-term value. Do not burden future generations with requirements for maintenance or vigilant administration of potential danger due to the careless creation of products, processes, or standards.

6 Eliminate the concept of waste. Evaluate and optimize the full life-cycle of products and processes, to approach the state of natural systems.

7 Rely on natural energy flows. Human designs should, like the living world, derive their creative forces from perpetual solar income. Incorporate this energy efficiently and safely for responsible use.

8 Understand the limitations of design. No human creation lasts forever and design does not solve all problems. Those who create and plan should practice humility in the face of nature. Treat nature as a model and mentor.

9 Seek constant improvement by the sharing of knowledge. Encourage direct and open communication between colleagues, patrons, manufacturers, and users to link long-term sustainable considerations with ethical responsibility, and re-establish the integral relationship between natural processes and human activity . . .

43 VICTOR PAPANEK

'Designing for a Safer Future'

from *The Green Imperative* (1995)

Ecology and the environmental equilibrium are the basic underpinnings of all human life on earth; there can be neither life nor human culture without it. Design is concerned with the development of products, tools, machines, artefacts and other devices, and this activity has a profound and direct influence on ecology. The design response must be positive and *unifying*. Design must be the bridge between human needs, culture and ecology.

This can be clearly demonstrated. The creation and manufacture of *any* product – both during its period of active use and its existence afterwards – fall into at least six separate cycles, each of which has the potential for ecological harm.

When we speak of pollution as related to products, we usually think of end results: the exhaust fumes from automobiles, the smoke from factory chimneys, chemical fertilizers or truck tyres in a dump poisoning the groundwater. But pollution falls into several phases.

PRODUCTION AND POLLUTION

1. The choice of materials

The materials chosen by designer and manufacturer are crucial. Mining metal for cars creates atmospheric pollution, and uses oil and petrol, thus wasting natural resources that cannot be replaced. The designer's decision to use foam plastics to make cheap, throw-away food containers damages the ozone layer. This is *not* a prescription for doing nothing at all, but an attempt to make designers aware that every choice and dilemma in their work can have far-reaching and long-term ecological consequences.

2. The manufacturing processes

The questions facing the designer are: Is there anything in the manufacturing process itself that might endanger the workplace or the workers, such as toxic fumes or radio-active materials? Are there air-pollutants from factory smoke-stacks, such as the gases that cause acid rain. Are liquid wastes from the factory leaking into the ground and destroying agricultural land or – worse still – entering the water supply?

3. Packaging the product

Further ecological choices face the designer when developing the package in which the product is transported, marketed and distributed. Foam plastics, which pose acute dangers to the ecological balance, are used by designers as a protection for fragile products. It is now known that propellants (such as CFCs) for lacquer sprays and other products are directly implicated in the depletion of the ozone layer. Considerations of materials and methods are therefore crucial in the packaging phase of ecologically aware design.

4. The finished product

There are too many different versions of the same item available in many cases. Since the manufacture of most industrial or consumer products uses up irreplaceable raw materials, the profusion of objects in the market-place constitutes a profound ecological threat. To give a typical example: in western Europe, Canada, Japan and the USA there are now more than 250 different video cameras available to consumers; the differences between them are minimal – in some cases they are identical but for the name-plate. The choice of consumer products in the West is highly artificial.

Other products threaten the ecological balance even more directly. Snowmobiles, which are largely sold as winter-sports and recreation equipment, are so noisy that when they go into roadless terrain they destroy breeding grounds and habitats. Yet, at the same time, they have assumed an important role in hunting and herding cycles and are now important tools for survival among the Inuit of Canada and Alaska. 'Off-road' vehicles and 'mountain bikes' affect the precious layer of topsoil and humus that can grow crops. 'Dune buggies' harm the sand-dune layers at the critical edge between ocean and land.

5. Transporting the product

The transporting of materials and products further contributes to pollution by the burning of fossil fuels, and by the necessity for a whole complex of roads, rails, airports and depots. There is transportation from the mill to the factory, the factory to the distribution centre, from there to the shops and, eventually, to the end-user.

6. Waste

Many products can have negative consequences *after the useful product life is over.* One only has to see the huge automobile graveyards in many countries to understand that these vast amounts of rusting metals, decaying paints and shellacs, deteriorating plastic upholstery, leaking oils and petrol are leaching directly into the ground, poisoning the soil, the water-supply and the wildlife, besides visually destroying the land-scape. It has been estimated that the average family in the technologically developed countries throws away some 16 to 20 tons of garbage and waste a year. This is not only an environmental hazard, but is also an enormous waste of materials that could be recycled responsibly. This is one area in which the so-called Third World countries are leading the way – because of material scarcities, recycling is an accepted way of life there and has been for generations.

PRODUCT ASSESSMENT

The relationship between design and ecology is a very close one, and makes for some unexpected complexities. The designed product goes, as shown, through at least *six* potentially eco-logically dangerous phases. Product Life Cycle Assessment is the evaluation incorporating all of them, from the original acquisition of raw materials, through the manufacturing process and assembly, the purchase of the complete product (which also includes shipping, packaging, advertising and the printing of instruction manuals), the use, the collection of the product after use, and finally the re-use or recycling and final disposal. It can best be understood through the hexagonal diagram, the six-sided 'Function Matrix' (Figure 1). At the moment Life Cycle Assessment is very new, and can be profoundly complicated, demanding a great deal of study, testing and experimentation.

Environmental issues in Life Cycle Assessment
- The exhaustion of scarce or finite resources
- The production of greenhouse gases
- The production of chlorofluorocarbons lead-ing to ozone depletion
- The production of acid rain
- Habitat destruction and species extinction
- Materials or processes that harm plants, animals and humans
- Air, soil and water pollution
- Noise pollution with its deleterious effect on the human psyche
- Visual pollution

PACKAGING AND SHROUDING

Most goods need to be packaged. The package protects the contents in transit and in store from spoilage, vermin, moisture and damage. It can serve as a powerful marketing tool through design, colour and texture. Furthermore, as explored in Chapter 7, it will frequently signify not only the contents, but also lend identity to the product-line. In terms of goods that are nearly identical – washing-up powders, breakfast cereals or cigarettes – it can be said that *the package is the product.*

It is clear that we routinely over-package things. In some cases this is to lend a visual

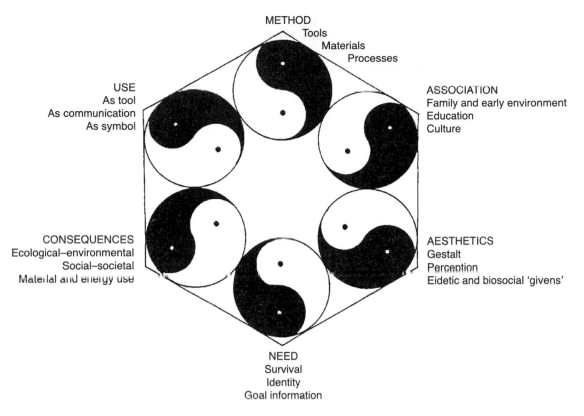

Figure 1 Six-sided function matrix

charisma to luxury goods such as perfumes that sell at enormously inflated prices. But the less luxurious package can be equally destructive of the environment. Fast-food suppliers have for decades used small coffins made of a plastic known as styrofoam in North America in which to serve their cheeseburgers and Big Macs. Some years ago, McDonald outlets in the American Midwest proudly proclaimed on an automatically changing neon sign: '*Seventy billion sold so far*' (italics supplied). More recently the McDonald corporation has been convinced of the ecological soundness of switching to paper containers.

Foam plastic is a very useful packaging material, yet profoundly damaging to the environment. After it has been discarded, it is doubtful whether it is possible to re-use it, and it continues to be an environmental and toxic hazard in spite of the optimistic assurances of the manufacturers relayed to the public by their public relations people. The advantages of foam plastic are that it makes an extremely lightweight protection for precision parts, is easily formed around delicate optical instruments or electronic assemblies, and is quite inexpensive. But there are alternative and organic ways of packaging.

It is a valuable concept that there is really nothing new in the world that needs to be packed and shipped. The immediate objection will be that this is sheer nonsense. After all, there were no computers, CD players or camcorders in the distant past. Yet Van Leuwenhoek had to ship his microscopes from the Netherlands to Padua in the 16th century, Galileo needed to send telescopes to the Danish astronomer Tycho Brahe on the island of Hven off southern Sweden, and forward 'philosophical instruments' and optics to various other parts of Europe. More recently, during the Civil War in the United States, delicate surgical instruments had to be shipped from northern factories to the front. The materials used to pack such early precision instruments were Spanish moss, other dried mosses, sand, sawdust, crushed and dried leaves or dried grasses, thin cotton bags filled with down or feathers, wood chips, and much else. The one thing that these materials have in common is that they can be recycled; they are all organic and will return to the natural environment.

My earliest introduction to this way of packing was my first job as a young boy in New York. I worked in the basement of the Museum of Modern Art packing small sculptures or ceramics to send to members of the museum who were renting art objects for a few months at a time. I remember that, in addition to shipping-boxes (which were made of wood or cardboard), we had two gigantic popcorn machines, and made popcorn – unsalted and without cheese, I may add – in which to pack the sculpture pieces; polystyrene 'worms' did not then exist. It was an intelligent and decent way of packing which in 1992, to my delight, began to be revived by some mail-order firms as an ecologically responsible way of dealing with fragile objects.

In 1989 I was hired by a Japanese corporation, specializing in computers, cameras, and other high-tech products, and spent three years conducting research and feasibility studies in the use of organic packaging materials. Research eventually concentrated on plants that, when maturing, surround their seeds with an enormous protective cradle of fluffy material. The specific seed we researched expands its bulk to more than forty times the original volume.

The package was for a professional precision 35mm camera and its lenses. Normally, expensive small cameras are cradled in a shaped foam-plastic cushion that has been covered with an equally plastic fake-velveteen fabric. This in turn is topped by another velvet-like foam-plastic lid on top, and both are bedded in a sarcophagus-like box, made of high-impact polystyrene. The box is held, or suspended, by two foam-plastic spacers within an outer (again plastic) case. Lenses are normally placed in plastic tubes that are upholstered with foam on the inside, and covered with a leather-like vinyl, called 'leatherette' (the very word makes one's flesh crawl), or a plastic called 'naughahide' on the exterior. A hideous example of over-packaging and transparent make-believe.

Eventually we created a small quilt, about 15 inches (37 cm) square. The 'shell' of the quilt is made of rice-paper, filled with fluffy plant fibres and then sewn into quilt squares with a hemp-derived thread. The quilt is wrapped around the camera body and inserted into a cardboard sleeve. Quilted pouches, made of 'green' cotton and filled with eiderdown, protect the lenses. In Japan this method of softly cradling precision parts is already in experimental use, and will probably soon be used for export models. The great advantages of this package are obvious. Reliance on oil-based plastics and the hazards of their manufacture are entirely eliminated. The new package is wholly organic, and will return to the soil. To exaggerate somewhat, theoretically it may be possible in a year or two for someone to buy a camera or CD player and literally dump the wrapping in the back garden where the recyclable, organic components of the package – augmented by trace amounts of nitrate boosters – will actually help the garden grow.

At the moment, packaging generally involves the use of plastics (discussed in detail later in this chapter), metal, wood, cardboard and paper. The use of paper has two major effects on the ecology. One of these is the cutting-down of trees and forests, the other the pollution that occurs in the paper production itself. Nine-tenths of paper products come from forests in northern temperate zones – Canada, the United States, and northern Europe. It is now widely known how to manage such forests commercially so that they can continue to function as renewable resources, but the timber industry generally refuses to engage in selective harvesting from multi-species mature forests, and continues to plant monocultural forests and to employ the clear-cutting of established woodlands.

Chlorines used in paper production as a bleach for wood fibres also pose an ecological hazard. Chlorine creates dioxins that are mutagenic, that is to say, they create genetic changes by bonding to the DNA structure of living cells. Furthermore the runoff of water tainted with dioxin and chlorine has endangered aquatic life (such as salmon in the Pacific north west of North America), as well as poisoning ground-water.

There are packaging items that are inherently impossible to recycle. Manufacturers can easily avoid using high-gloss papers, highly coated or plastic-coated papers, glues that are not water-soluble and plastic windows on envelopes. Instead designers could specify non-bleached papers or those whitened with new, bleach-free methods. More than three-quarters of all paper types can be recycled, but usually a percentage of new fibres are added. Recycling waste paper can be 50% more energy efficient than the use of virgin pulp. It is good practice to use paper with the highest percentage of recycled material. The Simpson Paper Company of San Francisco has emerged as one of the leaders in this field.

Lead, mercury, arsenic, chromium, cadmium, beryllium and vanadium are all carcinogenic and neurotoxic. They are frequently used in the composition of printing inks on packages, and pose a severe threat when they leach into the water supply from landfills. De-inking is difficult and costly. Vegetable-based inks, made from soya for instance, can be used effectively, and here again the Simpson Paper Company has led the way.

There are other materials that are dangerously poisonous to human beings and the environment. Some countries have already restricted polyvinyl chloride (PVC) since, unless it is burnt in special handling ovens, it releases dioxins and hydrochloric acids into the air. It has also been established that the making of PVC is directly linked to kidney cancers in workers. There are many other substances, such as cadmium-based pigments, certain flame-retardants and chlorinated solvents, that are still unrestricted in the United States and the United Kingdom, yet pose major health threats. A series of labels and symbols have been developed and accepted internationally, and more specific national or regional markers also exist.

Terms for hazardous materials, products or processes

- **Carcinogenic:** can cause cancer either in manufacture or in use
- **Mutagenic:** can cause genetic mutations in human beings or other organisms
- **Neurotoxic:** can attack the nervous system of human beings or other animals
- **Biocidal:** destructive to the environment and ecology

In January 1993 a comprehensive set of laws was introduced in Germany to deal with the reduction of packaging waste. These include requirements that producers and suppliers take back all sales and transport packaging. Manufacturers are also obliged to remove wrapping when selling to the end-users, and to inform them where and how to return the rest of the package. Furthermore, the law states specifically that manufacturers must take back, sort and recycle the following:

> Electrical or electronic appliances in the sense of this ordinance are household goods, entertainment electronics appliances and appliances and installations for office, information and communications technology, banking machines, electrical tools, measuring, steering and lighting technology, toys, clocks which contain electrical or electronic component parts ... component categories such as casing, screens, keyboards or plates.

Furthermore, these laws decree that packages are merely for the protection of the contents and are not to be used for advertising messages or point-of-sale graphics.

Packages come in various guises. With complex mechanisms and electronic parts forming a machine or device, the package in industrial design frequently turns into a 'shroud', that is, an external cover or shell that keeps dust from the working parts, protects them, and cuts down the visual confusion of a complicated working arrangement that can no longer be understood by the end-user.

'Urban Ecology: A Basis for Shaping Cities'

from *Cities and Natural Process* (1995)

[. . .]

THE CONTRADICTION OF VALUES

Towns and cities are perceived largely through their exterior environment. The average urban dweller going about the business of daily living will experience the city through its patterns of streets and pedestrian ways, shopping places, civic squares, parks and gardens and residential areas. However, there is another generally ignored landscape lying beneath the surface of the city's public places and thoroughfares. It is the landscape of industry, railways, public utilities, vacant lands, urban expressway interchanges, abandoned mining lands and waterfronts. Thus two landscapes exist side by side in cities. The first is the nurtured 'pedigreed' landscape of mown turf, flowerbeds, trees, fountains and planned places everywhere that have traditionally been the focus of civic design. Its basis for form rests in the formal design doctrine and aesthetic priorities of established convention. Its survival is dependent on high energy inputs, engineering and horticultural technology. Its image is that of the design solution independent of place: it can be found everywhere from Washington DC to Jakarta, Indonesia; from the city centre to the outlying suburbs. The second is the fortuitous landscape of naturalized urban plants and flooded places left after rain, that may be found in the forgotten places of the city. Urban 'weeds' emerge through cracks and gratings in the pavement, on rooftops, walls, poorly drained industrial sites or wherever a foothold can be gained. They provide shade and flowering groundcovers and wildlife habitat at no cost or care and against all the odds of gasoline fumes, sterile or contaminated soils, trampling and maintenance men. There is also a humanized landscape hidden away in the back alleys, rooftops and backyards of many an ethnic neighbourhood that can be described as the product of spontaneous cultural forces. It is here that one may find a rich variety of flourishing gardens and brightly painted houses. The turfed front yard of the well-to-do neighbourhood gives way to sunflowers, daisies, vegetable gardens, intricate fences, ornaments and religious icons of every conceivable variety, expressing rich cultural traditions, and the imperatives of necessity. The forces that shape the built vernacular, in fact, have remarkable similarities to the fortuitous landscape. Both have evolved in response to minimum interference from authority.

These two contrasting landscapes, the formalistic and the natural, the pedigreed and the vernacular, symbolize an inherent conflict of environmental values. The first has little connection with the dynamics of natural process. Yet it has traditionally been held in high public value as an expression of care, aesthetic value and civic spirit. The second represents the vitality of altered but none the less functioning natural and social processes at work in the city. Yet, it is regarded as a derelict wasteland in need of urban renewal, the disorderly shambles of the poorer parts of town. If we make the not unreasonable assumption that diversity is ecologically and socially necessary to the health and quality of urban life, then we must question the values that have determined the image of nature in cities. A comparison between the plants and animals present in a regenerating vacant lot, and those present in a landscaped residential front yard, or city park, reveals that the vacant lot generally has far greater floral and faunal diversity than the lawn or city park. Yet all efforts are directed towards nurturing the latter and suppressing the former. The reclamation of 'derelict' areas, or the creation of new development of the city's edge where the native and cultural landscape is replaced by a cultivated one, involves reducing diversity, rather than enhancing it. The question that arises, therefore, is this: which are the derelict sites in the city requiring rehabilitation? Those fortuitous and often ecologically diverse landscapes representing urban natural forces at work, or the formalized landscapes created by design?

It is my contention that the formal city landscape imposed over an original natural diversity is the one in need of rehabilitation. While it will be obvious that such a landscape has a time-honoured place in the city, its *universal* application for the making of urban places is the most persuasive argument for considering it as a derelict landscape. Other paradoxes become apparent when we apply ecological insights to our observation of the city environment.

- Attitudes and perceptions of the environment expressed in town planning since the Renaissance, have, with some exceptions, been more concerned with utopian ideals than with natural process as determinants of urban form. Examples of cities and institutions all over the world attest to the aesthetic and cultural baggage of a past era, transported to hostile climatic environments and wholly inappropriate to them. Cheap fossil fuel together with a misguided sense of civic pride, or expressions of power and wealth, has enabled the inorganic structure of planning theory to persist and maintain the illusion that the creation of benign outdoor climates has little relevance to urban development.

- Traditional storm drainage systems, the conventional method of solving the problem of keeping the city's paved surfaces free of water, have until recently been unquestioned. As the established vocabulary of engineering, water drains to the catchbasin. But the benefits of 'good design' – well-drained streets and civic spaces – is paid for by the environmental costs of eroded streams, flooding and impairment of water quality in downstream watercourses.

- Sewage disposal systems are seen as an engineering rather than a biological solution to the ultimate larger problem of eutrophication of water bodies and wasted resources. We have the paradox of the city as the centre for enormous concentrations of nutrient energy, while urban soils remain sterile and non-productive.

- The notions of humanity and nature have long been understood to be separate issues. Such a dichotomy has had profound influences on the way people have thought about themselves: the cities where people live and the non-urban regions beyond the city where nature lives. In the unique culture from which the disciplines of intervention spring – engineering, building, planning and design – this perceived separation has also had a profound effect on the control, not only of nature, but of human behaviour. Thus the nature of pedigreed design has had little time or understanding for the innate forces that shape human environments, or the peculiar needs of multi-cultural communities that are the norm in most cities today.

The resolution of these contradictions must be found in an ecological view that encompasses the total urban landscape and the people who live there. This includes the unstructured spatial and social environments that are not currently seen to contribute to the city's civic image as well as those that do.

PART 3

*N*egotiating Urban Cultures

SOCIAL JUSTICE

INTRODUCTION

Social justice sounds good. Like motherhood and apple pie it signifies a society which has attained a high degree of social development. It implies an 'other' of an unjust primitive, though its characterization as a virtue of civilization also renders it a kind of luxury, possibly to be gained after basic problems like money have been sorted out. Of course, it is not quite so easy. The texts in this section, which concern the contemporary city and its occupation, draw out some of the contradictions and some of the opportunities in the idea and actuality of social justice today. They can be read in conjunction with Sharon Zukin's essay in Section II on the symbolic economy of the city. Zukin raises a question which is equally relevant here: whose is the city, and who decides that?

Most liberal ideas correspond in some part to the ideal of liberty, which is not the same as freedom. Whilst the latter is associated with radical politics, and liberation movements in countries emerging from colonial rule, liberty is a bourgeois concept. Delacroix in his painting of *Liberty Leading the People* of 1831 depicts the crowd in contemporary dress. But Liberty is in Greek costume, one breast partially uncovered in a gendered gesture of classicism, like an animated statue. The disjuncture of two pictorial languages – realism and classicism – indicates, probably unconsciously for Delacroix, the difficulty in reconciling the conceptual realm of liberty with real struggle.

The precondition for struggle is injustice, and in the industrialized countries this is associated with capitalism. As Herbert Marcuse argues in his essay on 'The Affirmative Character of Culture', first published in 1937 and included in the collection *Negations* (1968, Penguin), bourgeois society proclaims liberty as a universal value, but cannot offer liberty to all. The ownership of the means of production by the few (capitalists, or entrepreneurs) and a mechanism of exchange whereby labour is sold at a market rate rather than rewarded with a share of profits, causes the labouring class to fall deeper into poverty and dependence. And alongside the growth of capital is the growth of the city in which, since wealth is no longer identified with land, capital is concentrated.

A history of western cities manifests a duality of exclusion and confinement. As Michel Foucault writes in *Madness and Civilisation* (1967, Tavistock), at the beginning of modernity, institutions such as the general hospital are places of confinement for the vagrant as well as the insane. So, in the birth of the modern city, certain publics, classified as dirt – that which is out of place – are excluded from the visible city. Ivan Illich (1986) writes of another aspect of this: the exclusion of smells. Not only the smells of the contents of the privy, but also the perceived smell of the dead threatened an idealized notion of city space. The city, since the Enlightenment, then, is both beautiful and of pure air. This relates to the idea of social justice because it lays an aesthetic foundation for injustice. To achieve the odourless utopia means also to clear the public spaces of a city, its sites of civic display, of the poor, who are collectively characterized as the smelly, whose foul odours spread contamination. Much of the marginalization of publics in contemporary urban development in the form of corporate or high-income enclaves follows the pattern set by the purging of the odour of the dead. Illich's analysis bears some comparison to that developed by David Sibley in his book *Geographies of Exclusion* (1995, Routledge) an extract from which is included in Section X.

Herman Hertzberger takes up the question of the emotional ownership of urban space, arguing that this is the precondition for conviviality. This goes against the grain of increasing control and surveillance, and the mechanistic thinking which governs attempts to solve everyday vandalism by repairing the things vandalized. Taking the amphitheatres of Arles and Lucca as cases, he deduces that forms can retain identity and accommodate changes in use. From this he urges flexibility in urban form, offering users a greater degree of freedom. Perhaps this is like the case for moveable chairs in urban parks made by W. H. Whyte in *The Social Life of Small Urban Spaces* (Conservation Foundation, Washington DC, 1980). When people move a chair they sit on, they own it for that time. If the chair is fixed, they do not. Hertzberger puts this as: 'Users project themselves onto the form.' He then argues that the users of buildings should be free to determine for themselves the use of space.

One form of asserting ownership of space is squatting. Direct action groups, such as 'The Land is Ours', which took over a vacant industrial site owned by the Guinness brewery in south London from May to October 1996, demonstrate the possibility for spontaneous settlement. Neil Smith, in an extract from *The New Urban Frontier* (1996, Routledge), describing parts of New York in the language of the old west, or frontier, discusses the occupation of Tompkins Square in the Lower East Side in the late 1980s. Smith sees an inversion of the pioneer myth in the re-coding of inner city neighbourhoods as frontier zones, both idyllic and dangerous. He links the gentrification of inner city areas with ideologically-rooted fashions for Tex-Mex food and cowboy boots; through the manufacture of such myths social conflict is displaced and class-specific and race-specific social norms are affirmed, so that the impact of development becomes naturalized. Smith's analysis of gentrification sees culture as a cover for the operations of capital. He illustrates this with a poster proclaiming 'the taming of the wild wild west' to advertise the Armory Building on 42nd Street. But such images belong to the 1980s. Smith concludes that, after the financial collapse of 1987, the new urban reality is a revanchist city of violent reaction on the part of gentrifiers marooned in areas which have not become zones of upward mobility, against minorities such as homeless people and the unemployed, as well as women, gays and lesbians.

Finally, Rajeev Patel (1997) considers urban violence. Like Hertzberger, he sees everyday acts of violence on city streets as denotive of deprivation. He notes research showing that wide differentials of wealth produce violence, and suggests that it is the poor who are the most common victims. Whilst a conventional view is that violence is dealt with by the forces, first of law, then of social work, Patel proposes responses through the provision of health, education, employment, housing and urban design, as well as law. Short-term and cosmetic measures, such as allowing youths to spray graffiti murals on designated walls do not solve the problem, which is systemic. Patel identifies political empowerment as a key element in any more enduring programme, together with a holistic approach involving all relevant agencies.

Perhaps, then, social justice is attainable, but only when it is redefined, not as a bourgeois concept within a society based on a contradiction, but as empowerment. This can easily become another rhetorical buzz-word. It can also be real.

'The Dirt of Cities', 'The Aura of Cities', 'The Smell of the Dead' and 'Utopia of an Odorless City'

from *H₂O and the Waters of Forgetfulness* (1986)

THE DIRT OF CITIES

When Aristotle drew up his rules for the siting of a city, he wanted the streets to be open to sunlight and prevailing winds. Complaints that cities can become dirty places go back to antiquity. In Rome special magistrates sat under their umbrellas in a corner of the Forum to adjudicate complaints from pedestrians soiled by the contents of chamberpots. Throughout classical antiquity, beginning with the palace at Knossos (1500 B.C.), the dwellings of the wealthy occasionally had a special room for bodily relief. In Rome, wealthy households owned a special slave to empty the night-chairs. Most homes had no designated place for bodily relief. Like the sewers beneath the Athenian agora, the sewers beneath the imperial fora and pay seats in marble latrines were restricted to city areas covered with marble. In popular two-story dwellings, Roman ordinances required a hole at the bottom of the staircase. Otherwise, the street was assumed to be the proper place for such disposals. Medieval cities were cleaned by pigs. There survive dozens of ordinances which regulate the right of burghers to own them and feed them on public waste. In Spain and the Islamic areas, ravens, kites, and even vultures were protected as sacred scavengers. These customs did not change significantly during the baroque period. Only during the last years of Louis XIV's reign was an ordinance passed that made the removal of fecal materials from the corridors of the palace in Versailles a weekly procedure. Underneath the windows of the ministry of finance, pigs were slaughtered for decades, and their encrusted blood caked the palace walls. Tanneries were operated within the city, even though their smell in the valley of

Ghinnom had become the symbol for hell (gehenna) in old Jerusalem. A survey carried out in Madrid in 1772 disclosed that the royal palace did not contain a single privy. These millennial city conditions prevailed in London when Harvey announced his discovery of the circulation of the blood. Only after the great London fire of 1660 and after Harvey's death were "lay-stalls" set up on London street-crossings for the disposal of waste, and an honorary scavenger was appointed for each ward to supervise the rakers – men and women willing to pay for the privilege of sweeping the streets so that they could sell the refuse for a profit. In 1817 the powers of these scavengers and rakers were codified in the London Metropolitan Paving Act, which remained the statute until 1855. By this time the houses of the well-to-do in London usually contained one privy, from which the night-soil was removed several times each week. But for the larger part of London, the collection of night-soil from the streets remained sporadic. In the late nineteenth century it was felt to interfere with rush hours. It was not until 1981 that the London County Council prescribed that privy cleaning had to be restricted in summertime to the hours between 4 A.M. and 10 A.M. Quite obviously, throughout history cities have been smelly places.

THE AURA OF CITIES

Nevertheless, the perception of the city as a place that must be constantly washed is of recent origin. It appears at the time of the Enlightenment. The reason most often given for this constant toilette is not the visually offensive features of waste or the residues that make people slip on the street but bad odors and their dangers. The city is suddenly perceived as an evil-smelling space. For the first time in history, the utopia of the odorless city appears. This new aversion to a traditional characteristic of city space seems due much less to its more intensive saturation with odors than to a transformation in olfactory perception.

The history of sense perception is not entirely new. Linguists have dealt with the changing semantics of colors, art historians with the style in which different epochs see. But only recently have some historians begun to pay closer attention to the evolution of the sense of smell. It was Robert Mandrou who, in 1961, first insisted on the primacy of touch, hearing, and smell in premodern European cultures. Complex non-visual sense perceptions gave way only slowly to the enlightened predominance of the eye that we take for granted when we "describe" a person or place. When Ronsard or Rabelais touched the lips of their love, they claimed to derive their pleasure from taste and smell, which could only be hinted at. Even the eighteenth-century writer does not yet describe the loved body; at best the publisher inserts into the text an etching that illustrates the scene, an etching which, during the early part of the century, effectively hides whatever is individual, personal, "touching" in the scene the author describes. But while it is easy to follow historically the ability of poets and novelists to perceive and then paint the flesh and landscape in their uniqueness, it is much more difficult to make statements about the perception of odors in the past. To write well about this past perception of odors would be a supreme achievement for a historian because the odors leave no objective trace against which their perception can be measured. When the historian describes how the past has smelled he is dependent on his source to know what was there and how it was perceived. The case is the same whether he deals with odors perceived by lovers or those that help physicians recognize the state of the ill or those with which devils or saints fill the spaces within which they dwell.

I still remember the traditional smell of cities. For two decades I spent much of my time in city slums between Rio de Janeiro and Lima, Karachi and Benares. It took me a long time to overcome my inbred revulsion to the odor of shit and stale urine which, with slight national variations, makes all unsewered industrial shantytowns smell alike. This smell is the characteristic for the early stage of industry; it is the stench of dwelling space that has begun to decay because it is threatened by imminent incorporation into the hygienic system of modern cities. It is distinct from the local atmosphere of a still vernacular town. A vernacular atmosphere is integral to dwelling space; according to traditional medicine, people waste away if they are sickened and repelled by the aura of a new place in which they are forced to live. Sensitivity to an aura and tolerance for it are requisites to enjoy being a guest. Many people today have lost the ability to imagine the geographic variety that once could be perceived through the nose. Because increasingly the whole world has come to smell alike: gasoline, detergents, plumbing, and junk foods coalesce into the catholic smog of our age. Where this smog mingles with the decay of vernacular atmosphere, as along the Rimac which carries Lima's sewage into the Pacific I learned to recognize the smell of development. It is there that I become sensitive to the difference between industrial pollution and the dense atmosphere of Paris between Louis XIV and Louis XVI. To describe it I shall draw heavily on Corbin.

THE SMELL OF THE DEAD

People then not only relieved themselves as a matter of course against the wall of any dwelling or church; the stench of shallow graves was evidence that the dead were present within its walls. This thick aura was taken so much for granted that it is rarely mentioned in contemporary sources. Universal olfactory nonchalance came to an end when a small number of citizens lost their tolerance for the smell of corpses. Since the Middle Ages, the corpses of clergy and benefactors had been entombed near the altar, and the procedures of opening and sealing these sarcophagi within the church had not changed over the centuries. Yet at the beginning of the eighteenth century, their miasma became objectionable. In 1737 the French parliament appointed a commission to study the danger that burial inside churches presented to public health. The presence of the dead was suddenly perceived as a physical danger to the living. Philosophical arguments were concocted to prove that the burial within churches was contrary to nature. An Abbé Charles Gabriel Porée, Fénelon's librarian, from Lyons argued in a book which went through several editions that, from a juridical point of view, the dead had a right to rest outside the walls. In his monumental history of attitudes toward death in the West since the Middle Ages, Philippe Ariès has shown that this new squeamishness in the presence of corpses was due to an equally new unwillingness to face death. Henceforth, the living refused to share their space with the dead. They demanded a special apartheid between live bodies and corpses at just the time when the innards of the live human body were beginning to be visualized as a machine whose elements were "prepared" for inspection on the dissecting table. Like the organs, the dead became more visible and less awesome; they also became increasingly more disgusting and physically dangerous for the living. Philosophical and juridical arguments calling for their exclusion from dwelling space went hand-in-hand with reported evidence of the deadly threat of their miasma. Corbin lists several instances of mass death among the members of a church congregation that occurred at the very moment when, during a funeral ceremony, miasma escaped from an opened grave. Burials within churches thereafter became rare – increasingly a privilege of bishops, heroes, and their like. The cemeteries were moved out of the cities. Though in 1760 the Cimetière des Innocents was still used for parties in the afternoon and for illicit love at night, it had been closed in 1780 by request of neighbors precisely because they objected to emanations from decomposing bodies. Yet even if the presence of the dead within the city was resented by rich and poor alike at the end of the ancien régime, it required almost two centuries to educate the lower classes to feel nausea from the odor of shit.

UTOPIA OF AN ODORLESS CITY

Both living and dead bodies have an aura. This aura takes up space and gives the body a presence beyond the confines of its skin. It mingles with the auras of other people; without losing its own personality, it blends into the atmosphere of a particular space. Odor is a trace that dwelling leaves on the environment. As fleeting as each person's aura might be, the atmosphere of a given space has its own kind of permanence, comparable to the building style characteristic of a neighborhood. This aura, when sensed by the nose, reveals the non-dimensional properties of a given space; just as the eyes perceive height and depth and the feet measure distance, the nose perceives the quality of an interior.

During the eighteenth century it became intolerable to let the dead contribute their aura to the city. The dead were either excluded from the city or their bodies were encased in airtight monuments celebrating hygienic disposal, for which Père Lachaise became the symbol in Paris. In the process of their removal, the dead were also transmogrified into the "remains of people who have been," subjects for modern history – but no more of myth. Disallowing them shared space with the living, their "existence" became a mere fiction and their relics became disposable remains. In this process western society has become the first to do without its dead.

The nineteenth century created a much more difficult task for deodorants. After removing the dead, a major effort was undertaken to deodorize the living by divesting them of their aura. This effort to deodorize utopian city space should be seen as one aspect of the architectural effort to "clear" city space for the construction of a modern capital. It can be interpreted as the repression of smelly persons who unite their separate auras to create a smelly crowd of commonfolk. Their "common" aura must be dissolved to make space for a new city through which clearly delineated individuals can circulate with unlimited freedom. For the nose a city without aura is literally a "Nowhere," a *u-topia*.

The clearing of city space coincides with a new stage of the professionalization of architects. Their profession had formerly been in charge of building palaces, squares, fountains, city walls, and perhaps bridges or channels. They were now empowered to condemn dwelling space and transform it into garages for people. Observing the course of Peruvian settlement thirty years ago, John Turner has described what happens when dwelling *by* people is transformed into housing *for* people. Housing is changed from an activity into a commodity. This transformation requires making dwelling activities impossible, so that persons become domesticated docile residents within shelters which they rent or buy. Each now needs a street address with a house number (and, in some cases, an apartment number too). People have lost the aura that allowed their whereabouts to be sniffed out in the old days. When the idea of the new city, made up of residents, began to register in the minds of the leaders of the Enlightenment, everything that smacked of quality in space came to be objectionable. Space had to be stripped of its aura once aura had been identified with stench. Unlike the architect who constructed a palace to suit the aura of his wealthy patron, the new architect constructed shelter for a yet unidentified resident who was supposed to be without odor.

46 HERMAN HERTZBERGER

'The Public Realm'

from *Architecture and Urbanism* (1991)

The more responsibility users have for an area – and consequently the more influence they can exert on it – the more care and love they will be prepared to invest in it. And the more suitable the area is for their own specific uses the more they will appropriate it. Thus *users* become *inhabitants*.

Strong affective relationships may thus arise, which help to turn the space into a (more) friendly environment.

The translation of the concepts 'public' and 'private' in terms of differentiated responsibilities thus makes it easier for the architect to decide in which areas user-participation in the design of the environment should be provided and where this is less relevant.

The point is to give public spaces form in such a way that the local community will feel personally responsible for them, so that each member of the community will contribute in his or her own way to an environment that he or she can relate to and can identify with.

It is the great paradox of the collective welfare concept as it has developed hand in hand with the ideals of socialism, that it actually makes people subordinate to the very system that has been set up to liberate them.

The services rendered by the Municipal Public Works departments are felt, by those for whose benefit those departments were created, as an overwhelming abstraction; it is as if the activities of Public Works are an imposition from above, the man in the street feels that they 'have nothing to do with him', and so the system produces a widespread feeling of alienation.

The public gardens and green belts around the blocks of flats in the new urban neighbourhoods are the responsibility of the Public Works departments, which do all they can to make these areas as attractive as possible – within the limits of the allocated budgets – on behalf of the community.

But the results that are achieved in this way cannot help being stark, impersonal and uneconomical, compared with what could have been achieved if all the flat-dwellers had been offered the opportunity of using a small plot of land (even if no bigger than a parking space) for their own purposes. What has now been collectively denied them could have become the contribution of each inhabitant to the community, while the space itself could have been used far more intensively if all that personal love and care had been lavished on it.

The reason why city dwellers become outsiders in their own living environment is either that the potential of collective initiative has been grossly overestimated, or because the organizational conditions to ensure true participation and involvement have been underestimated. The occupants of a house are not really concerned with the space outside their homes, but nor can they really ignore it. This opposition leads to alienation from your environment and – in so far as your relations with others are influenced by the environment – also to alienation from your fellow residents.

The mounting degree of control imposed from above is making the world around us increasingly inexorable: and this elicits aggression, which in turn leads to further tightening of the web of regulations. A vicious circle is the result, the lack of commitment and the exaggerated fear of chaos have a mutually escalating effect.

The incredible destruction of public property – which is on the rise in the world's major cities – can probably only partly be blamed on alienation from the living environment. The fact that public transport shelters and public telephones are completely destroyed week in week out is a truly alarming indictment of our society as a whole. What is almost as alarming, however, is that this trend – and its escalation – is dealt with as if it were a mere problem of organization: by undertaking periodical repairs as if they were a question of routine maintenance, and by applying extra reinforcements ('vandal-proofing') the situation appears to be accepted as 'just one of those things'.

The whole suppressive system of the established order is geared to avoiding conflict; to protecting the individual members of the community from incursions by other members of the same community, without the direct involvement of the individuals concerned. This explains why there is such a deep fear of disorder, chaos and the unexpected, and why impersonal, 'objective' regulations are always preferred to personal involvement. It seems as if everything must be regulated and quantifiable, so as to permit total control; to create the conditions in which the suppressive system of order can make us all into lessees instead of co-owners, into subordinates instead of participants. Thus the system itself creates the alienation and, by claiming to represent the people, obstructs the development of conditions that could lead to a more hospitable environment.

Just as important as the disposition of the residential units vis à vis each other is the fenestration, the placement of bay windows, balconies, terraces, landings, doorsteps, porches – whether they have the correct proportions and how they are spatially organized, i.e. adequately separated but certainly not too much so. It is always a question of finding the right balance to enable the residents to withdraw into privacy when they want to but also to seek contact with others.

Having departed from the traditional block siting principle, architects have endeavoured, inspired especially by Team X and Forum, to invent a stream of new dwelling forms. This often gave rise to spectacular results, but whether they function properly is only partially dependent on the quality of the dwellings themselves. What is at least as important is whether the architect can find a way, using the dwellings as his construction material to form a street that

functions adequately. The quality of each is dependent on that of the other: *houses and streets are complementary*.

If the houses are private domains, then the street is the public domain. Paying equal attention to housing and street alike means treating the street not merely as the residual space between housing blocks, but rather as a fundamentally complementary element, spatially organized with just as much care so that a situation is created in which the street can serve more purposes besides motorized traffic. If the street as a collection of building blocks is basically the expression of the plurality of individual, mostly private, components, the sequence of streets and squares as a whole potentially constitutes the space where a dialogue between inhabitants can take place. The street was, originally, the space for actions, revolutions, celebrations, and throughout history you can follow from one period to the next how architects designed the public space on behalf of the community which they in fact served. So this is a plea for more emphasis on the enhancement of the public domain in order that it might better serve both to nurture and to reflect life in the community. With respect to every urban space we should ask ourselves how it functions: for whom, by whom and for what purpose. Are we merely impressed by its sound proportions or does it perhaps also serve to stimulate improved relations between people.

The break away from the closed perimeter block siting in twentieth-century urbanism, meant the disintegration of the clear-cut spatial definition given by the street pattern. As the autonomy of the buildings grew, their interrelationship diminished, so that they now stand devoid of alignment as it were, like an irregular scattering of megaliths far away from each other in an excessively large open space. The 'rue corridor' has degenerated into an 'espace corridor'.

This new open type of siting, so innovative for the 'physical' conditions of housing construction in particular, has had a disastrous effect on the cohesion of the whole – a fate that has befallen most cities.

The more buildings stand apart as autonomous building volumes with individualized façades and unambiguously private entrances not only the less cohesion there is, but also and especially the greater the opposition between public and private space – even though the housing blocks may be designed, for instance, with access galleries or interior covered streets, and even though, conversely, the buildings are surrounded by private space which is often not accessible to others besides residents.

Urbanism with buildings as autonomous freely dispersed monuments has given rise to a huge exterior environment – at best a pleasant park landscape where nature is never far away.

While modern architects and town planners already started breaking open the city before the Second World War, the demolition work was continued by the war; later on the traffic mania dealt this fragmentation the 'coup de grâce' wherever it could.

So all of us are by now convinced of the need for reconstruction of the interior of the city and for a revival of interest and concern for the street area, and hence for the exterior of housing. But that must not be allowed to lead to an architecture of street walls with the actual dwellings as mere punctuation marks or props to support the decor.

We must not forget that the 'Nieuwe Bouwen' aimed specifically at the improvement of buildings, and notably at the improvement of the dwellings by means of better siting to ensure more sunlight, wider views, more satisfactory exterior spaces etc.

The face of a city is half the truth – satisfactory housing is the other, complementary half.

The many examples of open urbanism, as designed in the 1920s and 1930s, are indeed still of great relevance, at least if each is judged according to its own specific qualities.

We must consider the quality of street-space and of buildings in relation to each other. A mosaic of interrelationships – as we imagine urban life to be – calls for a spatial organization in which built form and exterior space (which we call street) are not only complementary in the spatial sense and therefore reciprocate in forming each other, but also and especially – for that is what we are primarily concerned with here – in which built form and exterior space offer maximal accessibility vis à vis each other. The point is therefore to permit 'building' and 'street' as spaces with different degrees of public accessibility to penetrate each other in such a way that not only the borderlines between outside and inside become less explicit, but also that the sharp division between private and public domain becomes softened.

If you enter a place gradually, the front door is divested of its significance as a single and abrupt moment; it is extended, as it were, to form a step-by-step sequence of areas which are not yet explicitly inside but also less explicitly public. The most obvious expression of such a mechanism of accessibility was to be seen in the arcades, and it is indeed not surprising therefore that the arcade-idea serves as an example today.

47 NEIL SMITH

'"Class Struggle On Avenue B": The Lower East Side as Wild Wild West'

from *The New Urban Frontier* (1996)

On the evening of August 6, 1988, a riot erupted along the edges of Tompkins Square Park, a small green in New York City's Lower East Side. It raged through the night with police on one side and a diverse mix of antigentrification protestors, punks, housing activists, park inhabitants, artists. Saturday night revelers and Lower East Side residents on the other. The battle followed the city's attempt to enforce a 1:00 A.M. curfew in the Park on the pretext of clearing out the growing numbers of homeless people living or sleeping there, kids playing boom boxes late into the night, buyers and sellers of drugs using it for business. But many local residents and park users saw the action differently. The City was seeking to tame and domesticate the park to facilitate the already rampant gentrification on the Lower East Side. "GENTRIFICATION IS CLASS WAR!" read the largest banner at the Saturday night demonstration aimed at keeping the park open. "Class war, class war, die yuppie scum!" went the chant. "Yuppies and real estate magnates have declared war on the people of Tompkins Square Park," announced one speaker. "Whose fucking park? It's our fucking park," became the recurrent slogan. Even the habitually restrained *New York Times* echoed the theme in its August 10 headline: "Class War Erupts along Avenue B."

In fact it was a *police* riot that ignited the park on August 6, 1988. Clad in space-alien riot gear and concealing their badge numbers, the police forcibly evicted everyone from the park before midnight, then mounted repeated baton charges and "Cossacklike" rampages against demonstrators and locals along the park's edge:

> The cops seemed bizarrely out of control, levitating with some hatred I didn't understand.

They'd taken a relatively small protest and fanned it out over the neighbourhood, inflaming hundreds of people who'd never gone near the park to begin with. They'd called in a chopper. And they would eventually call 450 officers ... The policemen were radiating hysteria. One galloped up to a taxi stopped at a traffic light and screamed, "Get the fuck out of here, fuckface ..." [There were] cavalry charges down East Village street, a chopper circling overhead, people out for a Sunday paper running in terror down First Avenue.

Finally, a little after 4:00 A.M. the police withdrew in "ignominious retreat," and jubilant demonstrators reentered the park, dancing, shouting and celebrating their victory. Several protestors used a police barricade to ram the glass-and-brass doors of the Christodora condominium, which borders on the park on Avenue B and which became a hated symbol of the neighborhood's gentrification.

In the days following the riot, the protestors quickly adopted a much more ambitious political geography of revolt. Their slogan became "Tompkins Square everywhere" as they taunted the police and celebrated their liberation of the park. Mayor Edward Koch, meanwhile, took to describing Tompkins Square Park as a "cesspool" and blamed the riot on "anarchists." Defending his police clients, the president of the Patrolmen's Benevolent Association enthusiastically elaborated: "social parasites, druggies, skinheads and communists" – an "insipid conglomeration of human misfits" – were the cause of the riot, he said. In the following days, the city's Civilian Complaint Review Board received 121 complaints of police brutality, and, largely on the evidence of a four-hour videotape made by local video artist Clayton Patterson, seventeen officers were cited for "misconduct." Six officers were eventually indicted but none was ever convicted. The police commissioner only ever conceded that a few officers may have become a little "overenthusiastic" owing to "inexperience," but he clung to the official policy of blaming the victims.

Prior to the riot of August 1988, more than fifty homeless people, evictees from the private and public spaces of the official housing market, had begun to use the park regularly as a place to sleep. In the months following, the number of evictees settling in the park grew, as the loosely organized antigentrification and squatters' movements began to connect with other local housing groups. And some of the evictees attracted to the newly "liberated space" of Tompkins Square Park also began to organize. But the City also slowly regrouped. City-wide park curfews (abandoned after the riot) were gradually reinstated; new regulations governing the use of Tompkins Square Park were slowly implemented; several Lower East Side buildings occupied by squatters were demolished in May 1989, and in July a police raid destroyed tents, shanties and the belongings of park residents. By now there were on average some 300 evictees in the park on any given night, at least three-quarters men, the majority African-American, many white, some Latino, Native Americans, Caribbean. On December 14, 1989, on the coldest day of the winter, the park's entire homeless population was evicted from the park, their belongings and fifty shanties hauled away into a queue of Sanitation Department garbage trucks.

It would be "irresponsible to allow the homeless to sleep outdoors" in such cold weather, explained a disingenous parks commissioner, Henry J. Stern, who did not mention that the city shelter system has beds for only a quarter of the city's homeless people. In fact, the city's provision for the evicted ran only to a "help center" that, by one account, "proved to be little more than a dispensary for baloney sandwiches." Many evictees from the park were taken in by local squats, others set up encampments in the neighborhood, but quickly they filtered back to Tompkins Square. In January 1990 the administration of supposedly progressive mayor David Dinkins felt sufficiently confident of the park's eventual recapture that it announced a "reconstruction plan." In the next summer the basketball courts at the north end were dismantled and rebuilt with tighter control of access; wire fences closed off newly constructed children's playgrounds; and park regulations began to be more strictly enforced. In an effort to force evictions, City agencies also heightened their harassment of squatters who now spearheaded the antigentrification movement. As the next winter closed in, though, more and more of the city's evictees came back to the park and began again to construct semi-permanent structures.

In May 1991, the park hosted a Memorial Day concert organized under the slogan "Housing is a human right" and, in what was becoming an annual May ritual, a further clash with park users ensued. It was now nearly three years since protestors had taken the park, and, with almost a hundred shanties, tents and other structures now in Tompkins Square, the Dinkins administration decided to move. The authorities finally closed the park at 5.00 A.M. on June 3, 1991, evicting between 200 and 300 park dwellers. Alleging that Tompkins Square had been "stolen" from the community by "the homeless," Mayor Dinkins declared: "The park is a park. It is not a place to live." An eight-foot-high chain-link fence was erected, a posse of more than fifty uniformed and plainclothes police was delegated to guard the park permanently – its numbers augmented to several hundred in the first days and during demonstrations – and a $2.3 million reconstruction was begun almost immediately. In fact, three park entrances were kept open and heavily guarded: two provided access to the playgrounds for children only (and accompanying adults); the other, opposite the Christodora condominium, provided access to the dog run. The closure of the park, commented *Village Voice* reporter Sarah Ferguson, marked the "death knell" of an occupation that "had come to symbolize the failure of the city to cope with its homeless population." No alternative housing was offered evictees from the park; people again moved into local squats, or filtered out into the city. On vacant lots to the east of the park, a series of shantytown communities were erected and they quickly took the name "Dinkinsville," linking the present mayor with the "Hoovervilles" of the Depression. Dinkinsville was less a single place than a collection of communities, with a similar impossible geography to that of Bophuthatswana. Existing collections of shanties under the Brooklyn, Manhatten and Williamsburg Bridges expanded.

As the site of the most militant antigentrification struggle in the United States, the ten acres of Tompkins Square Park quickly became a symbol of a new urbanism being etched on the urban "frontier." Largely abandoned to the working class amid postwar suburban expansion, relinquished to the poor and unemployed as reservations for racial and ethnic minorities, the terrain of the inner city is suddenly valuable again, perversely profitable. This new urbanism embodies a widespread and drastic repolarization of the city along political, economic, cultural and geographical lines since the 1970s, and is integral with larger global shifts. Systematic gentrification since the 1960s and 1970s is simultaneously a response and contributor to a series of wider global trans-formations: global economic expansion in the 1980s; the restructuring of national and urban economies in advanced capitalist countries toward services, recreation and consumption; and the emergence of a global hierarchy of world, national and regional cities. These shifts have propelled gentrification from a com-paratively marginal preoccupation in a certain niche of the real estate industry to the cutting edge of urban change.

Nowhere are these forces more evident than in the Lower East Side. Even the neighborhood's different names radiate the conflicts. Referred to as *Loisaida* in local Puerto Rican Spanish, the Lower East Side name is dropped altogether by real estate agents and art world gentrifiers who, anxious to distance themselves from the historical association with the poor immigrants who dominated this community at the turn of the century, prefer "East Village" as the name for the neighborhood above Houston Street. Squeezed between the Wall Street financial district and Chinatown to the south, the Village and SoHo to the west, Gramercy Park to the north and the East River to the east, the Lower East Side feels the pressure of this political polarization more acutely than anywhere else in the city.

Highly diverse but increasingly Latino since the 1950s, the neighborhood was routinely described in the 1980s as a "new frontier." It mixes spectacular opportunity for real estate investors with an edge of daily danger on the streets. In the words of local writers, the Lower East Side is variously a "frontier where the urban fabric is wearing thin and splitting open" or else "Indian country, the land of murder and cocaine." Not just supporters but antagonists have found this frontier imagery irresistible. "As the neighborhood slowly, inexorably gentrifies," wrote one reporter, in the wake of the 1988 police riot, "the park is a holdout, the place for one last metaphorical stand." Several weeks later, "Saturday Night Live" made this Custer imagery explicit in a skit cast in a frontier fort. Custer (as Mayor Koch) welcomes the belligerent warrior Chief Soaring Eagle into his office and inquires: "So how are things down on the Lower East Side?"

The social, political and economic polarization of "Indian country" is drastic and fast becoming more so. Apartment rents soared throughout the 1980s and with them the numbers of homeless: record levels of luxury condo construction are matched by a retrenchment in public housing provisions: a nearby Wall Street boom generated seven- and eight-figure salaries while unemployment rose among the unskilled; poverty is increasingly concentrated among women, Latinos and African-Americans while social services are axed; and the conservatism of the 1980s spewed a recrudescence of racist violence throughout the city. With the emergence of deep recession in the early 1990s, rents have stabilized, but unemployment has soared. In the late 1990s the resurgence of gentrification and development is destined to magnify the polarization of the 1980s.

Tompkins Square lies deep in the heart of the Lower East Side. On its southern edge along Seventh Street a long slab of residential buildings overlooks the park, mostly late-nineteenth-century five- and six-storey walk-up tenements adorned with precariously affixed fire escapes, but also including a larger building with a dreary, modern, off-white facade. To the west, the tenements along Avenue A are barely more interesting, but many cross streets and the mix of smoke shops, Ukrainian and Polish restaurants, upscale cafes and hip bars, groceries, candy stores and night clubs make this the liveliest side of the park. Along Tenth Street on the northern edge stands a stately row of 1840s and 1850s townhouses, gentrified as far back as the early 1970s. To the east, Avenue B presents a more broken frontage: tenements, St Brigid's Church from the mid-nineteenth century, and the infamous Christodora building – a sixteen-storey brick monolith built in 1928 that dominates the local skyline.

"One day," laments the tony, habitually understated AIA guide to New York architecture, "when this area is rebuilt, the mature park will be a godsend." Actually, the park itself is rather unexceptional. An oval rosette of curving, crisscross walkways, it is shaded by large plane trees and a few surviving elms. The walkways were lined by long rows of cement benches, replaced in the park reconstruction by wooden benches sectioned into individual seats by wrought iron bars designed to prevent homeless people from sleeping. Wide grassy patches, often bare, made up the body of the park and these were fenced off in the reconstruction. At the north end of the park are handball and basketball courts, playgrounds and the dog run, and at the south end a bandshell, which hosted everyone from the Fugs to the Grateful Dead in the 1960s to May Day demonstrations and the annual Wigstock Parade in the late 1980s. By day, before its reconstruction, the park would be filled with Ukrainian men playing chess, young guys selling drugs, yuppies walking to and from work, a few remaining punks with boom boxes. Puerto Rican women strolling babies, residents walking dogs, kids in the playgrounds. After 1988, there were also cops in cruisers, and photographers, and a growing population of evictees attracted to the relative safety of this "liberated" if still contested space. The encampments burgeoned before June 1991, and were made from tents, cardboard, wood, bright blue tarpaulins, and all sorts of scavenged material that could provide shelter. Hard drug users traditionally congregated in "crack alley" on the southern edge; a group of mostly working people clustered to the east, and Jamaican Rastafarians hung out by the temperance fountain closer to Avenue A. Political activists and squatters congregated closer to the bandshell, which also provided shelter during the rain. The bandshell was demolished in the reconstruction.

Variously scruffy and relaxing, free-flowing and energetic, but rarely dangerous unless the police are on maneuvers, Tompkins Square exemplifies the kind of neighborhood park that Jane Jacobs adopted as a *cause célèbre* in her famous antimodernist tract, *The Death and Life of Great American Cities* (1961). If it hardly has the physical features of a frontier, neither class conflict nor police riots are new to Tompkins Square Park. Originally a swampy "wilderness," its first evictees may have been the Manhattoes whose acceptance of some rags and beads in 1626 led to their loss of Manhattan Island. Donated to the city by the fur trader and capitalist John Jacob Astor, the swamp was drained, a park was constructed in 1834, and it was named after Daniel Tompkins, an ex-governor of New York State and US vice-president from 1817 to 1825. Immediately the park became a traditional venue for mass meetings of workers and the unemployed, although, to the apparent consternation of the populace, it was commandeered for use as a military parade ground in the 1850s and throughout the Civil War.

The symbolic power of the park as a space of resistance crystallized after 1873 when a catastrophic financial collapse threw unprecedented numbers of workers and families out of job and home. The city's charitable institutions were overwhelmed and at the urging of the business classes the city government refused to provide relief. "There was in any case a strong ideological objection to the concept of relief itself and a belief that the rigors of unemployment were a necessary and salutary discipline for the working classes." A protest march was organized for January 13, 1874 in Tompkins Square, and the following account is reconstructed by labor historian Philip Foner:

> By the time the first marchers entered the Square, New Yorkers were witnessing the largest labour demonstration ever held in the city. The Mayor, who was expected to address the demonstration, changed his mind and, at the last minute, the police prohibited the meeting. No warning, however, had been given to the workers, and the men, women and children marched to Tompkins Square expecting to hear mayor Havemeyer present a program for the relief of the unemployed. When the demonstrators had filled the Square, they were attacked by the police. "Police clubs," went one

account, "rose and fell. Women and children went screaming in all directions. Many of them were trampled underfoot in the stampede for the gates. In the street bystanders were ridden down and mercilessly clubbed by mounted officers."

Within an hour of the first baton charges, a special edition of the *New York Graphic* appeared in the streets with the headline: "A Riot Is Now in Progress in Tompkins Square Park."

Following the police riot the New York press provided a script that would have gratified the 1988 mayor. Decrying the marchers as "communists," and evoking the "red spectre of the commune," the New York *World* consistently built an analogy between the repression of the urban hordes in Tompkins Square and Colonel Custer's heroic Black Hills expedition against the savage Sioux of South Dakota. What began in 1874 as an outlandish juxtaposition between the park and the frontier had by the 1980s become an evocative but seemingly natural description.

The destiny of the Lower East Side has always been bound up with international events. The immigration of hundreds of thousands of European workers and peasants in the following decades only intensified the political struggles in the Lower East Side and its depiction in the press as a depraved environment. By 1910, some 540,000 people were crammed into the area's tenements, all competing for work and homes: garment workers, dockers, printers, laborers, craftsmen, shopkeepers, servants, public workers, writers, and a vital ferment of communists. Trotskyists, anarchists, suffragists and activist intellectuals devoted to politics and struggle. Successive economic recessions forced many into unemployment: tyrannical bosses, dangerous work conditions and a lack of workers' rights

elicited large-scale union organizing. And landlords proved ever adept at rent gouging. The decade that began with the Triangle fire of 1911 – the fire engulfed 146 women garment workers from the Lower East Side, imprisoned behind locked sweatshop doors, forcing them to jump to their death in the street below – ended with the Palmer Raids of 1919 in which a wave of state-sponsored political terror was unleashed against the now notorious Lower East Side. In the 1920s as the suburbs burgeoned, landlords throughout the neighborhood allowed their buildings to fall into dilapidation, and many residents who could were following capital out to the suburbs.

Like other parks, Tompkins Square came to be viewed by middle-class reformers as a necessary "escape valve" for this dense settlement and volatile social environment. Following the 1874 riot, it was redesigned explicitly to create a more easily controllable space, and in the last decade of the century the reform and temperance movements constructed a playground and a fountain. The contest for the park ebbed and flowed, but took another surge during the Depression when Robert Moses redesigned the park, and again two decades later when the Parks Department tried unsuccessfully to usurp park land with a baseball diamond. Local demonstrations diverted this redesign. A hangout for Beat poets in the 1950s and the so-called counterculture in the 1960s, the park and its surroundings were again the scene of battles in 1967 when police waded into hippies sprawled out in the park in defiance of the "Keep off the Grass" signs.

This explosive history of the park belies its unremarkable form, making it a fitting locale for a "last stand" against gentrification.

'Urban Violence: An Overview'

from J. Beall (ed.), *A City for All* (1997)

Violence has started to receive considerable attention as a significant urban phenomenon. Its recent acknowledgement as a cause for concern accompanies both its increasing severity and the increasing recourse to social development as a means of tackling its root causes. Social development is a broad church which positively encourages the multidisciplinary approach necessary to understand and tackle the causes of urban violence. In particular, methods such as incarceration are being re-evaluated in recognition of the fact that any durable solution to urban violence must involve the marginalised urban youth often responsible for its perpetration. This chapter examines the multidimensional nature of violence and initiatives to deal with the problem. Diverse experiences of violence are recognised. It is argued that a broader conception of violence than that of 'violence as criminal activity' is necessary so that attention is drawn to groups other than young urban men. The chapter concludes by noting that schemes which have successfully dealt with urban violence have included *all* members of a community.

Dictionary definitions tend to stress the physical nature of violence; the *Collins English Dictionary* defines violence as 'the exercise of physical force, usually effecting or intending to effect injuries, destruction, etc.' To the extent that violence is a physical phenomenon, this is a correct definition. Violence may be actual harm or damage, against either people or property. This harm may comprise either physical or sexual assault. However, policy initiatives that have been constructed around this definition have missed out other important dimensions of urban violence that are highlighted below.

CHARACTERISTICS OF URBAN VIOLENCE

Acts of violence in the urban context have some characteristic features. First, from a purely statistical perspective, the *frequency* of violence is greater in urban than in rural areas. Consider, for instance, that urban violence in São Paulo is the single most common cause of death, with homicide accounting for 5% of all deaths in the city in 1986, or that the most significant cause of death in Colombia in 1995 was violence.

Second, the *intensity* of violence is greater in cities than in rural areas. In the domestic context, particularly concerning violence against women and children, there are high levels of physical, sexual and non-physical violence. These are increased by the heightened incidence of urban alcoholism and drug abuse. At a neighbourhood level, violence tends not to escalate to the same extent in rural areas because social structures exist to prevent and manage pathologies as and when they arise ... Within urban communities, reciprocal relations and bonds of trust are particularly compromised ... to the point where social capital stocks have, in some urban areas, been entirely consumed.

Third, there exists the fundamentally urban phenomenon of *street violence*. In addition to 'individualist' physical or sexual urban violence where victim and perpetrator are not known to each other, this category includes rioting and gang or organised violence. The incidence of gang violence is increasing. The breakdown of family units makes the gang an attractive social support network, especially to young males – often, gangs provide a collective identity or 'street family'. Violence under such circum-stances of anomie and marginalisation becomes a means of self-expression. Gangs themselves are not entirely negative phenomena. Through long--term community measures, gangs have been 'reintegrated' into communities. Successful urban regeneration schemes in Columbia and the United States have had gang cooperation as a prerequisite for success.

Significantly, it is poor families who dis-proportionately suffer the consequences of urban violence. It has been suggested that the existence of vast inequalities in wealth and their frequent juxtaposition in the urban context may be the cause of a decline in psycho-social health. If this is true, it is a further argument to buttress the already compelling case for a reduction in income disparities in the cities of developing countries. Within poor communities, it is men who suffer most direct harm from, and are most likely to be perpetrators of, urban violence. In the US, for instance, black men are more likely to die of homicide than of any other cause, and male life expectancy in the Bronx, New York, is 30 years – far lower than that in low-income countries such as Bangladesh.

RESPONDING TO URBAN VIOLENCE

Previous developmental paradigms held urban violence as a pathology to be dealt with by the police in the short term, and by social workers in the medium term. This view has been replaced by an understanding that the causes of urban violence are systemic and multidimensional. Examples of sectoral responses to urban violence can be found in respect of health, education, employment, the law, housing and urban design.

Recently violence has come to be seen as a health threat. A postmodern analysis of health has brought to the fore the issues of psy-chological and social health and well-being. The threats to health posed by urban violence have been incorporated into strategic public health initiatives in the US and South Africa, for example. Further, a link between psycho-social health, income and urban violence has recently been suggested . . . [by some writers].

Children in particularly poor families may often be prevented from attending school. The influences on children outside the education system are likely to be less benign than those within it. In the absence of parental supervision due to employment commitments, other influences such as street gangs (and television in more developed countries) loom large in formative education. For (predominantly male) adolescents, gangs provide the only oppor-tunity for esteem-building and, to an extent, skill-building (where the skills may be antisocial ones such as burglary). Invariably, formal education programmes are not designed to cope with such problems, while successful alternatives include informal participatory approaches.

The existence of urban unemployment and the lack of opportunities for work create stresses to which poor people are particularly susceptible. For young men, this stress is often compounded by social norms which demand that they provide the main income for the household. The social consequences of urban unemployment are well documented, and the failure of government to provide opportunities to harness these human resources is another factor contributing to urban violence. The lack of income lowers the opportunity cost of crime as a means to an end, and hence increases the incidence of violence in poor areas. A vicious circle then ensues as a region becomes known for violence. In Jamaica, for example, employers are unwilling to hire people from certain areas for fear of the violence they will bring with them.

The law, in certain communities, is seen as antagonistic to the interests of the community, as just another means by which the elite suppress the majority. This is not polemicism – these were the terms used by rioters in Los Angeles after the Rodney King incident in 1992, in which a young black man was gratuitously and severely beaten by police. This image of the judiciary and law enforcement agencies as an 'army' of a certain class (and often race) is not limited to the United States. There is abundant evidence that such a view is held by communities in India, Brazil and, most obviously, South Africa. Major reform of the methods of policing urban communities is necessary. But law enforcement agencies have argued that while it is true that they may need to reconsider their techniques, their duties are largely residual – they are forced to face and deal with the consequences of failures in other agencies in the urban context. In short, the police argue, with some justification, that they are the scapegoats for urban violence.

Housing and urban design can influence urban violence in two ways. First, the degree to which housing and work environments create comfortable environments to live in is a parameter of psycho-social health. Poorly designed buildings can contribute to poor health (consider 'sick building syndrome'), physical isolation and depression, the latter particularly affecting women. Moreover, the built environment can provide circumstances conducive to violence. The labyrinthine nature of urban housing areas, in both developed and developing countries, makes street crime (e.g. mugging, pickpocketing, etc.) easy and fosters fear.

Second, the degree to which a community is able to influence, plan and transform its environment is a measure of its ability to control itself. Violence is often a response to a lack of precisely this kind of autonomy. Evidence need be sought only in the relatively high levels of violence against property in poor urban areas. Writing graffiti, for instance, is an act predominantly carried out by marginalised young men and boys. A consideration of this action supports an interpretation in which the artists are disenfranchised from their environment and feel a need to shape it, and to achieve recognition for it.

In an attempt to respond to this, and to reduce the incidence of inner-city vandalism, a London city council made available several walls for graffiti artists to use. The walls were filled up quickly in an initiative involving artists and local schoolchildren in the production of several often beautiful murals. But despite a brief reduction in vandalism, the council did not provide similarly attractive follow-up opportunities. Within weeks, incidences of graffiti returned to their previous levels. This has been observed in parallel schemes in both developed and developing countries.

If physical violence is an assault on physical health, non-physical violence is an assault on psycho-social health and is an important component of urban violence. This sort of harm may include such phenomena as verbal harassment on the basis of sex, race or religion, and psychological stress resulting from environmental factors. Most countries acknowledge the existence of non-physical violence such as racial hatred, sexual harassment and harassment such as 'stalking'. However, it is not always acted on, while it is difficult to legislate against some aspects of non-physical violence, such as the *fear* of rape or of domestic violence.

WHO ARE THE VICTIMS OF UBAN VIOLENCE?

Women in particular bear the brunt of urban violence, both as victims, and as those forced to deal with the long-term consequences of it (as mothers whose sons kill or have been killed, for instance). Domestic violence is an important component of urban violence . . . [and] persistent sexual harassment is one of the long-term problems faced by women in developing country cities. Bombay, the centre of the Indian film industry, produces 800 films per year in which women are portrayed as sex objects . . . as a consequence there are increased sexual expectations of women, and non-fulfilment of sexual demands has even led to murder. Rinka Patel, for instance, did not want to marry immediately after her college graduation and was dubbed 'too modern' by young men in her neighbourhood (and indeed, subsequently by the media). The son of a community leader, together with other youths, gang-raped her before killing her and burning her body. Because of the perpetrators' status, the incident was initially registered as a suicide.

The gendered effect of urban violence has an impact on women's choice of housing . . . Women are prepared to pay at least 50% over the market rate to live in accommodation that is secure.

Finally, evidence that the long-term stresses of violence rest disproportionately on the shoulders of women comes from the World Bank (1993). Psycho-social disorders are, according to the *World Development Report 1993*, among the top ten diseases in the disease profile of Third World cities. Further, within cities, women and the young seem to be more affected than men. The *elderly and disabled*, although statistically unlikely to suffer direct physical violence, face an increased fear of vulnerability to attack. This can often severely affect their quality of life.

In some areas, *ethnic, religious, sexual or other minorities* are subject to open targeted violence. South Africa has been the scene of institutionalised and targeted ethnic violence, but any city where minorities exist is liable to communal violence. The UNHCR estimates that there are currently around 30 million internally displaced people. In times of conflict, these people tend to head for the cities. This leads to tensions, and conflicts which increase the chance of escalating violence.

Street children face particular difficulties, in terms of both vulnerability to violence and exposure to the factors that normalise it. Exposure to violence in urban areas has also led to children accepting and promoting violence as acceptable behaviour. They risk death by vigilante groups 'cleaning the streets', as in Rio. Both girls and boys face being forced into prostitution – in Nairobi, there have been instances of street girls as young as eight being involved. Children are also at risk in violent homes . . . [facing] particular difficulties . . . with respect to psychological and domestic violence.

BETTER PRACTICE

A common feature of the better practice solutions is that they all have a political empowerment component. It is often the case that the poorest members of a community are also among those least politically represented. Violence is not only a response to bad governance, however; it is a symptom of it, a sign that voices are not being heard and that groups of people are marginalised from the political process. Given the chance, communities worst affected by urban violence have shown themselves able to begin to tackle it at community and municipal levels, by trying to change the factors that are partly the result of, partly a contributing factor to, urban violence. An exceptional example of this is a scheme in Cali, Colombia. In 1992, murder was the most frequent cause of death for the general population, and Cali had a murder rate of 87 per 100,000 per year. In June 1992, a public initiative called DESEPAZ (development, security and peace), involving local NGOs, CBOs and municipal authorities, began. It acknowledged the multidisciplinary nature of the problems, and responded in kind with a three-tier policy solution.

At the level of *law enforcement*, the court system was modernised to increase local involvement, protection for the police in dealing with violence was increased, and public safety councils were created to deal with community violence issues. At the intermediate level, *micro-enterprise schemes* were started and the creation of 25,000 sites and service plots was authorised. It was hoped that the security of tenure provided through these would decrease incentives for criminal income, and decrease psycho-social stress. At the long-term level, DESEPAZ had a strong *community education* component: young people, often in gangs, surrendered their weapons and persuaded children to give up their war toys in exchange for passes that allowed access to play areas citywide. This was mirrored by a gun amnesty for gang members, which resulted in the disarmament of four entire gangs. Similar moves, fundamentally from a community level, have been encouraged in the US.

Another example of successful urban violence management comes from Uganda. 'Resistance Councils', similar to the Public Safety Councils of Cali, have a range of responsibilities to the small urban populations that elect members to them. Councils must guarantee security and respect for the law within their jurisdiction. Significantly, they are also links between government and the community, and as such aim to promote community development within their constituency. By all accounts, they have proved successful. Not only have they contributed to the socio-economic status of their areas, but crime rates have been reduced, relations with the police improved and, most tellingly, the 'abuse of power by the military' has been avoided. Their success has been based on an effective involvement of local residents, and on genuinely inclusive practice.

CONCLUSION

Urban violence has diverse causes and effects. Moves to include urban youth in programmes to combat urban violence, are, of course, to be welcomed, but urban violence is more than just street crime. The disempowering psychological and domestic dimensions tend to be underplayed in policy initiatives, and thus those people excluded from voicing their experiences are ignored. To remedy this, two courses of action suggest themselves. The first is to enhance municipal and government cooperation across the sectors mentioned above, in order that a coordinated solution might be brokered. But, as the DESEPAZ and Ugandan Council cases show, institutional multi-sectoralism is insufficient by itself. Inclusive community level involvement is also necessary. Resources also need to be directed at these community-level initiatives. The initial results from DESEPAZ appear promising, and if they continue perhaps governments can be persuaded that these measures are a cost-effective solution to an increasingly costly, isolating and dangerous problem.

CULTURES OF RESISTANCE AND TRANSGRESSION

INTRODUCTION

One of the most important traditions of the city is that it has never been a place of uniform, homogeneous culture. It has always been a place of difference and of variation. And, of course, such heterogeneity often leads to competition and contestation among different groups. For the culture of the city, important questions then arise. If there is contestation in the city, whose will is to dominate and, conversely, how are people to resist and transgress that domination? What are the cultural sites on which this resistance should take place? What are the most appropriate cultural themes to challenge? And what form should these challenges adopt?

In the 'Border Crossings' extract from *Geographies of Exclusion* (1995, Routledge), urban geographer David Sibley considers the spatial and social entity at which resistance and transgression most often occurs: the boundary or border. As Sibley points out, it is crossing boundaries that provides moments of cultural meaning and personal exhilaration, and this is, therefore, a perceived and relative culture as much as a quantitative and finite occurrence. Border crossings can then occur in relation to the home, the person, age, familiarity, social beliefs and morality. Boundaries are therefore things to be transgressed as well as reinforced, inverted as well as upheld. Resistance and transgression, for their part, are revealed as spatial as well as social, cultural as well as economic, actively performed as well as passively recorded. They occur everywhere rules and order are established.

One of the most powerful examples of this kind of performative resistance, playing on all manner of cultural and political assumptions, is provided by Susana Torre in her essay 'Claiming the Public Space: The Mothers of Plaza de Mayo' (1996). In seven years of rule by a military junta, Argentina experienced the 'disappearance' – the systematic abduction, torturing and execution – of a large sector of its civilian population. In response, a protest was mounted by the 'Mothers' of those who disappeared, meeting regularly in the highly symbolic Plaza de Mayo as an overt resistance to the oppressors. Here, several actions were resisted and boundaries transgressed, involving not only highly significant political acts and places, but also the age and status of the protesters, the way in which they moved and acted, and the frequency of their actions. More than a simple protest march, the very actions of the Mothers confronted the authority and tactics of the male military.

Protests against state actions have increasingly tended from the later years of the twentieth century onward to use a multiplicity of tactics and rationales by which to act. Greatly influenced by many of these resistances have been the ideas of the Situationist International, a highly politicized group of European artists, film-makers, writers, architects and activists operating in the late 1950s and 1960s. 'Introduction to a Critique of Urban Geography' (1955) (in Knabb (ed.) *Situationist International Anthology*, 1981, Bureau of Public Secrets) (see Suggestions for further reading, p. 8 above), written by the Situationists' main theorist Guy Debord, set out many tenets of the group, notably the idea of 'psychogeography' where participants respond to the emotional and behavioural aspects of the city. By turning resistance into a kind of game, the Situationists attacked some of the fundamental principles of the capitalist city – not only functionality and efficiency but the very idea of work, rationality, commodification – thus 'flooding the market with a mass of desires'.

Many of the Situationists' ideas and tactics are echoed, explicitly or implicitly, in more recent environmental protests against the building of roads. George McKay, in an extract from *Senseless Acts of Beauty* (1996, Verso), details road protests in the UK. Arguing against urban blight and pollution as well as class privilege, these activities involved a highly spatial form of cultural protest, comprising not only the occupation of land and trees, and the squatting of houses, but the setting up of whole new communities ranging from a tree-as-home to a whole street. Alternative publications, graffiti, cafés, public art and forms of defensive construction were all used not only to communicate their cause but to promote the alternative forms of cultural reasoning that underlay the protesters' actions.

Not all such protests and resistances, however, are treated as wholly oppositional by city authorities. In 'The Stimulus of a Little Confusion' (1996), Edward W. Soja considers the rather different cultural context of Amsterdam where the squatters and alternative communities in various pieces of canal-side architecture have been treated with 'repressive tolerance' and 'flexible inflexibility'. Although maintaining some of the cultural rhetoric and tactics of the road protesters in McKay's essay, here the squatters are seen to exist side-by-side with the rest of the city, socially and spatially absorbed into an urban heterogeneity.

49 DAVID SIBLEY

'Border Crossings'

from *Geographies of Exclusion* (1995)

The sense of border between self and other is echoed in both social and spatial boundaries. The boundary question, a traditional but very much undertheorized concern in human geography, is one that I will explore in this chapter from the point of view of groups and individuals who erect boundaries but also of those who suffer or whose lives are constrained as a result of their existence. Crossing boundaries, from a familiar space to an alien one which is under the control of somebody else, can provide anxious moments; in some circumstances it could be fatal, or it might be an exhilarating experience – the thrill of transgression. Not being able to cross boundaries is the common fate of many would-be migrants, refugees, or children in the home or at school. Boundaries in other circumstances provide security and comfort. I will start by examining some general characteristics of boundary zones and then describe some of the diverse ways in which boundaries are constructed, demolished and energized.

In a rather formalistic treatment of the boundary problem, Edmund Leach played with the idea of separating and combining categories, focusing particularly on the intersections of sets ... The need to make sense of the world by categorizing things on the basis of crisp sets – A, not-A, and so on – is evident in most cultures, although I do not think it is a universal need, as Leach suggests. However, it is a good place to start because problems associated with this mode of categorization are readily identifiable.

Problems arise when the separation of things into unlike categories is unattainable. The mixing of categories ... by the intersections of sets, creates liminal zones or spaces of ambiguity and discontinuity. As Leach recognized, 'There is always some uncertainty about where the edge of Category A turns into the edge of Category not-A.' For the individual or group socialized into believing that the separation of categories is necessary or desirable, the liminal zone is a source of anxiety. It is a zone of abjection, one which should be eliminated in order to reduce anxiety, but this is not always possible. Individuals lack the power or organize their world into crisp sets and so eliminate spaces of ambiguity.

To move from Leach's abstraction to the concrete, we might consider the case of the home. To the occupier, the home may represent a space clearly separated from the outside. Inside the home, the owner or tenant may feel that space is ordered according to his or her values. However, problems can be created by entrance, breaches in the boundaries of the home. The entrance, the hallway or passage provides a link between the private and the public, but it constitutes an ambiguous zone where the private/public boundary is unclear and in need of definition and regulation in order to remove the anxiety of the occupier. If you admit strangers to the house, are they confined to an entrance area or allowed to enter a living space? How do you cope with the Jehovah's Witnesses or with the person selling double-glazing? The response will depend on where the householder locates the boundary, but this may be variable, depending on how the outsider is perceived in relation to the occupier's conception of privacy.

A second example, child/adult illustrates a similarly contested boundary. The limits of the category 'child' vary between cultures and have changed considerably through history within western, capitalist societies. The boundary separating child and adult is a decidedly fuzzy one. Adolescence is an ambiguous zone within which the child/adult boundary can be variously located according to who is doing the categorizing. Thus, adolescents are denied access to the adult world, but they attempt to distance themselves from the world of the child. At the same time, they retain some links with childhood. Adolescents may be threatening to adults because they transgress the adult/child boundary and appear discrepant in 'adult' spaces. While they may be chased off the equipment in the children's playground, they may also be thrown out of a public house for under-age drinking. These problems encountered by teenagers demonstrate that the act of drawing the line in the construction of discrete categories interrupts what is naturally continuous. It is by definition an arbitrary act and thus may be seen as unjust by those who suffer the consequences of the division.

BOUNDARY MAINTENANCE AND SOCIAL ORGANIZATION

In using these two examples, I am suggesting that liminality presents as many problems for highly developed capitalist societies as for the relatively simple agrarian and hunter-gatherer collectivities which have been the primary focus of anthropological research and where much of the theory of boundary dynamics has been developed. Dichotomies like traditional/modern or simple/complex do not seem to have much relevance to the questions of boundary drawing, inclusions and exclusions. Perhaps a meaningful distinction could be made between what Davis and Anderson term *high-density* and *low-density* social networks. Albeit crude, this dichotomy does suggest varying attitudes to difference which might be attributed to the density of social interaction. Davis and Anderson suggest that in high-density networks 'most links are strong and one is likely to know and have direct ties to most people affected by the misbehavior [*sic*] of a member of one's network'. Conversely, they argue that in networks of low density, difference is less visible because there is less shared knowledge of individuals within the community. In pre-industrial societies, people are enmeshed, involved in each other's lives through extended family, kin connections or clan membership coupled, in many cases, with simple physical

propinquity. Gypsies and other semi-nomadic minorities demonstrated the characteristics of shared knowledge of members of the community and of physical nearness very clearly. An outsider in such a community is very exposed. However, similar forms of social organization are found within developed societies, both within traditional working-class and suburban neighbourhoods.

A division based on the density of social networks is fairly close to Durkheim's schema in which he distinguishes between societies exhibiting *mechanical* and *organic* solidarity. Where social identity is based on mechanical solidarity, Durkheim argues that

> the society is dominated by the existence of a strongly formed set of sentiments and beliefs shared by all members of the community [so] it follows that there is little scope for differentiation between individuals: each individual is a microcosm of the whole.

With shared beliefs, we can talk of a *conscience collective* which 'completely envelopes individual consciousness'. By contrast, individualism is consistent with social solidarity in developed industrial societies because solidarity is organic, that is, deriving from contractual relationships which develop with an increasing division of labour. Durkheim assumed that these contracts were governed by norms which comprised the glue holding society together, but norms did not preclude individual difference. This dichotomy does considerable violence to reality and it was probably a fairly crude representation of varying forms of social organization when Durkheim was writing in the late nineteenth century. Like Davis and Anderson's view of social networks, it does provide some ideas about the way people might collectively react to difference, but we need knowledge of the social, political and geographical contexts of community responses to 'others' in order to say anything useful about conflicts based on difference. With the globalization of culture and the almost total penetration of capitalist forms of consumption, it certainly does not make sense to characterize societies in Durkheim's terms. Yet, there is something approaching a conscience collective in some middle-class North American suburbs and on some local authority housing estates in Britain, manifest in reactions to the mentally disabled, Gypsies and Bangladeshis, for example.

Some useful ideas on this kind of hostility to others comes from studies of small groups by social anthropologists. Here, the work of Mary Douglas and her critics is particularly illuminating.

POLLUTION, DISCREPANCY AND SMALL GROUP BOUNDARIES

At the social level, as at the individual level, an awareness of group boundaries can be expressed in the opposition between purity and defilement. In Mary Douglas' *Purity and Danger* and subsequent writing, she developed this thesis, gathering support for her argument largely from fieldwork with tribal societies and ancient texts, particularly the Old Testament of the Bible as a record of Judaic ritual. Her key argument is that 'Uncleanness or dirt is that which must not be included if a pattern is to be maintained . . . in the primitive culture, the rule of patterning works with greater force and more total comprehensiveness [*sic*] . . . [than in a modern industrial society]'. By patterning, Douglas means the imposition of a symbolic order 'whose keystone, boundaries, margins and internal lines are held in relation by rituals of separation'. Separation is a part of the process of purification – it is the means by which defilement or pollution is avoided – but to separate presumes a categorization of things as pure or defiled . . .

From such observations, Douglas proposes a rule for categorization in ancient Israel which forms the basis for a general rule, namely, that 'the underlying principles of cleanliness in animals is that they shall conform fully to their class. Those species are unclean which are imperfect members of their class or whose class itself confounds the general scheme of the world'. Thus, it is those animals, people or things that are discrepant, that do not fit in a group's classification scheme, which are polluting. The evidence for this from records of the practices current in ancient Israel are quite convincing and Neusner has compiled a long list of polluting activities, conditions of the body, animals, and so on, which lends support to Douglas's thesis. However, to generalize from one ancient culture, where there were strong rules of exclusion laid down by the rabbis who also wrote the texts that constitute the evidence,

to all small groups and tribal societies is dangerous. Murray, in particular, questioned the empirical validity of her argument, suggesting that people were not really that concerned about defilement and happily mixed discrepant categories in their daily struggle for survival. He asserted that: 'if there is any psychological reality to the 'horror' purportedly inspired by such classification difficulties, it is confined to anthropologists intent on eliciting complete and exhaustive contrast sets.' Accordingly to Murray, the people who were supposedly engaged in these boundary rituals would be driven to anxiety 'by proscribing and even attempting to annihilate what is not readily classifiable from the world'. This criticism was hardly fair because Mary Douglas had herself recognized the limitations of her original thesis, long before Murray's assault.

I think we can conclude that Douglas's argument about purification and defilement needs to be qualified in regard to time and place, but I would also argue that it has wider application than she recognized. In *gemeinschaft*-like groups, that is, closed, tightly knit communities with something approaching a conscience collective, it may be that adherence to the rules is more likely in times of crisis, when the identity of the community is threatened. However, my observations in English Gypsy communities suggest that poverty and family size are also factors affecting observance of pollution taboos. In order to ensure that things are not polluted (*mochadi* or *marime*), numerous separations are required, including among utensils which are used for washing food, clothes, and the body. In poor families with several small children, however, it is often impractical to comply with all these rules. There may be a shortage of water or not enough washing bowls and the mother may be too tired to meet the ritual requirements all the time. In affluent families, particularly with grown-up children, pollution taboos are much more likely to be observed. There is some support here for Murray's argument, but he overstates his case.

Boundary consciousness is also a characteristic of the mainstream in modern, western society, or, at least, it is in some kinds of locales and at certain times. The North American suburb has been represented as a particular kind of *gemeinschaft* within the swathe of individual anonymous worlds that are supposed to constitute the modern metropolis. The suburb was first described as an exclusionary, purified social space by Richard Sennett, in *The Uses of Disorder*; the anatomy of the North American variety has been examined in some detail by Constance Perin; and Mike Davis describes the enclosed communities, socially purified and defended fortress-like, against the supposed threat posed by the poor, which are an increasingly prominent feature of the geography of Los Angeles. Affluent suburbs in Britain are similarly coming to resemble these closed communities where the discrepant is clearly identified and expelled, like a suburb in Bristol which hired a private security firm to patrol its leafy avenues and eject what the security officers described as 'hostiles', including young men in baseball caps.

In these suburbs, there is a concern with order, conformity and social homogeneity, which are secured by strengthening the external boundary, but, as Davis recognizes, 'the greater the search for conformity, the greater the search for deviance; for without deviance, there is no self-consciousness of conformity and *vice-versa*'. This process is seen by the members of the community as a virtuous one – it brings into being a morally superior condition to one where there is mixing because mixing (of social groups and of diverse activities in space) carries the threat of contamination and a challenge to hegemonic values. Thus, spatial boundaries are in part moral boundaries. Spatial separations symbolize a moral order as much in these closed suburban communities as in Douglas's tribal societies.

BOUNDARY ENFORCEMENT

Generalizations like Sennett's 'purified suburb' have to be qualified. While there is plenty of evidence that purified suburbs exist, with damaging consequences for the welfare of the rest of the population in metropolitan areas, not all suburbs are like this. Apart from an increase in racially mixed suburbs in the United States, it has also been argued that (British) suburbs can provide a refuge for eccentrics. A concern with privacy, minding your own business, is also characteristic suburban behaviour. However,

communities which much of the time appear to be indifferent to others do occasionally turn against outsiders, particularly when antagonism is fuelled by moral panics. Moral panics heighten boundary consciousness but they are, by definition, episodic. Fears die down and people subsequently rub along with each other. Often, but not invariably, panics concern contested spaces, liminal zones which hostile communities are intent on eliminating by appropriating such spaces for themselves and excluding the offending 'other' . . .

One of the most remarkable features of moral panics is their recurrence in different guises with no obvious connection with economic crises or periods of social upheaval, as if societies frequently need to define their boundaries . . .

The resonance of historical panics in modern crises is worth noting because it demonstrates the continuing need to define the contours of normality and to eliminate difference. This is as evident in what are claimed to be post-modern western societies, post-modern in the sense that they embrace difference, as it was in Nazi Germany or England in the seventeenth century. I will describe one case from seventeenth-century England because of its remarkable similarity with modern panics, particularly in the way in which difference is represented by the popular media.

In the early modern period, before the industrial revolution in Europe and North America, Christian religion was an important source of conflict. The line between conformity and dissent, good and evil, light and darkness was of great concern for an established church attempting to isolate or eliminate religious minorities which appeared to threaten its theological hegemony. The Ranters were one such group who appeared threatening despite their very small numbers. Their dissent from the established church was not unique but their views were publicized at a critical juncture in English history, just after the Revolution of 1649 and the establishment of the Commonwealth, when there was general political uncertainty. At this time, religion was politics and the collapse of religious authority was portrayed in sensationalist literature as 'a prelude to unbridled immorality and social chaos'.

In this context, the Ranters represented a subversive and threatening group and their difference and deviance were amplified in a sustained press campaign which had little regard for the truth. Connections were suggested between religious dissent, atheism and immorality. Interestingly, as in witch crazes, Ranters were associated with inversions which threatened moral values, specifically, devil worship and promiscuity. While women generally had little scope for sexual relations outside marriage, Ranter women were portrayed as sexually unbounded: 'They were free to copulate with any man and did so enthusiastically and openly'.

[. . .]

The Ranter panic is comparable to modern instances in the sense that the group is represented as a threat to core values. Core values in the seventeenth century were to a great extent religious values, whereas in modern secular societies, core values are embodied in the family, the home and the national, and thus they have implications for 'deviant' youth, other sexualities and racial minorities. Although the folk devil comes in different guises, panics do not introduce a succession of new characters bearing no resemblance to each other. Rather, they are manifestations of deep antagonisms within society, for example, between adults and teenagers, blacks and whites, heterosexuals and homosexuals. The alterity personified in the folk devil is not any kind of difference but the kind of difference which has a long-standing association with oppression – racism, homophobia, and so on. The moral panic will be accompanied by demands for more control of the threatening minority, for the state to provide stronger defences for, say, white, heterosexual values. This may include a call for the stronger bounding of space to counter the perceived threat, as in attempts by the British government to exclude New Age Travellers from the countryside and secure this terrain for the middle classes. An account of two recent sources of moral panics in North America and Britain – AIDS and muggings – demonstrates most of these points.

[. . .]

Two things came together in the reporting of street crime in the 1970s. The first was the notion of the British inner city as a *black* inner city, characterized by lawlessness and vice, so that inner city became a coded term for black deviance. The idea of a black inner city bore little relation to demographic or geographical reality, but the myth is more important than the reality. Thus, in a typical example of place labelling, Weaver claimed that the local press 'portrayed north central Birmingham as a violent, crime-ridden area, beset by problems rooted in the nature of its coloured [*sic*] residents rather than in the district's disadvantaged position in British urban space'. Second, there was a rise in recorded street crime but a perceived rise, particularly in 1973, amounting to 'a national mugging scare', and this fixed the idea of black youth as an inherently criminal minority and inner cities as inherently criminal localities. Susan Smith's quotations from Birmingham newspapers, like 'Society at limit of leniency' and 'Angry suburb', indicate a panic which was generated through the stereotyping of minority group and locality. This required a silence about policing, unemployment, the population composition of the district and comparative crime statistics for the city, which would have put a different complexion on the issue. The labels attached to the inner city during this panic strengthened the boundary separating the 'respectable white suburbs' and 'black inner areas' and decreased the likelihood of white people gaining knowledge of Afro-Caribbean and Asian communities through experience. A panic surrounding inner-city riots in the 1980s again confirmed the boundary, with material consequences for inner-city residents, such as the withdrawal of financial services. Different moral panics with slightly different scripts signalled the continuing presence of racism.

'Family', 'suburb' and 'society' all have the particular connotation of stability and order for the relatively affluent, and attachment to the system which depends for its continued success on the belief in core values is reinforced by the manufacture of folk devils, which are negative stereotypes of various 'others'. Moral panics articulate beliefs about belonging and not belonging, about the sanctity of territory and the fear of transgression. Since panics cannot be sustained for long, however, new ones have to be invented (but they always refer to an old script).

INVERSIONS AND REVERSALS

Moral panics bring boundaries into focus by accentuating the differences between the agitated guardians of mainstream values and excluded others. Occasionally, these social cleavages are marked by inversions – those who are usually on the outside occupy the centre and the dominant majority are cast in the role of spectators. Inversions can have a role in political protest in the sense that they expose power relations by reversing them and, in the process, raise consciousness of oppression. They energize boundaries by parodying established power relations.

In early modern Europe, inversions constituted a popular genre known as World Upside Down. Broadsheets illustrating such a world were widely distributed among the illiterate, the themes being virtually unchanged for several hundred years. Illustrations showed, for example, the blind leading the sighted, sheep eating wolf, child punishing father, beggar giving alms to the rich. Their popularity with the oppressed could be accounted for by the fact that they fantasized about the existing order. Sometimes, reversals could serve as a symbol of actual revolt.

[. . .]

The occasions when inversions assume a centre–periphery form, when the dominant society is relegated to the spatial margins and oppressed minorities command the centre, may represent a challenge to established power relations and, thus, be subject to the attentions of the state. There may be attempts to control or suppress such events because they harness the energies of groups which challenge mainstream values ... This is particularly the case with carnivalesque events which are licensed but have contested spatial and temporal bounds. For example, Caribbean carnivals in British cities have been grudgingly accepted by the state as legitimate celebrations of black culture in an avowedly pluralist society. But, in the past, they have been heavily policed and contained. The appeal of the exotic for the white majority mixes uneasily with the images of black criminal stereotypes which have informed the responses of the control agencies. Similar conflicts occurred over the carnivalesque centre on festivals in rural England which attract New Age Travellers. With the Criminal Justice and Public Order Bill, the British government is attempting to seriously limit or ban festivals which are seen as a threat to the cherished values of rural England. Inversions of this kind are thus important indicators of marginality. Responses to carnivalesque events demonstrate how the majority constructs the 'other'.

Other reversals may have less political currency although they can still be symbolically potent. One such case is the Gypsy pilgrimage to Saintes Maries de la Mer on the Camargue coast in the south of France. Gypsies have been relegated to the margins of French society for centuries, and being a Gypsy in the seventeenth and eighteenth centuries was under most regimes a crime warranting execution, mutilation, transportation or a life sentence in the galleys. Since 1935, however, French and other European Gypsies have taken over the small town of Saintes Maries de la Mer on 24 and 25 May and again in October, for a ritual which inverts the practice of the established church in that the object of reverence is Sara, a black madonna. It was only in 1935 that French Gypsies gained ecclesiastical authority to venerate Sara, who has not been canonized by Rome. Although the pilgrimage has now been given a tourist gloss and the Gypsy veneration includes the other Saintes Maries, Salome and Jacobe, it is still a subversive event which expresses the collective but highly circumscribed power of European Gypsies and expresses the long history of racism to which they have been subject.

CONCLUSION

The propositions of object relations theory – the bounding of the self, the role of good and bad objects as stereotypical representations of others, as well as their representation as material things and places – can be projected onto the social plane. The construction of community and the bounding of social groups are a part of the same problem as the separation of self and other. Collective expressions of a fear of others, for example, call on images which constitute bad objects for the self and thus contribute to the definition of the self.

The symbolic construction of boundaries in small groups which have been studied by social anthropologists has its counterpart in the marking off of communities in developed western societies. Consciousness of purity and defilement and intolerance of difference secure some groups within the larger spaces of the modern metropolis. The outside is populated by a different kind of people who threaten disorder, so it is important to keep them at a distance. These fears, however, are fuelled by the exaggerated accounts of some sections of the media and the state who represent the claims of others for space, or simply for the right to dissent, as a threat to core values. Social and spatial boundaries in these circumstances become charged and energized. The defence of institutions like the family and spaces like the suburb becomes a more urgent undertaking during a moral panic. The oppressed, however, have their own strategies which challenge the domination of space by the majority, if only briefly and in prescribed locales. Ultimately, carnivalesque events confirm their subordination.

The problems that I have been discussing here concern, in part, territoriality, the defence of spaces and transgressions. Space is implicated in many cases of social exclusion, and in the next chapter I will try to identify in more detail the characteristics of exclusive social spaces and to relate these spaces to questions of power and social control.

'Eco-Rads on the Road'

from *Senseless Acts of Beauty* (1996)

[. . .]

'WE ARE MORE POSSIBLE THAN YOU CAN POWERFULLY IMAGINE': WANSTONIA, LEYTONSTONIA AND OTHER INDEPENDENT FREE STATES OF BRITAIN

> At Twyford they weren't really connecting it. It was nice fluffy landscapes and not about houses and people and their communities. The M11 has made roads into the issue it should have been.
>
> No M11 Link protester

In the case of the direct action protest against the construction through London of a link to the M11 motorway, homes, roads and the environment are all connected. The etymology of the word *ecology* is apt here: one of its roots is the Greek word *oikos*, meaning 'habitat' or 'house'. From the pastoral beauty of the South Downs, road protest spread to embrace urban concerns. Already in the early seventies, André Gorz describes the impact of the car on urban space: 'The car has made the big city uninhabitable. It has made it stinking, noisy, suffocating, dusty . . . cars have killed the city.' But Gorz also began to feel (will?) an alternative: now, ironically, 'After killing the city, the car is killing the car.' Gorz sees the possibility of the car turning in on itself, consuming itself, as its rhetoric of individual freedom and speed is replaced by its social reality of pollution, death and worst of all, for we allow the car industry to cope with these first two awful facts with surprising ease, the nightmare of contemporary existence, *slowness*. (The point of the M11 link roads is to cut seven or eight minutes from commuters' driving time into central London.)

Planning blight, the gradual decay of an area proposed for development of demolition, has a long history in east London. Patrick Field writes that 'Long-term residents of Leyton and Leytonstone will tell you about shops that closed in the fifties because the Link Road was coming.' Such blight had an up side, too: gentrification bypassed the blighted streets, and 'As well as architectural authenticity the side roads in the Link Road corridor became a preservation area for legendary East London life.' An article in the Earth First! magazine *Do or Die* called 'News From the Autonomous Zones' describes the situation:

> The area threatened by the Link Road comprises two very different localities. At the eastern end of the route is Wanstead, a reasonably affluent, conservative leafy-green London suburban. To the west are Leyton and Leytonstone, areas of high-density urban housing, built at the turn of the century, but badly neglected ever since the proposal for the Link Road first blighted the area forty years ago.

As at Twyford Down, when the latest campaign began there was already strongly established opposition to the road, in the form of local committees, presentations to official inquiries, and so on. Some of this opposition had employed direct action, too, such as the disruption of public inquiries by a long-term vociferous campaign group. However, in September 1993 contractors moved in to Wanstead with their bulldozers to begin clearing trees along the proposed route. This led to the first sustained direct action of the No M11 Link Campaign: the first squat in Wanstead, in a roofless house at 110 Eastern Avenue, and six weeks of obstruction of the tree-clearing operation by local protesters and a small number of experienced eco-rads. A chestnut tree (later capitalized and given a definite article) at George Green in Wanstead suddenly became the focus for protesters and increasing numbers of locals when they realized that it was to be cut down to make way for a cut-and-cover road tunnel under the park of George Green. The protection of the Chestnut Tree in the park came quickly to symbolize what was under threat from the road: residents' health, their past, nature more generally, even simply an effective local voice. Through November and early December 1993 the Chestnut Tree functioned as new impetus and the focal point – a tree house and a bender appeared, and each evening locals, squatters, schoolchildren, the odd Donga from Twyford would gather round the campfire for chat and campaign. After a letter was delivered by the local postman to the tree itself, campaign solicitors argued in court that 'the tree-house should be formally recognized as a "dwelling"'. This duly happened, with the result that the authorities 'now had to apply for a court order to evict the tree dwellers from their new "home"' – causing further delay. However, in the early morning of Tuesday, 7 December the familiar and inevitable happened: several hundred police arrived with full equipment to claim back the land and the Chestnut Tree. This may have been the first time a cherry-picker – a high-hoist hydraulic platform – was used to reach and lower protesters from on high. The sight of such an operation – and its widespread reporting in the media – reflected badly on the Department of Transport, signalled its seemingly inflexible approach to the community. The local lollipop woman (a worker whose place of work is a road) said after being sacked for attending a ceremony at the ancient Chestnut Tree of Wanstonia while wearing her uniform: 'All life is politics – if you step out of line.'

A number of large Edwardian houses were next in line for demolition, on Cambridge Park Road. Two were still lived in by long-term residents, and these became the heart of the Independent Free Area of Wanstonai, born on 9 January 1994, the inspired result of a long session in a pub between a campaign solicitor and a Donga. Wanstonia was picked up by the media as self-consciously constructed and was declared in part to refer back to the anti-bureaucracy little England of the 1950s Ealing comedy *Passport to Pimlico*. Wanstonia printed its own passports, its flag was the Union Jill. This seriously parodic unilateral declaration of independence set the tone for much of the rest of the No M11 Link Campaign: Situationist-inspired humour and art and an emphasis on alternative English social and cultural practices (as with the politicized echo of Ealing) became paramount features of daily life for protesters. Wanstonia was evicted on Ash Wednesday 1994, renamed Bash Wednesday by some of those on the receiving end of things.

'The M11 campaign as a whole has largely been this sequence of evictions, each one of which has broadened the campaign and been a spur forcing us to think up fresh ideas,' explains protester Phil McLeish. The focus of the campaign moved farther into London along the line of the planned route. A bender site in which protesters lived was set up in old gardens and waste ground at the rear of Fillebrooke Road, signalling a nomadic tradition and the continuing concern with the destruction of living spaces all around. The seemingly temporary living space of the bender showed a surprising solidarity with the previously considered permanence of bricks and mortar. The bender site, and the old yew tree in it, were the short-lived heart of Leytonstonia, decorated with a kerbhenge, a replica of Stonehenge made with torn-up kerbstones. (Some talked of the significance of the names Leyton and Leytonstone, of their relation to ley lines more generally, as though this might explain the energy of the protest sites. A brickhenge was made at Claremont Road, halfway between Leyton and Leytonstone; again art and countercultural reference were created from the waste products of destruction.) 'Particularly after the bender site was trashed, Claremont Road became the centre of the campaign … In concentrating on Claremont Road, we gradually moved into a siege situation where defence became the overriding priority, although irregular actions on site continue[d].'

The idea of direct action as performance was seen at its most vibrant in the squatted row of terraces at Claremont Road, which gained much of its impetus from the artwork it produced. The relative density of artists in the local population was, ironically, partly due to the link road: houses that had been compulsorily purchased over the years were released to housing associations which encouraged artist communities by providing short-term low-rent accommodation. This short street was squatted and blocked off to prevent any vehicles entering it; it was an effort to reclaim a safe public place. Armchairs on the road blurred the distinction of internal and external space, and more: the project was 'the turning of the street itself into a "living room" by using the furniture, carpets, fittings and other objects from some of the houses . . . to make actual rooms on the street'. There were two cafés, an information centre and an exhibition area, the Art House. In the Art House coinage was painted onto the enamel of a sink: money going down the drain. The street itself was full of public art, a chessboard painted on the road, hubcaps as pawns, traffic cones and broken hoovers as other pieces. Sculptures in the road were also barricades. A car was 'pedestrianized': chopped in two with each half placed on opposite sides of the road, painted and joined by the black and white stripes of a pedestrian crossing. Another car sprouted plants, had RUST IN PEACE on its side. Walls were painted with bright slogans and murals, including a daisy chain the length of the terrace. Installations vied with rubbish for attention. Above the street there was strong netting, an aerial walkway from tree to roof. A scaffolding tower built on a roof reached up into the air, described by Claremont Road residents variously as a critical parody of the Canary Wharf tower, an update of Tatlin's unbuilt monument to the Russian Revolution, a NASA rocket launcher for other campaigns. It was also a landmark visible for miles and an effective obstruction to bailiffs when the eviction finally took place – in all these interpretations of a single piece of resistance culture can be seen the wealth of these new versions of what John Jordan calls 'activist aesthetics'.

Campaigners and sympathetic visitors to Claremont Road – like much of the culture of resistance before the nineties, though possibly more so – view and respect alternative histories and possibilities of Britain to those put forward by majority culture. Utopian space created in the present reaches back and forward in time; the temporary autonomous zone puts down roots . . . Regular Sunday parties began to attract people other than strict environmentalists, including, following the July 1994 national demonstration in London against the Criminal Justice Bill, ravers, and a less welcome element of 'lunch outs' which severely tested the tolerance praised above.

> Although some people in the No M11 campaign have argued that the new people didn't seem to be here because of the fight against the road, it was already apparent to many of us that our struggle was about far more than the road anyway. The road itself obviously raised issues that didn't fit neatly into the 'environmental issues' category: housing, the issue of protest itself, etc. But in taking over Claremont Road it became increasingly clear that the struggle was about a whole way (or ways) of life. Thus for many of the new people coming onto the street, the issue was the Criminal Justice Bill – an attack on non-mainstream ways of life. Claremont Road was a free space . . .

In an uncomfortable irony, the protest which had done so much to dodge eviction had itself on occasion to fall back on a similar tactic: some 'lunch outs' were evicted from the free space of Claremont Road. The effort talked about here – to take a single-issue campaign beyond its self-imposed limitations in a sustained manner – was termed by one protester 'polysemic dissent', by another a 'life of permanent struggle'. By this stage there was a definite split, both geographical and political or tactical, between the No M11 Link campaigns of Wanstead and Leytonstone. As one protester wrote at the time, 'Some of the Wanstead residents feel disappointed that the perceived Donga ethos of earlier times has been replaced by the dominance of a harder edged attitude . . . The campaign has become more proletarianized in many ways. This conflicts with the middle-class pacifism, nature-loving environmentalism and educated-dropout characters that used to predominate.' In fairness, Wanstead residents did attempt to organize their own direct action site to protect Bush Wood, but contractors moved in in September 1994 and prevented it.

At the end of November 1994, the eviction of Claremont Road that all had been waiting and preparing for finally came. Tip-offs led to protesters having time to finalize what they had been working towards for months, and a phone tree whereby each protester alerted by phone called others spread the word and brought people onto the street, or onto streets as near as possible to the officially closed-off areas. 'Knowing that the future is rubble gave us the strength to approach the eviction as a game . . . Barricades, bunkers, towers, nets, tree houses, lock-ons, the tunnel . . . all energy stored and stockpiled for that final moment, everything tuned for maximum intensity during that one hundred hour explosion.' Two hundred bailiffs, seven hundred police and four hundred security guards arrived to carry out the eviction, or just to stand around and intimidate. For four more days Claremont Road obstructed the M11 link road, from the first protesters encountered, apparently nonchalantly lying on mattresses in the road until police tried to remove them and found arms buried into the tarmac and concrete, to the very last protester on the greasy-poled scaffolding tower one hundred feet in the air.

51 EDWARD W. SOJA

'The Stimulus of a Little Confusion: On Spuistraat, Amsterdam'

from I. Borden, J. Kerr, A. Pivaro and J. Rendell (eds), *Strangely Familiar: Narratives of Architecture in the City* (1996)

In Amsterdam in 1990, I dwelt for a time in a traditional canal house on Spuistraat, a border street on the western flank of the oldest part of the Centrum, the largest and most successfully reproduced historic inner city in Europe. To live in one of these artfully preserved spaces is to immediately and precipitously encounter Amsterdam's past and presence.

With its narrow nooks and odd-angled passageways, its flower-potted corners and unscreened windows that both open and close to the views outside, everyday life inside becomes a crowded reminder of at least three rich centuries of urban history and urban geography being preserved on a scale and scope that is unique to Amsterdam. At home, one is invited daily into the creative spatiality of the city's social life and culture, an invitation that is at the same time embracingly tolerant and carefully guarded. Not everyone can become a true Amsterdammer, but everyone must at least be given the chance to try.

The view through my front window was a wonderful illustration of this peculiar urban genius. Immediately opposite, in a building very much like mine, each floor was a separate flat and each storey told a vertical story of subtle and creative citybuilding. It was almost surely a squatter-occupied house in the past and probably still was one now, for Spuistraat has long been an active scene of the squatter movement. On the first floor of the house across the way were the most obviously elegant living quarters, occupied by a woman who had probably squatted there as a student but had since comfortably entered the job market. She spent a great deal of time in the front room, frequently had guests in for candlelit dinners, and would occasionally wave to us across the street. On the floor above there was a young

couple. They were probably still students and still poor, although the young man may have been working at least part time for he was rarely seen, except in the morning and late at night. The woman was obviously pregnant and spent most of her time at home. Except when the sun was bright and warm, they tended to remain away from the front window and never acknowledged anyone outside, for their orientation was decidedly inward. The small top floor, little more than an attic, still had plastic sheeting covering the roof. A single male student lived there and nearly always ate his lunch leaning out the front window alone.

This vertical transect through the current status of the squatter movement was matched by an even more dramatic horizontal panorama along the east side of Spuistraat. To my left, looking north, was an informative sequence of symbolic structure, beginning with a comfortable two-storey corner house that was recently rehabilitated with near squatter rentals above and below, a series of shops also run by the same group of rehabilitated and socially absorbed squatter-renters, another of those creative contradictions in terms that characterize Amsterdam. There was a well-stocked fruit and vegetable market-grocery selling basic staples at excellent prices, a small beer-tasting store stocked with dozens of imported (mainly Belgian) brews and their distinctively matching drinking glasses and mugs, a tiny bookstore and gift shop specializing in primarily Black, gay and lesbian literature, a used household furnishings shop with dozens of chairs and tables set out on the front sidewalk and, finally, closest to my view, a small hand-crafted woman's cloth hat shop.

This remarkably successful example of the gentrification by the youthful poor is just a stone's throw away from the Royal Palace on the Dam, the focal point for the most demonstrative peaking of the radical squatter movement that blossomed city-wide in conjunction with the coronation of Queen Beatrix in 1980. A more immediate explanation of origins, however, is found just next door on Spuistraat, where a new office-construction site has replaced former squatter dwellings in an accomplished give and take trade-off with the urban authorities. And just next door to this site, even closer to my window, was still another paradoxical juxta-positioning, one which signalled the continued life of the radical squatter movement in its old anarchic colours.

A privately owned building had been recently occupied by contemporary squatters and its façade was brightly repainted, graffitoed, and festooned with political banners and symbolic bric-a-brac announcing the particular form, function and focus of the occupation. The absentee owner was caricatured as a fat tourist obviously beached somewhere with sunglasses and tropical drink in hand, while a white-sheet headline banner bridged the road to connect with a similar squat on my side of the street, also bedecked with startling colors and slogans and blaring with music from an established squatter pub. I was told early in my stay that this was the most provocative ongoing squatter settlement in the Centrum.

These vertical and horizontal panoramas concentrate and distill the spectrum of forces that have creatively rejuvenated the residential life of the Centrum and preserved its anxiety-inducing superabundance of urban riches. At the center of this rejuvenation has been the squatter movement, which has probably etched itself more deeply into the urban built environment of Amsterdam than in any other inner city in the world. To many of its most radical leaders, the movement today seems to be in retreat, deflected if not co-opted entirely by an embracing civic tolerance that extends even to distributing official pamphlets on "How to be a Squatter". But it has been this ever-so-slightly repressive tolerance that has kept open the competitive channels for alternative housing, counter-cultural lifestyles, and that most vital of rights to the city: the right to be different.

From my vantage point on Spuistraat a moving picture of contemporary life in the vital center of Amsterdam visually unfolded, opening my eyes to much more than I ever expected to see. The view from the window affirmed what I continue to believe is the most extraordinary quality of this city, its unheralded achievement of highly regulated urban anarchism, another of those creative paradoxes like "repressive tolerance", "flexible inflexibility", "squatter-renters" and indeed the "strangely familiar" that two-sidedly filter through the geohistory of Amsterdam in ways that seem to defy comparison with almost any other *polis*, past or present. This deep and enduring commitment to libertarian socialist values and participatory spatial democracy is openly apparent throughout the urban built environment and in the social practices of urban planning, popular culture and everyday life. One senses that Amsterdam is not just preserving its own Golden Age but is actively keeping alive the very possibility of a socially just and humanely scaled urbanism. And it is doing so by adding to our regular habits of thought that entertaining stimulus of a little confusion.

'Claiming the Public Space: The Mothers of Plaza de Mayo'

from Diane Agrest *et al.* (eds), *The Sex of Architecture* (1996)

The role of women in the transformation of cities remains theoretically problematic. While women's leadership in organizations rebuilding communities and neighborhoods and their creation of new paradigms for monumentality are sometimes noted in the press, these interventions have yet to inform cultural discourse in the design disciplines or in the history and theory of art and architecture.

The largest body of current feminist scholarship on women in urban settings is concerned with the construction of bourgeois femininity in nineteenth-century European capitals. Within this framework, women are seen as extensions of the male gaze and as instruments of the emerging consumer society and its transformative powers at the dawn of modernity. In other words, they are described as passive agents rather than engaged subjects. When women have assumed transformative roles, feminist critics and biographers have seen them as exceptional individuals or female bohemians, publicly flaunting class and gender distinctions; in contrast, women in general, and working-class women in particular, are presented as unintentional agents of a collective social project, acting out assigned scripts. As a class, women share the problematic status of politically or culturally colonized populations. Both are seen as passively transformed by forced modernization rather than as appropriating modernity on their own and, through this appropriation, being able to change the world that is transforming them.

From this perspective it is difficult to see the current individual and collective struggle of women to transform urban environments as anything of cultural significance, or to reevaluate the enduring influence of traditional female enclaves originated in the premodern city. Many of these enclaves continue to serve their traditional functional and social roles, like the public washing basins in major Indian cities or the markets in African villages, while others have persisted as symbolic urban markings, like the forest of decorated steel poles that once held clotheslines in Glasgow's most central park. Some of these enclaves have even become a city's most important open space, like River Walk in San Antonio, Texas, where women once congregated to wash laundry and socialize.

A literature is now emerging, focused on the participation by marginalized populations in the transformation of postmodern cities and establishing the critical connection between power and spatiality, particularly within the disciplines of art and architectural history and architectural and urban design. To these contributions, which have revealed previously unmarked urban sites as well as the social consequences of repressive urban planning ideologies, should be added feminist analyses of women's traditional urban enclaves and of women's appropriations of public sites that symbolized their exclusion or restricted status. These appropriations, whether in the form of one of the largest mass demonstrations ever held on the Washington Mall (in favor of abortion rights) or in the display of intimacy in very public settings (such as the private offerings and mementos that complete Maya Lin's Vietnam Memorial and compose the monumental Names Quilt commemorating AIDS victims), continue to establish women's rights not merely to inhabit but also to transform the public realm of the city. It is in such situations that women have been most effective in constructing themselves as transformative subjects, altering society's perception of public space and inscribing their own stories into the urban palimpsest.

As in all instances where the topic of discussion is as complex as the transformative presence of women in the city – and particularly when this topic does not yet operate within an established theoretical framework – the main difficulty is to establish a point of entry. In the present essay I propose entering this territory through the examination of one dramatic case of a successful, enduring appropriation: the Mothers of the Plaza de Mayo in Argentina. This small but persistent band of women protesters first captured international attention in the mid-1970s with their sustained presence in the nation's principal "space of public appearance," as Hanna Arendt has called the symbolic realm of social representation, which is controlled by the dominant political or economic structures of society. This case illustrates the process that leads from the embodiment of traditional roles and assigned scripts as wives and mothers to the emergence of the active, transformative subject, in spite of – or perhaps because of – the threat or actuality of physical violence that acts of protest attract in autocratic societies. As we will see, this case is also emblematic of architecture's complicity with power in creating a symbolic system of representation, usually of power hierarchies. The hegemony of this system has been threatened ever since the invention of the printing press and is now claimed by electronic media and its virtual space of communication. Finally, the Mothers of Plaza de Mayo's appropriation of the public square as a stage for the enactment of their plea is a manifestation of *public space* as social production. Their redefinition of that space suggests that the public realm neither resides nor can be represented by buildings and spaces but rather is summoned into existence by social actions.

THE MOTHERS OF THE PLAZA DE MAYO

In March 1976, after a chaotic period following Juan Perón's death, a military junta wrested power from Perón's widow, Isabel, in order (as the junta claimed) to restore order and peace to the country. The first measures toward achieving this goal were similar to those of General Pinochet in Chile three years earlier, and included the suspension of all civil rights, the dissolution of all political parties, and the placement of labor unions and universities under government control. It would take seven long, dark years for a democratically elected government to be restored to Argentina, which at last permitted an evaluation of the extent of open kidnappings, torture, and executions of civilians tolerated by the military. Because of the clandestine, unrecorded activities of the paramilitary groups charged with these deeds, and because many burial sites still remain undisclosed, agreement as to the exact number of "disappeared" may never be achieved, but estimates range from nine thousand to thirty thousand. Inquiries to the police about the fate of detainees went unanswered. Luis Puenzo's 1985 film, *The Official Story*, offers glimpses into the torture and degradation endured by thousands of men, women, and even babies, born in detention, some of whom were adopted by the torturers' families.

"Disappearances" were very effective in creating complicitous fear: many kidnappings were conducted in broad daylight, and the victims had not necessarily demonstrated open defiance of the military. In fact, later statistics show that almost half of the kidnappings involved witnesses, including children, relatives, and friends of those suspected of subversion. Given the effectiveness of arbitrary terror in imposing silence, it is astonishing that the public demands of less than a score of bereaved women who wanted to know what had happened to their children contributed so much to the military's fall from power. Their silent protest, opposed to the silence of the authorities, eventually had international resonance, prompting a harsh denunciation of the Argentinean military, which led, finally, to the demise of state terrorism and the election of a democratic government.

The actions of the "Mothers," as they came to be known, exemplified a kind of spatial and urban appropriation that originates in private acts that acquire public significance, thus questioning the boundaries of these two commonly opposed concepts. Gender issues, too, were not unimportant. The Mothers' appropriation of the plaza was nothing like a heroic final assault on a citadel. Instead, it succeeded because of its endurance over a protracted period, which could only happen because the

Mothers were conspicuously ignored by the police, the public, and the national press. As older women they were no longer sexually desirable, and as working-class women they were of an inferior ilk. Nevertheless, their motherhood status demanded conventional respect. Communicating neither attraction nor threat, they were characterized by the government as "madwomen." The result of their public tenacity, which started with the body exposed to violence, eventually evolved into a powerful architecture of political resistance.

Plaza de Mayo is Argentina's symbolic equivalent of the Washington Mall. It is, however, a much smaller and very different kind of space: an urban square that evolved from the Spanish Plaza de Armas, a space that has stood for national unity since Creoles gathered there to demand independence from Spain in May of 1810. The national and international visibility of Plaza de Mayo as *the* space of public appearance for Argentineans is unchallenged. Originally, as mandated by the planning ordinances of the Law of the Indies, its sides were occupied by the colonial Cabildo, or city council, and the Catholic Cathedral. Today the most distinctive structure is the pink, neoclassical Casa Rosada, the seat of government.

Military exercises, executions, and public market commingled in the plaza until 1884, when Torcuato de Alvear, the aristocratic mayor, embarked on a Haussmannian remodeling of the center of Buenos Aires shortly after important civic structures – such as Congress and the Ministries of Finance and Social Welfare – had been completed. A major element of Alvear's plan was Avenida de Mayo, an east–west axis that put Congress and the Casa Rosada in full view of each other. Such a potent urban representation of the checks and balances of the modern, democratic state was achieved through selective demolition, including the removal of the plaza's market stalls and the shortening of the historic Cabildo's wings by half their original length. Currently, the plaza's immediate area includes several government offices, the financial district, and the city's most famous commercial street, Florida. This densely populated pedestrian thoroughfare links Avenida de Mayo to Plaza San Martin, another major urban square. A plastered masonry obelisk, the May Pyramid, erected on the square in 1811 to mark the first anniversary of the popular uprising for independence, was rebuilt as a taller, more ornate structure and placed on the axis between Congress and the Casa Rosada. In this new position, it became a metaphorical fulcrum in the balance of powers.

The now well-known image of a ring of women with heads clad in white kerchiefs circling the May Pyramid evolved from earlier spontaneous attempts at communication with government officials. At first, thirteen wives and mothers of the "disappeared" met one another at the Ministry of the Interior, having exhausted all sources of information about their missing children and husbands. There a small office had been opened to "process" cases brought by those who had filed writs of *habeas corpus*. One woman well in her sixties, Azucena Villaflor de Vicente, rallied the others: "It is not here that we ought to be," she said. "It's the Plaza de Mayo. And when there are enough of us, we'll go to the Casa Rosada and see the president about our children who are missing." At the time, popular demonstrations at the plaza, frequently convened by the unions as a show of support during Juan Perón's tenure, were strictly forbidden, and gatherings of more than two people were promptly dispersed by the ever-present security forces. The original group of thirteen women came to the plaza wearing white kerchiefs initially to identify themselves to one another. They agreed to return every Thursday at the end of the business day in order to call their presence to the attention of similarly aggrieved women. The Mothers moved about in pairs, switching companions so that they could exchange information while still observing the rule against demonstrations. Eventually they attracted the interest of the international press and human rights organizations, one of which provided an office where the women could congregate privately. Despite this incentive to abandon the plaza for a safer location, the Mothers sustained a symbolic presence in the form of a silent march encircling the May Pyramid. That form, so loaded with cultural and sexual associations, became the symbolic focus of what started as a literal response to the police's demand that the women "circulate."

The white kerchiefs were the first elements of a common architecture evolved from the body. They were adopted from the cloth diapers a few of the Mothers had worn on their heads in a pilgrimage to the Virgin of Luján's sanctuary. The diapers were those of their own missing children, whose names were embroidered on them, and formed a headgear that differentiated the Mothers from the multitude of other women in kerchiefs on that religious march. In later demonstrations the Mothers constructed full-size cardboard silhouettes representing their missing children and husbands, and shielded their bodies with the ghostly blanks of the "disappeared."

By 1982, the military had proven itself unable to govern the country or control runaway inflation of more than 1000 percent per year. The provision of basic services was frequently disrupted by the still powerful Peronista labor unions, and many local industries had gone bankrupt due to the comparative cheapness of imported goods under an economic policy that eliminated most import taxes. Then, in the same year, the military government embarked on an ultimately ruinous war with Great Britain over the sovereignty of the Falkland/Malvinas Islands. With the help of the United States satellite intelligence and far superior naval might, Great Britain won with few casualties, while Argentina lost thousands of ill-equipped and ill-trained soldiers. The military government, which had broadcast a fake victory on television using old movie reels rather than current film footage, was forced to step down in shame by the popular outcry that followed. Following the collapse of the military government, the Mothers were a prominent presence at the festivities in Plaza de Mayo, their kerchiefs joyously joined as bunting to create a city-sized tent over the celebrants. They have continued their circular march to this day, as a kind of living memorial and to promote their demands for full accountability and punishment for those responsible for the disappearance of their husbands and children.

After the election of a democratic government, the military leadership was prosecuted in civil rather than military court, resulting in jail sentences for a few generals and amnesty for other military personnel. Although the amnesty was forcefully contested by the Mothers and other organizations, the protest was seen by many as divisive. Nevertheless, the Mothers and a related organization of grandmothers pressed on with attempts to find records about disappearances and fought in the courts to recover their children

and grandchildren. Then, early in 1995, more than a decade after the restoration of democratic government, a retired lieutenant publicly confessed to having dumped scores of drugged but still living people from a helicopter into the open ocean, and he invited other military men on similar assignments to come forth. The Mothers were present to demonstrate this time as well, but now the bunting had become a gigantic sheet that was waved overhead as an angry, agitated sea.

The Mothers were able to sustain control of an important urban space much as actors, dancers, or magicians control the stage by their ability to establish a presence that both opposes and activates the void represented by the audience. To paraphrase Henri Lefebvre, bodies produce space by introducing direction, rotation, orientation, occupation, and by organizing a *topos* through gestures, traces, and marks. The formal structure of these actions, their ability to refunctionalize existing urban spaces, and the visual power of the supporting props contribute to the creation of public space.

What is missing from the current debate about the demise of public space is an awareness of the loss of architecture's power to represent the *public*, as a living, acting, and self-determining community. Instead, the debate focuses almost exclusively on the *physical space* of public appearance, without regard for the social action that can make that environment come alive or change its meaning. The debate appears to be mired in regrets over the replacement of squares (for which Americans never had much use) with shopping malls, theme parks, and virtual space. But this focus on physical space – and this ideological potential to encompass the public appearance of all people, regardless of color, class, age, or sex – loses credibility when specific classes of people are denouncing their exclusion and asserting their presence and influence in public life. The claims of these excluded people underscore the roles of *access* and *appearance* in the production and representation of public space, regardless of how it is physically or virtually constituted. They also suggest that public space is produced through public discourse, and its representation is not the exclusive territory of architecture, but is the product of the inextricable relationship between social action and physical space.

UTOPIA AND DYSTOPIA

INTRODUCTION

Utopia is most commonly understood as an ideal place and state, a condition of perfection where all issues are resolved to the benefit of all those who live there. Utopia is, therefore, thought of as a place in the past or future, and/or far away from the contemporary world. There are, however, at least two other ways of thinking about utopia. The first is in the sense of a utopian impulse, a desire always to move towards such a state even if it is not yet seen to be immediately attainable. A second alternative follows on from this, in that if we wish to enact a utopia does that not automatically imply a sense of dissatisfaction with the present state of affairs? Utopia, in this sense, becomes less a goal to move towards as a position from which to critique the present – a political perspective directed towards contemporary aims. It should be noted that utopia also sometimes carries with it a pejorative sense, often being used by those trying to emphasize the futility and impossibility of an idea.

The concept of dystopia is also pejorative, but in a rather different sense – the notion that certain conditions represent an extreme version of all that is bad, all that is unjust, all that is iniquitous. Although seemingly negative, the image of dystopia can be equally helpful as that of utopia for those interested in social change and progress, offering a way of representing conditions to be avoided in the present and in the future. In this sense, dystopia may be no different to utopia – both are conditions to be constantly born in mind when engaging in politicized cultural activity.

Judith Shklar, in 'The Political Theory of Utopia: From Melancholy to Nostalgia' (1973), notes the relative lack of notions of utopia today. Shklar rejects Karl Mannheim's categorization of absolutely all activities which seek to overthrow contemporary society as 'utopian' (whereas activities supportive of the status quo were deemed to be 'ideological'), preferring the more historically accurate version offered by Marx and Engels, distinguishing older, idealist 'classical' utopians and true, socialist utopians who properly understood the contradictions of capitalism. Shklar also notes the existence of utopias of condemnation, such as those of Swift and Diderot, and, alternatively, utopias of those such as Bellamy who used utopia as a call to action. Whatever their practicality, and whatever their nostalgic character, Shklar suggests that the creativity of thought in such formulations is not to be dismissed. It is not, then, that there should be more utopias today, but that we should not hark to the past for political models of our present and future.

A similarly intense attack on the relevance of utopia today is provided by the Italian Marxist architectural historian Manfredo Tafuri. In his polemic *Architecture and Utopia* (1976, MIT) (see suggestions for further reading), Tafuri focuses his attentions on the modernist architect and city planner Le Corbusier and his work at Algiers. More generally, however, Tafuri is at pains to explain not only the failure of Le Corbusier to get his huge experiment actually built, but the failure of the whole of modernist architecture. Underlying Tafuri's history here is an argument that capitalism initially turned critical utopias to its own purpose, using them to destroy old and unwanted ideologies and forms, while simultaneously suggesting new directions and spaces for action. Unfortunately for modernism, its model of utopian action was related to the needs of nineteenth-, not twentieth-century capitalism, and so was doomed to failure because it no longer served the needs of capitalism.

In a rather different interpretation of architecture, the designer and architectural theorist Lebbeus Woods considers in 'Everyday War' (1995, Princeton Architectural Press) that architecture is an act of war. As an attack on Aristotelian logic and Descartes' duality of opposites, Woods contends that identities are 'transformational, sliding and shifting in an ongoing complex stream of becoming'. In this light, architecture can be at once construction and destruction, as both are necessary to create a building. A building is 'by its very nature an aggressive, even warlike act', involving the clearing of sites, an indifference to existing culture and the naked exercise of power. 'War', then, is everywhere, ever present in the architecture of the city, thus destroying normality and routine while emphasizing consciousness and independence. The dystopian nature of architecture is, ultimately, at once terrorizing and liberating.

A far less ambiguous interpretation of an architectural dystopia is presented by Marshall Berman. In 'Robert Moses: The Expressway World', a section of *All That Is Solid Melts Into Air* (1982, Simon and Schuster), Berman castigates a single figure, Robert Moses (a figure admired by Woods), for his authoritarian and dictatorial acts of planning violence in New York. The devastation of the Bronx (the scene of Berman's childhood) by the Cross-Bronx Expressway, along with the construction of innumerable other large infrastructural projects across New York, are seen by Berman to be self-glorifying acts that frequently resulted in the erasure of local neighbourhoods and communities. Architecture and planning may have been a way of dealing with problems of congestion, chaos and economic decline, but they were also, in the Moses' version, examples of brutal power wielded idiosyncratically and indifferently.

Robert Moses' exercise of planning power took place in the 1920s, 1930s and immediate post-war years. Some forty or more years later, other planning dystopias are still being constructed, but of a different kind. In 'Dystopia on the Thames' (1993, Routledge), cultural historian Jon Bird discusses the development of the Canary Wharf and Docklands area of London – an area with great similarities to other sites worldwide, including New York's Battery Park City. Running against the trend of protest elsewhere in contemporary London, Bird sees Docklands as an attempt to recreate the (presumed) tranquillity and homogeneity of the past. Docklands emerges as a place of enterprise over communities, business over pleasure, monetarism over socialism, private over public wealth, imaged harmony over social difference. But Bird also notes that this very dystopia can give rise to new forms of social and artistic action – cultural forms ranging from publications and large-scale posters to exhibitions and the 'Art of Change' consultancy. Dystopia is then not just something to be regretted – it is something to be fought against.

53 JUDITH SHKLAR

'The Political Theory of Utopia: From Melancholy to Nostalgia'

from F. E. Manuel (ed.), *Utopias and Utopian Thought* (1973)

What does the plaintive question, "why are there no utopias today?" mean? Does it merely express the nostalgia of those who were young and socialist in the thirties? Is it just that they resent the lack of sympathy among younger people? Do some of the latter, perhaps, long to re-experience the alleged political excitements of the romanticized thirties, but find that they cannot do so? For it is pre-eminently a question about states of mind, and intellectual attitudes, not about social movements. And it says something about the historical obtuseness of those who ask, "why no good radicals?" that they do not usually consider the obvious concomitant of their question, "why no Nazism, no fascism, no imperialism and no bourbonism?" If the absence of utopian feeling mattered only to that relatively small number of intellectuals who are distressed by their inability to dream as they once did, then it might concern a social psychologist, but it would scarcely interest the historian.

There is, however, more to the question than the temporary malaise of a few relics of the inter-war period. The questions "after socialism, what?" and "can we go on without utopias?" were already being asked before 1930, specifically by Karl Mannheim. Here, an entire theory of history and of the historical function of utopian thought was involved. Mannheim's now celebrated proposition was that all the political thought of the past could be divided into two classes, the utopian and the ideological. The former was the "orientation" of those aspiring classes that aimed at the complete or partial overthrow of the social structure prevailing at the time. Ideology, on the other hand, was the typical outlook of the dominant classes, intent upon preserving the established order. It is, of course, more than questionable whether the vast variety of Europe's intellectual past can be squeezed into this Manichean strait-jacket. And, in fact, it was a perfectly deliberate falsification of history on Mannheim's part. As he blandly admitted, the historian's concern with actual differences, contrasts, and nuances was a mere nuisance to one who sought to uncover the "real," though hidden, patterns beneath the actual men and events of the past. As seen through the spectacles of the "sociology of knowledge," history *had* to show successive waves of revolutionary fervor as the chief constant feature of European intellectual and social life. This meant, among other things, that so marginal a figure as the "chiliastic" Thomas Müntzer had to be pushed to the very front ranks of intellectual luminaries. He was the first in a series that included such first-class thinkers as Condorcet and Marx. Karl Kautsky had indeed allowed Sir Thomas More to share with Müntzer the honor of being the first socialist, but he saw More rather as a unique intellectual prophet of the socialist future than as a mere class manifestation. Mannheim, however, rejected Sir Thomas More summarily as a figure of no sociological significance in the "real" history of utopian thought. This entirely Marxian view of the past as dominated by incidents of revolutionary conduct and its reflected thought is of considerable importance, because it is what makes the contemporary absence of such zeal appear so entirely new, unique, and catastrophic and thus gives the question, "why no utopias?" its tense historical urgency. Certainly it had that effect upon Mannheim. If "art, culture and philosophy are nothing but the expression of the central utopia of the age, as shaped by contemporary social, and political forces," then indeed the disappearance of utopia might well mean the end of civilization. And since Mannheim assumed that the classless society was at hand, and that no challenging, utopia-inspiring classes would again appear, the new "matter-of-factness" seemed threatening and ominous indeed. The disappearance of "reality–transcending doctrines" brings about "a static state of affairs in which man himself becomes no more than a thing" and in relinquishing utopia men lose the will to shape history and so the ability to understand it. What, above all, is to become of the heirs of Müntzer, Condorcet, and Marx, of the intellectual élite who, until now, have been the producers of utopias? Mannheim's response, natural under the circumstances, was to provide a blueprint of a future society to be run by an intellectual élite trained in the sociology of knowledge, capable of both trans-forming and controlling history in the interests of freedom, democracy, and rationality.

Since the social role and ideas of the intellectuals are the central concern of the sociology of knowledge, and since Mannheim, unlike Marx, seemed to believe that this élite was of supreme importance in shaping the pattern of history, it is not at all surprising that their notions of utopia should have differed so much. While Mannheim accepted Marxian ideas about ideology, utopia was for him the intellectuals' vehicle of self-expression and it was they, not the voiceless classes, who ultimately shaped the ages. Yet Marx's and Engels' views on utopian thought were historically far more sound in at least one respect. The classical utopia, the critical utopia inspired by universal, rational morality and ideals of justice, the Spartan and ascetic utopia was already dead after the French Revolution. Doomed to impracticability, since the material conditions necessary for this realization had not been prevalent, the classical utopia could be admired even though it had lost its intellectual function with the rise of "scientific socialism." This judgment, according to Marx and Engels, was also applicable to such socialist precursors as Owen, Fourier, and Saint-Simon. their successors, and indeed all non-Marxist socialists, were, however, utopians in a very different sense. For these rivals *were* in a position to understand the true course and future of bourgeois society and to act accordingly. *They* had the benefit of Marx's theories of surplus value and of dialectical materialism. They could recognize both his "scientific" truths and the necessity for revolutionary activity. Instead they produced "duodecimo editions of the New Jerusalem," preached the brotherhood of man to the bourgeoisie, and ignored the Eleventh Thesis against Feuerbach. Here, "utopian" clearly becomes a mere term of opprobrium for un-Marxian, "unscientific" socialists. One wishes that Marx and Engels might have chosen another epithet. Certainly many useless verbal wrangles over the "true" meaning of the adjective "utopian" might have been avoided. What remains relevant in their views, however, is the serious importance they attached to the classical utopia and their recognition that it was a thing of the past because socialism had replaced it in their own age. To this, one must add that it was not only Marxian socialism, but all forms of socialism and, indeed, all the social belief systems (especially Social Darwinism) which prevailed in the nineteenth century, that joined in this task. Moreover, all of these, in spite of Mannheim's idiosyncratic vocabulary, are now called ideologies. In short, it was ideology that undid utopia after the French Revolution.

To understand why the classical utopia declined, not yesterday, but almost two hundred years ago, demands a more detailed analysis of its character than either Marx, Engels, or Mannheim offered. It also requires a return to that historical way of looking at the past which they despised, because it does not try to uncover "real" patterns, nor to establish laws. Instead of concentrating on paradigmatic, even if obscure, figures which fit a preconceived scheme, it looks at the acknowledged masters of utopian liter-ature: at Sir Thomas More and his successors. Paradoxically, this utopia is a form of political literature that cannot possibly be fitted into either one of Mannheim's categories, for it is in no sense either revolutionary and future directed or designed to support the ruling classes. All the utopian writers who followed More's model were critical in two ways. In one way or another all were critical of some specific social insti-tutions of their own time and place. But far more importantly, utopia was a way of rejecting that notion of "original sin" which regarded natural human virtue and reason as feeble and fatally impaired faculties. Whatever else the classical utopias might say or fail to say, all were attacks on the radical theory of original sin. Utopia is always a picture and a measure of the moral heights man could attain using only his natural powers, "purely by the natural light." As one writer put it, utopia is meant "to confound those who, calling themselves Christians, live worse than animals, although they are specially favored with grace, while pagans, relying on the light of nature manifest more virtue than the Reformed Church claims to uphold." No one doubts the intensity of Sir Thomas More's Christian faith, but the fact remains that his Utopians are not Christians, "define virtue as living according to nature," pursue joy and pleasure and are all the better for it – which is, of course, the main point.

The utopian rejection of original sin was, however, in *no sense* a declaration of historical hopefulness – quite the contrary. Utopia was, as Sir Thomas More put it, something "I wish rather than expect to see followed." It is a vision not of the probable but of the "not-impossible." It was not concerned with the historically likely at all. Utopia is nowhere, not only geographically, but historically as well. It exists neither in the past nor in the future. Indeed, its esthetic and intellectual tension arises precisely from the melancholy contrast between what might be and what will be. And all utopian writers heightened this tension by describing in minute detail the institutions and daily lives of the citizens of utopia while their realization is scarcely mentioned. "Utopus" simply appears one day and crates utopia. This is very much in keeping with the Platonic metaphysics which inspired More and his imitators as late as Fénelon. For them, utopia was a model, an ideal pattern that invited contemplation and judgment but did not entail any other activity. It is a perfection that the mind's eye recognizes as true and which is described as such, and so serves as a standard of moral judgment. As Miss Arendt has said, "in (Platonic contemplation) the beholding of the model, which ... no longer is to guide any doing, is prolonged and enjoyed for its own sake." As such it is an expression of the craftsman's desire for perfection and permanence. That is why utopia, the moralist's artifact, is of necessity a changeless harmonious whole, in which a shared recognition of truth units all the citizens. Truth is single and only error is multiple. In utopia, there cannot, by definition,

be any room for eccentricity. It is also profoundly radical, as Plato was; for all historical actuality is here brought to judgment before the bar of trans-historical values and is found utterly wanting.

If history can be said to play any part at all in the classical utopia, it does so only in the form of an anguished recollection of antiquity, of the polis and of the Roman Republic of virtuous memory. This is a marked feature also of utopias not indebted to Platonic metaphysics, even those of libertine inspiration. The institutional arrangements of Plato's *Laws*, Plutarch's *Lycurgus*, and Roman history also served as powerful inspirations to the utopian imagination. Thus, to the melancholy contrast between the possible and the probable was added the sad confrontation between a crude and dissolute Europe and the virtue and unity of classical antiquity. It is this, far more than the prevalence of Platonic guardians, perfected and effective education, and rationalist asceticism, ubiquitous as these are, that marks utopia as an intellectualist fantasy. Until relatively recently nothing separated the educated classes from all others more definitely than the possession of laboriously gained classical learning. It might be more correct to say classicism possessed them. They identified themselves more deeply and genuinely with the dead of Athens and Rome than with their own despised and uncouth contemporaries. And inasmuch as utopia was built on classical lines it expressed the values and concerns of the intellectuals. It was to them, not to unlettered lords or peasants, that it was addressed. As such, it was the work of a socially isolated sensibility, again not a hope-inspiring condition. But it survived even the literary and scientific victory of the Moderns over the Ancients. Nothing seemed to shake the long-absorbed sense of the moral and political superiority of classical man. That is why wistful Spartan utopias were still being written in the second half of the eighteenth century.

Of course, the political utopia, with its rational city-planning, eugenics, education, and institutions, is by no means the only vision of a perfect life. The golden age of popular imagination has always been known, its main joy being food – and lots of it – without any work. Its refined poetic counterpart, the age of innocence, in which men are good without conscious virtue, has an equally long history. The state of innocence can exist, moreover, side by side with a philosophic utopia and illuminate the significance of the latter. Plato's Age of Kronos and Fénelon's Boetica, in which wisdom is spontaneous, are set beside rational, Spartan-style models. The state of innocence is what moral reason must consciously recreate to give form and coherence to what all men can feel and imagine as a part, however remote, of their natural endowment. Both utopias, in different ways, try to represent a timeless "ought" that never "is."

Among the utopias that do not owe anything to classical antiquity at least one deserves mention here: the utopia of pure condemnation. Of this genre Swift is the unchallenged master, with Diderot as a worthy heir. The king of Brobdingnag, the city of giants, of supermen, that is, notes, after he hears Gulliver's account of European civilization, that its natives must be "the most pernicious Race of little odious Vermin that Nature ever suffered to crawl upon the Surface of the Earth." A comparison of his utopian supra-human kingdom with those of Europe could yield no other conclusion. Gulliver, then, tastes the delights of a non-human society of horses, an experience which leaves him, like his author, with an insurmountable loathing for his fellow-Yahoos. Here, utopia serves only to condemn not merely Europe, whether ancient or modern, but the human louse as such. To the extent that Diderot's account of Tahiti slaps only at European civilization, it can be said to be more gentle. However, after observing the superiority of primitive life, his European travelers return home wiser in recognizing the horrors of their religion, customs, and institutions, but in no way capable or hopeful of doing anything about them. The aim, as in Swift, is to expose absurdity and squalor simply for the sake of bringing them into full view.

This all too brief review of classical utopia should suffice to show how little "activism" or revolutionary optimism or future-directed hope there is in this literature. It is neither ideology nor utopia in Mannheim's sense, but then neither is most of the great critical political literature before the end of the eighteenth century. Machiavelli, Bodin, Hobbes, Rousseau: were they "reactionary ideologues" because they were not "revolutionary utopists"? Significantly, it

is only during the course of the English Civil War that action-minded utopists appear. However, even the two most notable among them, Harrington and Hartlib, were concerned with constitutional and educational reform, respectively, rather than with full-scale utopias. Nevertheless, for once imminent realization was envisaged. As for poor Winstanley and his little band of Diggers, they have merely been forced to play the English Müntzer in Marxian historiography in search of precursors and paradigms. These were all voices in the wilderness, part of a unique revolutionary situation. It is only as partial exceptions to the rule that they are really illuminating. They seem only to show how unrevolutionary the general course of utopian thought and political thinking was before the age inaugurated by the French Revolution. The end of utopian literature did not mark the end of hope; on the contrary, it coincided with the birth of historical optimism.

Utopia was not the only casualty of the revolution in political thought. Plutarchian great-man historiography and purely critical political philosophy were never the same again either. Nor was it solely a matter of the new theory of historical progress. As Condorcet, one of the first and most astute of its authors, observed, the real novelties of the future were democracy and science, and they demanded entirely new ways of looking at politics. If a democratic society was to understand itself it would need a new history: "the history of man," of all the inconspicuous and voiceless little people who constitute humanity and who have now replaced the star actors on the historical stage. The various historical systems of the nineteenth century, with their "laws" – whether progressive, evolutionist, dialectical, positivist, or not – were all, in spite of their endless deficiencies, efforts to cope with this new history. To write the history of the inarticulate majority, of those groups in society which do not stand out and therefore must be discovered, was a task so new and so difficult that it is scarcely surprising that it should not have succeeded. After all, contemporary sociology is, in a wiser and sadder mood, still plodding laboriously to accomplish it. As for science, Condorcet recognized not only that technology, that is, accumulated and applied knowledge, would transform material and social life, but also that science was not just an acquisition, but an entirely new outlook. With its openmindedness and experimentalism it had to replace older modes of thought which were incompatible with it. As such it was not only the vehicle of progress, but also the sole way in which the new society-in-change could be understood and guided. Scientific thought was inherently a call to action. The new world, as Condorcet saw it, would be so unlike the old that its experiences could be grasped, expressed, and formed only by those who adopted the openness of scientific attitudes. It was this that his systematizing successors, the victims of classical habits of thought, did not appreciate in the slightest. Whatever one may think of Condorcet's own historiography with its simple challenge-and-response ladder of improvement, he, at least, had the great merit of understanding why classical history and critical political theory had to be replaced by more democratic, dynamic, and activist social ideas.

Given the revolutionary changes of nineteenth-century Europe, the preoccupations of the classical utopists were no longer relevant. Original sin and the critical model were not vital interests. Marx, in spite of his protestations, was not the only one to take over the critical functions of the old utopists and to expand them into a relentlessly future-directed activism. All his rivals were just as intent upon action as he was. It was merely that some of them thought, as Saint-Simon had, that philosophers would exercise the most significant and effective authority by ruling over public opinion, rather than by participating directly in political action. Thus "the philosopher places himself at the summit of thought. From there he views the world as it has been and as it must become. *He is not only an observer. He is an actor* [Italics added.] He is an actor of the first rank in the moral world because it is his opinions on what the world must become which regulate human society." This bit of intellectualist megalomania could be illustrated by endless quotations not only from socialist sources, but also from liberal writings and from the distressed conservative deprecators of both.

The activism of the age was, moreover, not a random one. The future was all plotted out, and beckoning. "The Golden Age lies before us and not behind us, and is not far away." The

inevitable had only to be hastened on. Certainly there was no point in contemplating the classical past. However, the impact of the polis was not quite gone. Especially in socialist thought, even if not so openly as in Jacobin rhetoric, the ancient republic was still an inspiration. The ideal of its unity, of its homogeneous order, colored all their visions of the future. Certainly Marx was no stranger to the nostalgia for the cohesive city or its medieval communal counterpart. It was the liberal Benjamin Constant who noted that efforts to impose the political values of classical antiquity upon the totally dissimilar modern world could lead only to forms of despotism, which, far from being classical, would be entirely new. John Stuart Mill, following him, found this notion, "that all perfection consists in unity," to be precisely the most repellent aspect of Comte's philosophy. Indeed, the engineered community, whose perfect order springs not from a rational perception of truth, but from a pursuit of social unity as a material necessity, provides neither ancient nor modern liberty. The imagery is revealing. Cabet delighted in the vision of factory workers who displayed "so much order and discipline that they looked like an army." Bellamy's "industrial army" speaks for itself.

The form of such works as Cabet's *Voyage to Icaria* and Bellamy's *Looking Backward* should not lead one to think that these pictures of perfected societies are in any sense utopias. Precisely because they affect the external format of the classical utopia they demonstrate most effectively the enormous differences between the old and new ways of thought. The nineteenth-century imaginary society is not "nowhere" historically. It is a future society. And, it too is a summons to action. The purpose of Cabet's expedition to set up Icaria in America was not simply to establish a small island of perfection; it was to be a nucleus from which a world of Icarias would eventually spring. No sooner had Bellamy's work appeared than Bellamy societies, often (and not surprisingly) sponsored by retired army officers, appeared to promote his ideas. Theodore Hertzka's *Freeland* led to similar organized efforts, as its author had hoped it would. He declared frankly that the imaginary society was merely a device to popularize social ideas which he regarded as practical and scientifically sound. This, in itself, suffices to account for the literary feebleness of virtually all nineteenth-century quasi-utopias. There was nothing in them that could not have been better presented in a political manifesto or in a systematic treatise. They were all vulgarizations and were devised solely to reach the largest possible audience. The form of the classical utopia was inseparable from its content. Both were part of a single conception. The social aspirations of the nineteenth century found their literary form in the realistic novel, not in the crude and unstylish fiction of social theorists turned amateur romancers. Even the "utopias" based on scientific, rather than social, predictions were either childish or tedious. Either their fancies displayed no insight into the real potentialities of technology or, if they were well-informed, they were rendered obsolete by the actual developments of technology. Not even the last and most talented of latter-day contrivers of imaginary societies, H. G. Wells, could save the genre. He at least saw that his utopia had nothing in common with the classical works of that name. Now the perfect model is in the future, that is, it has a time and a place and is, indeed, already immanent in the present. It must be world-wide, devoted to science, to progress, to change, and it must allow for individuality. Only the intellectual ruling class of "samurai" is left to remind one of the classical past. As a liberal socialist, Wells was, no doubt, especially aware of the need to root out the remnants of that illiberal, self-absorbed, and closed social order which classicism had left as its least worthy gift to the democratic imagination. However, the novel of the happy future did not prosper in Wells' or in any other hands. For it was simply superfluous. Its message could be presented in many more suitable ways. Certainly it was in no way a continuation of the classical utopian tradition.

It has of late been suggested that the radicalism of the last century was a form of "messianism," of "millennialism," or of a transplanted eschatological consciousness. Psychologically this may be quite true in the sense that for many of the people who participated in radical movements social ideologies fed religious longings that traditional religions could not satisfy. These people may even have been responding to the same urges as the members of the medieval revolutionary millennialist sects. In this sense one may well regard radical ideology

as a surrogate for unconventional religiosity. It should, however, not be forgotten that millennialism always involves an element of eternal salvation. And this was entirely absent in the message of even those social prophets who called for new religions as a means of bringing social discipline to Europe. For them it was only a matter of social policy, not of supra-terrestrial truth. Marxism and social Darwinism, moreover, did not even involve this degree of "new" religiosity. Whatever they did for the fanaticized consciousness that eagerly responded to them, the intellectual structure of radical doctrines was not a prophetic heresy either in form or in intent. It represented an entirely new chapter in European thought. As Condorcet had clearly seen, it was a matter of new responses to a new social world; and the aspirations, methods of argument, and categories of thought of these historical systems were correspondingly unique, however primordial the human yearnings that they could satisfy might be. One ought not to forget the rational element, the effort of intellectual understanding that is perfectly evident in the writings of Saint-Simon, Marx, Comte, and all the rest. The various political revolutions after 1789 gave more than a semblance of reality to a vision of social history as a perpetual combat between the forces of progress and of conservatism. Conservative and liberal social observers no less than socialists took that view of historical dynamics. Even John Stuart Mill, who recognized that the categories of order and progress were inadequate concepts for a deeper analysis of politics, could see the past as a sequence of struggles between freedom and repression. The theory of class war was by no means the only one that, in a projection of nineteenth-century experience into the past, saw the history of Europe as a series of duels. Some saw it as progress, some as doom, but all perceived the same pattern. Looking back, of course, the century before the First World War appears infinitely more complicated than that, and so do the eras preceding it. However, if it is quite understandable why one should be sensitive to pluralistic social complexities today, it is also not difficult to see why dualistic patterns tended to dominate the historical imagination of the nineteenth century. Nor is it totally irrational that the experience of these rapid changes, so unlike those of the past, should lead men to entertain great expectations of the future. Neither the view of history as a dualistic combat of impersonal social forces nor the confident belief in a better future which would at last bring rest to mankind was a "millennial" fancy, nor was either really akin to the chiliastic religious visions that inspired that apocalyptic sects. If they were not utopias, neither were they New Jerusalem. The desire to stress similarities, to find continuities everywhere, is not always helpful, especially in the history of ideas, where the drawing of distinctions is apt to lead one more nearly to the truth.

The reason that ideology has been represented so often as a type of religiosity is, of course, a response to the terrifying fervor expressed by the members of modern mass-movements. It is the emotional element in Nazism, communism, and other revolutionary movements all over the world that is so reminiscent of many of the old popular heresies. The dynamism of mass parties, however, is really at stake here, not the actual systems of ideas which were produced in such great quantity by nineteenth-century Europe. Ideology, however, when it refers to those systems of ideas which were capable of replacing all the inherited forms of social thought, utopia among them, was clearly more significant intellectually than the brutish "isms" that animate both the leaders and the led of these movements. The latter should not be confused with either ideology, or utopia, or even with the religious extravagances of other ages. Nor is the question, "why no utopias?" really concerned with the organization of mass-parties. Indeed, even in Mannheim's theory, utopia and ideology refer to highly developed modes of thought and not to quasi-instinctive mental reactions. If, unlike Mannheim, one does not identify utopia with the charted mission of the intellectual class, one can recognize that "the end of utopia" involves not sociological, but philosophical, issues. It is the concern of political theory – of the high culture of social thought. What is really at stake is the realization that the disintegration of nineteenth-century ideology has not made it possible simply to return to classical-critical theory, of which utopia was a part. The post-ideological state of mind is not a classical one, any more than an ex-Christian is a pagan. on the contrary, the end of the great ideological systems may well also mark the exhaustion of

the last echo of classicism in political theory, even if, occasionally, a nostalgic appreciation for the integral classicism of the more distant past can still be heard.

The occasional contemporary efforts to construct pictures of perfect communities illustrate the point. They are compromises between the old utopia and the newer historical consciousness. Thus, for example, Martin Buber and Paul Goodman argue only for the historical non-impossibility of their plans, not for their inevitability. *Kvuzas*, or perfectly planned cities, are feasible, and certainly their admirers hope for their realization. These are still calls to action, but modest ones. Their scope, moreover, is limited, and their very essence is a revival of that dream of the polis, of the "authentic" small community that truly absorbs and directs the lives of its inhabitants. These relatively mild and moderate proposals, and the more general concern with *real* community life, do show, among other things, the lingering power of classical values. Here the longing for utopia and nostalgia for antiquity are inseparable. And indeed the question, "why is there no utopia?" expresses not only an urge to return to antiquity, but also, and far more importantly, a sense of frustration at our inability to think as creatively as the ancients apparently did.

Classicism, in one form or another, was, as we have seen, an integral part not only of utopia, but also of most political thinking. Hobbes and Bentham in their firm rejection of the conventional classical model were intellectually far more radical than the later ideologists. For in spite of occasional liberal protests, socialist doctrines were by no means the only ones that contrived to perpetuate classical notions throughout the nineteenth century. Long after Platonic metaphysics and the critical-contemplative mode of thought had been abandoned, classical imagery and values retained their hold on the political imagination, and classical methods of description and argument continued to mold the expression of political ideas in a social context in which classicism had ceased to be relevant. In this respect, all the ideologies served to retard political thinking. Their decline now has left political theory without any clear orientation and so with a sense of uneasiness. It is not that political theory is dead, as has often been claimed, but that so much of it consists of an incantation of clichés which seem to have no relation to social experiences whose character is more sensed than expressed. Could it be that classicism, not only as a set of political values and memories, but as a legacy of words, conceptions, and images, acts as a chain upon our imagination? Is it not, perhaps, that language, mental habits, and categories of thought organically related to a social world completely unlike our own are entirely unsuitable for expressing our experiences? May this not be the cause of our inability to articulate what we feel and see, and to bring order into what we know? Certainly a vocabulary and notions dependent upon Greek and Latin can no longer be adequate to discuss our social life-situation. Nor will the continual addition of implausible neologisms composed of more Greek and Latin words help, for they do not affect the structure of thinking. The malaise induced by this state of affairs is responsible for much of the ill-tempered and ill-informed hostility of many humanists toward the natural sciences which do not share these inherited difficulties. It also accounts for many ill-considered efforts to "imitate" science by the metaphorical or analogical use of words drawn from biology or physics. Nor is analytical philosophy of much use, for it does not address itself to concerns which are more nearly felt than spoken and which involve not so much what can be said as the difficulty of saying anything at all. To be sure, nostalgia is the least adequate response of all to these discomforts. And that is just what the question, "why is there no utopia?" does express in this context.

With these considerations, the question, "why is there no utopia today?" has, hopefully, been reduced to its proper proportions, which are not very great. To the extent that it depends on an erroneous and dated view of the European past, it is simply irrelevant. As a psychological problem its interest is great, but of a clinical nature. Lastly, it is only one item in the far more complex range of questions that concern the possibilities of contemporary political philosophy. Here, however, it does at least have genuine significance, even though it does not ask for an answer. For it is more a comment upon an intellectual situation than a real query. That is why a journey, however quick, through the utopian and ideological past seemed a fitting response, since it might show what the question implies, even if it does not offer any solutions.

54 MARSHALL BERMAN

'Robert Moses: The Expressway World'

from *All That Is Solid Melts Into Air* (1982)

Among the many images and symbols that New York has contributed to modern culture, one of the most striking in recent years has been an image of modern ruin and devastation. The Bronx, where I grew up, has even become an international code word for our epoch's accumulated urban nightmares: drugs, gangs, arson, murder, terror, thousands of buildings abandoned, neighborhoods transformed into garbage- and brick-strewn wilderness. The Bronx's dreadful fate is experienced, though probably not understood, by hundreds of thousands of motorists every day, as they negotiate the Cross-Bronx Expressway, which cuts through the borough's center. This road, although jammed with heavy traffic day and night, is fast, deadly fast; speed limits are routinely transgressed, even at the dangerously curved and graded entrance and exit ramps; constant convoys of huge trucks, with grimly aggressive drivers, dominate the sight lines; cars weave wildly in and out among the trucks: it is as if everyone on this road is seized with a desperate, uncontrollable urge to get out of the Bronx as fast as wheels can take him. A glance at the cityscape to the north and south – it is hard to get more than quick glances, because much of the road is below ground level and bounded by brick walls ten feet high – will suggest why: hundreds of boarded-up abandoned buildings and charred and burnt-out hulks of buildings; dozens of blocks covered with nothing at all but shattered bricks and waste.

Ten minutes on this road, an ordeal for anyone, is especially dreadful for people who remember the Bronx as it used to be: who remember these neighborhoods as they once lived and thrived, until this road itself cut through their heart and made the Bronx, above all, a place to get out of. For children of the Bronx like myself, this road bears a load of special irony: as we race through our childhood world, rushing to get out, relieved to see the end in sight, we are not merely spectators but active participants in the process of destruction that tears our hearts. We fight back the tears, and step on the gas.

Robert Moses is the man who made all this possible. When I heard Allen Ginsberg ask at the end of the 1950s, "Who was that sphinx of cement and aluminum," I felt sure at once that, even if the poet didn't know it, Moses was his man. Like Ginsberg's "Moloch, who entered my soul early," Robert Moses and his public works had come into my life just before my Bar Mitzvah, and helped bring my childhood to an end. He had been present all along, in a vague subliminal way. Everything big that got built in or around New York seemed somehow to be his work: the Triborough Bridge, the West Side Highway, dozens of parkways in Westchester and Long Island, Jones and Orchard beaches, innumerable parks, housing developments, Idlewild (now Kennedy) Airport, a network of enormous dams and power plants near Niagara Falls; the list seemed to go on forever. He had generated an event that had special magic for me: the 1939–40 World's Fair, which I had attended in my mother's womb, and whose elegant logo, the trylon and perisphere, adorned our apartment in many forms – programs, banners, postcards, ashtrays – and symbolized human adventure, progress, faith in the future, all the heroic ideals of the age into which I was born.

But then, in the spring and fall of 1953, Moses began to loom over my life in a new way: he proclaimed that he was about to ram an immense expressway, unprecedented in scale, expense and difficulty of construction, through our neighborhood's heart. At first we couldn't believe it; it seemed to come from another world. First of all, hardly any of us owned cars: the neighborhood itself, and the subways leading downtown, defined the flow of our lives. Besides, even if the city needed the road – or was it the state that needed the road? (in Moses' operations, the location of power and authority was never clear, except for Moses himself) – they surely couldn't mean what the stories seemed to say: that the road would be blasted directly through a dozen solid, settled, densely populated neighborhoods like our own; that something like 60,000 working- and lower-middle-class people, mostly Jews, but with many Italians, Irish and Blacks thrown in, would be thrown out of their homes. The Jews of the Bronx were nonplussed: Could a fellow-Jew really want to do this to us? (We had little idea of what kind of Jew he was, or of how much we were all an obstruction in his path.) And even if he did want to do it, we were sure it couldn't happen here, not in America. We were still basking in the afterglow of the New Deal: the government was *our* government, and it would come through to protect us in the end. And yet, before we knew it, steam shovels and bulldozers were there, and people were getting notice that they had better clear out fast. They looked numbly at the wreckers, at the disappearing streets, at each other, and they went. Moses was coming through, and no temporal or spiritual power could block his way.

For ten years, through the late 1950s and early 1960s, the center of the Bronx was pounded and blasted and smashed. My friends and I would stand on the parapet of the Grand Concourse, where 174th Street had been, and survey the work's progress – the immense steam shovels and bulldozers and timber and steel beams, the hundreds of workers in their variously colored hard hats, the giant cranes reaching far above the Bronx's tallest roofs, the dynamite blasts and tremors, the wild, jagged crags of rock newly torn, the vistas of devastation stretching for miles to the east and west as far as the eye could see – and marvel to see our ordinary nice neighborhood transformed into sublime, spectacular ruins.

[. . .]

Moses seemed to glory in the devastation. When he was asked, shortly after the Cross-Bronx road's completion, if urban expressways like this didn't pose special human problems, he replied impatiently that "there's very little hardship in the thing. There's a little discomfort and even that is exaggerated." Compared with his earlier, rural and suburban highways, the only difference here was that "There are more houses in the way . . . more people in the way – that's all." He boasted that "When you operate in an overbuilt metropolis, you have to hack your way with a meat ax." The subconscious equation here – animals' corpses to be chopped up and eaten, and "people in the way" – is enough to take one's breath away. Had Allen Ginsberg put such metaphors into his Moloch's mouth, he would have never been allowed to get away with it: it would have seemed, simply, too much. Moses' flair for extravagant cruelty, along with his visionary brilliance, obsessive energy and megalomaniac ambition, enabled him to build, over the years, a quasi-mythological reputation. He appeared as the latest in a long line of titanic builders and destroyers, in history and in cultural mythology: Louis XIV, Peter the Great, Baron Haussmann, Joseph Stalin (although fanatically anti-communist, Moses loved to quote the Stalinist maxim "You can't make an omelette without breaking eggs"), Bugsy Siegel (master builder of the mob, creator of Las Vegas), "Kingfish" Huey Long; Marlowe's Tamburlaine, Goethe's Faust, Captain Ahab, Mr. Kurtz, Citizen Kane. Moses did his best to raise himself to gigantic stature, and even came to enjoy his increasing reputation as a monster, which he believed would intimidate the public and keep potential opponents out of the way.

In the end, however – after forty years – the legend he cultivated helped to do him in: it brought him thousands of personal enemies, some eventually as resolute and resourceful as Moses himself, obsessed with him, passionately dedicated to bringing the man and his machines to a stop. In the late 1960s they finally succeeded, and he was stopped and deprived of his power to build. But his works still surround us, and his spirit continues to haunt our public and private lives.

It is easy to dwell endlessly on Moses' personal power and style. But this emphasis tends to obscure one of the primary sources of his vast authority: his ability to convince a mass public that he was the vehicle of impersonal world-historical forces, the moving spirit of modernity. For forty years, he was able to pre-empt the vision of the modern. To oppose his bridges, tunnels, expressways, housing developments, power dams, stadia, cultural centers, was – or so it seemed – to oppose history, progress, modernity itself. And few people, especially in New York, were prepared to do that. "There are people who like things as they are. I can't hold out any hope to them. They have to keep moving further away. This is a great big state, and there are other states. Let them go to the Rockies." Moses struck a chord that for more than a century has been vital to the sensibility of New Yorkers: our identification with progress, with renewal and reform, with the perpetual trans-formation of our world and ourselves – Harold Rosenberg called it "the tradition of the New." How many of the Jews of the Bronx, hotbed of every form of radicalism, were willing to fight for the sanctity of "things as they are?" Moses was destroying our world, yet he seemed to be working in the name of values that we ourselves embraced.

I can remember standing above the con-struction site for the Cross-Bronx Expressway, weeping for my neighborhood (whose fate I foresaw with nightmarish precision), vowing remembrance and revenge, but also wrestling with some of the troubling ambiguities and contradictions that Moses' work expressed. The Grand Concourse, from whose heights I watched and thought, was our borough's closest thing to a Parisian boulevard. Among its most striking features were rows of large, splendid 1930s apartment houses: simple and clear in their architectural forms, whether geometrically sharp or biomorphically curved; brightly colored in contrasting brick, offset with chrome, beautifully interplayed with large areas of glass; open to light and air, as if to proclaim a good life that was open not just to the elite residents but to us all. The style of these buildings, known as Art Deco today, was called "modern" in their prime. For my parents, who described our family proudly as a "modern" family, the Concourse buildings represented a pinnacle of modernity. We couldn't afford to live in them – though we did live in a small, modest, but still proudly

"modern" building, far down the hill – but they could be admired for free, like the rows of glamorous ocean liners in port downtown. (The buildings look like shell-shocked battleships in drydock today, while the ocean liners themselves are all but extinct.)

As I saw one of the loveliest of these buildings being wrecked for the road, I felt a grief that, I can see now, is endemic to modern life. So often the price of ongoing and expanding modernity is the destruction not merely of "traditional" and "pre-modern" institutions and environments but – and here is the real tragedy – of everything most vital and beautiful in the modern world itself. Here in the Bronx, thanks to Robert Moses, the modernity of the urban boulevard was being condemned as obsolete and blown to pieces, by the modernity of the interstate highway. *Sic transit!* To be modern turned out to be far more problematical, and more perilous, than I had been taught.

What were the roads that led to the Cross-Bronx Expressway? The public works that Moses organized from the 1920s onward expressed a vision – or rather a series of visions – of what modern life could and should be. I want to articulate the distinctive forms of modernism that Moses defined and realized, to suggest their inner contradictions, their ominous under-currents – which burst to the surface in the Bronx – and their lasting meaning and value for modern mankind.

Moses' first great achievement, at the end of the 1920s, was the creation of a public space radically different from anything that had existed anywhere before: Jones Beach State Park on Long Island, just beyond the bounds of New York City along the Atlantic. This beach, which opened in the summer of 1929, and recently celebrated its fiftieth anniversary, is so immense that it can easily hold a half million people on a hot Sunday in July without any sense of con-gestion. Its most striking feature as a landscape is its amazing clarity of space and form: abso-lutely flat, blindingly white expanses of sand, stretching forth to the horizon in a straight wide band, cut on one side by the clear, pure, endless blue of the sea, and on the other by the boardwalk's sharp unbroken line of brown. The great horizontal sweep of the whole is punctuated by two elegant Art Deco bathhouses of wood, brick and stone, and half-way between them at the park's dead center by a monumental columnar water tower, visible from everywhere, rising up like a skyscraper, evoking the grandeur of the twentieth-century urban forms that this park at once complements and denies. Jones Beach offers a spectacular display of the primary forms of nature – earth, sun, water, sky – but nature here appears with an abstract horizontal purity and a luminous clarity that only culture can create.

We can appreciate Moses' creation even more when we realize (as Caro explains vividly) how much of this space had been swamp and wasteland, inaccessible and unmapped, until Moses got there, and what a spectacular meta-morphosis he brought about in barely two years. There is another kind of purity that is crucial to Jones Beach. There is no intrusion of modern business or commerce here: no hotels, casinos, ferris wheels, roller coasters, parachute jumps, pinball machines, honky-tonks, loudspeakers, hot-dog stands, neon signs; no dirt, random noise or disarray. Hence, even when Jones Beach is filled with a crowd the size of Pittsburgh, its ambience manages to be remarkably serene. It contrasts radically with Coney Island, only a few miles to the west, whose middle-class constituency it immediately captured on its opening. All the density and intensity, the anarchic noise and motion, the seedy vitality that is expressed in Weegee's photographs and Reginal Marsh's etchings, and celebrated symbolically in Lawrence Ferlinghetti's "A Coney Island of the Mind," is wiped off the map in the visionary landscape of Jones Beach.

[. . .]

Moses' Northern and Southern State parkways, leading from Queens out to Jones Beach and beyond, opened up another dimension of modern pastoral. These gently flowing, artfully landscaped roads, although a little frayed after half a century, are still among the world's most beautiful. But their beauty does not (like that of, say, California's Coast Highway or the Appalachian Trail) emanate from the natural environment around the roads: it springs from the artificially created environment of the roads themselves. Even if these parkways adjoined nothing and led nowhere, they would still constitute an adventure in their own right. This is especially true of the Northern State Parkway, which ran through the country of palatial estates that Scott Fitzgerald had just immortalized in *The Great Gatsby* (1925). Moses' first Long Island roadscapes represent a modern attempt to recreate what Fitzgerald's narrator, on the novel's last page, described as "the old island here that flowered once for Dutch sailors' eyes – a fresh, green breast of the new world." But Moses made this breast available only through the mediation of that other symbol so dear to Gatsby: the green light. His parkways could be experienced only in cars: their underpasses were purposely built too low for buses to clear them, so that public transit could not bring masses of people out from the city to the beach. This was a distinctively techno-pastoral garden, open only to those who possessed the latest modern machines – this was, remember, the age of the Model T – and a uniquely privatized form of public space. Moses used physical design as a means of social screening, screening out all those without wheels of their own. Moses, who never learned to drive, was becoming Detroit's man in New York. For the great majority of New Yorkers, however, his green new world offered only a red light.

[. . .]

Where did it all go wrong? How did the modern visions of the 1930s turn sour in the process of their realization? The whole story would require far more time to unravel, and far more space to tell, than I have here and now. But we can rephrase these questions in a more limited way that will fit into the orbit of this book: How did Moses – and New York and America – move from the destruction of a Valley of Ashes in 1939 to the development of far more dreadful and intractable modern wastelands a generation later only a few miles away? We need to seek out the shadows within the luminous visions of the 1930s themselves.

[. . .]

Part of Moses' tragedy is that he was not only corrupted but in the end undermined by one of his greatest achievements. This was a triumph that, unlike Moses' public works, was for the most part invisible: it was only in the late 1950s that investigative reporters began to perceive it. It was the creation of a network of enormous, interlocking "public authorities," capable of raising virtually unlimited sums of money to build with, and accountable to no executive, legislative or judicial power.

[. . .]

Moses' projects of the 1950s and 60s had virtually none of the beauty of design and human sensitivity that had distinguished his early works. Drive twenty miles or so on the Northern State Parkway (1920s), then turn around and cover those same twenty miles on the parallel Long Island Expressway (1950s/60s), and wonder and weep. Nearly all he built after the war was built in an indifferently brutal style, made to overawe and overwhelm: monoliths of steel and cement, devoid of vision or nuance or play, sealed off from the surrounding city by great moats of stark empty space, stamped on the landscape with a ferocious contempt for all natural and human life. Now Moses seemed scornfully indifferent to the human quality of what he did: sheer quantity – of moving vehicles, tons of cement, dollars received and spent – seemed to be all that drove him now. There are sad ironies in this, Moses' last, worst phase.

The cruel works that cracked open the Bronx ("more people in the way – that's all") were part of a social process whose dimensions dwarfed even Moses' own megalomaniac will to power. By the 1950s he was no longer building in accord with his own visions; rather, he was fitting enormous blocks into a pre-existing pattern of national reconstruction and social integration that he had not made and could not have substantially changed. Moses at his best had been a true creator of new material and social possibilities. At his worst, he would become not so much a destroyer – though he destroyed plenty – as an executioner of directives and imperatives not his own. He had gained power and glory by opening up new forms and media in which modernity could be experienced as an adventure; he used that power and glory to institutionalize modernity into a system of grim, inexorable necessities and crushing routines. Ironically, he became a focus for mass personal obsession and hatred, including my own, just when he had lost personal vision and initiative and become an Organization Man; we came to know him as New York's Captain Ahab at a point when, although still at the wheel, he had lost control of the ship.

The evolution of Moses and his works in the 1950s underscores another important fact about the postwar evolution of culture and society: the radical splitting-off of modernism from modernization. Throughout this book I have tried to show a dialectical interplay between unfolding modernization of the environment – particularly the urban environment – and the development of modernist art and thought. This dialectic, crucial all through the nineteenth century, remained vital to the modernism of the 1920s and 1930s: it is central in Joyce's *Ulysses* and Eliot's *Waste Lane* and Doblin's *Berlin, Alexanderplatz* and Mandelstam's *Egyptian Stamp*, in Léger and Tatlin and Eisenstein, in William Carlos Williams and Hart Crane, in the art of John Marin and Joseph Stella and Stuart Davis and Edward Hopper, in the fiction of Henry Roth and Nathanael West. By the 1950s, however, in the wake of Auschwitz and Hiroshima, this process of dialogue had stopped dead.

It is not that culture itself stagnated or regressed: there were plenty of brilliant artists and writers around, working at or near the peak of their powers. The difference is that the modernists of the 1950s drew no energy or inspiration from the modern environment around them. From the triumphs of the abstract expressionists to the radical initiatives of Davis, Mingus and Monk in jazz, to Camus' *The Fall*, Beckett's *Waiting for Godot*, Malamud's *The Magic Barrel*, Laing's *The Divided Self*, the most exciting work of this era is marked by radical distance from any shared environment. The environment is not attacked, as it was in so many previous modernisms: it is simply not there.

This absence is dramatized obliquely in what are probably the two richest and deepest novels of the 1950s, Ralph Ellison's *Invisible Man* (1952) and Günter Grass's *The Tin Drum* (1959): both these books contained brilliant realizations of spiritual and political life as it had been lived in the cities of the recent past –

Harlem and Danzig in the 1930s – but although both writers moved chronologically forward, neither one was able to imagine or engage the present, the life of the postwar cities and societies in which their books came out. This absence itself may be the most striking proof of the spiritual poverty of the new postwar environment. Ironically, that poverty may have actually nourished the development of modernism by forcing artists and thinkers to fall back on their own resources and open up new depths of inner space. At the same time, it subtly ate away at the roots of modernism by sealing off its imaginative life from the everyday modern world in which actual men and women had to move and live.

55 JON BIRD

'Dystopia on the Thames'

from Jon Bird *et al.* (eds), *Mapping the Futures* (1993)

The tidal current runs to and fro in its unceasing service, crowded with memories of men and ships it had borne to the rest of home or to the battles of the sea. It had known and served all the men of whom the nation is proud, from Sir Francis Drake to Sir John Franklin, knights all, titled and untitled – the great knight-errants of the sea. It had borne all the ships whose names are like jewels flashing in the night of time, from the Golden Hind returning with her round flanks full of treasure, to be visited by the Queen's Highness and thus pass out of the gigantic tale, to the Erobas and Terror, bound on other conquests – and that never returned. It had known the ships and the men. They had sailed from Deptford, from Greenwich, from Erith – the adventurers and the settlers; king's ships and the ships of men on 'charge'; captains, admirals, the dark 'interlopers' of the Eastern trade, and the commissioned 'generals' of East India fleets. Hunters for gold or pursuers of fame, they all had gone out on that stream, bearing the sword, and often the torch, messengers of the might within the land, bearers of a spark from the sacred fire. What greatness had not floated on the ebb of that river into the mystery of an unknown earth! ... The dreams of men, the need of commonwealths, the germs of empires.

(Joseph Conrad, *Heart of Darkness*, 1902)

THE HERITAGE: London's bustling docks and busy river were the very evidence of the nation's development as the Empire grew and flourished; jewels in Britain's commercial crown. Products of engineering and industrial genius despatched around the world; exotic cargoes and evocative aromas from far-flung corners in return. But eventually new trading patterns emerged and modern cargo-handling techniques moved the tide of shipping away from those docks. Already, as the docks declined in commercial importance, so the neighbouring City of London was increasing in stature and importance – as *the* international centre for financial services.

Strategically placed between the New and Old Worlds, spanning the globe, creating new opportunities, linking all kinds of trade. New technology, new communications, new demands and new services reaffirmed the importance of London, with the greatest concentration of markets in Europe – and the innovative skills and experience that go with them.

(London Docklands Development Corporation, 1989)

City life in Britain has never conveyed the alluring resonances of the great centres of European modernism – Paris, Barcelona, Madrid, Milan, Hamburg, or the glittering but brittle spectacles of American urbanization. Neither the Left nor the Right has really laid claim to the city as a site for the construction of subjectivity and political identity other than as the backdrop for the enactment of ritual and tradition: the ceremonial commemoration of privilege, national identity, or loss. During the 1970s and early 1980s, various social and democratic movements, notably the Labour-controlled Metropolitan Councils, challenged the dominant meanings of city spaces and activities, only to be abolished by Conservative government legislation replacing the social infrastructure with privatized services and facilities. This process did not go unchallenged and the 1980s were marked by eruptions of the inner cities in anarchic refusal of the patterns of unemployment, repressive policing, and authoritarian government. Starting with violent clashes between police and rioters in Brixton in southeast London in 1981, this rapidly spread to depressed areas throughout the country, frequently as a result of years of racial harassment and social and economic neglect. Many of the tactics learnt by the police in combating urban unrest during this period were later successfully deployed against a fantasized 'enemy within' – the miners in their year-long strike of 1984/85, and against the print workers in the Wapping struggles as Murdoch shifted his production to East London.

These moments of civic upheaval have occurred against a continuing nostalgia for a mythical past of tranquillity and order running deep within the political and social fabric of Britain as a nation state, and manifested in the proliferating cultures of 'heritage' and 'enterprise' that so clearly characterized the last decade.

In London, the most public and visual expression of 1980s aggressive monetarism has been the changing skyline of East London's riverside redevelopment from the Tower of London to the Thames Barrier, spear-headed by the Toronto conglomerate Olympia & York's Canary Wharf scheme on the Isle of Dogs. This represents the single largest building project ever undertaken – a $7bn investment into 12 million square feet of office space, apartments, restaurants, shopping malls and riverside facilities, which includes the country's (and until recently, Europe's) highest single building at over 800 feet, designed by Cesar Peli, and on a scale totally disproportionate to the scale of the City, casting a permanent shadow over the immediate vicinity.

Docklands has been made possible by the Thatcher government's establishment in 1981 of Urban Development Corporations (UDCs) – consortia of business and property interests run by boards directly appointed by the Secretary of State and able to grant planning permission free from local authority investigation or sanction. Under the ideology of 'regeneration of the inner cities', the UDCs can create companies, subsidize existing businesses, and develop or dispose of land and buildings. Although UDCs have also been established in several areas of the north-east, it is the high profile London Docklands Development Corporation (LDDC) that has become the testing ground for free-market intervention into what had previously been the concerns and responsibilities of democratically elected local councils and boroughs – in Docklands the three boroughs of Tower Hamlets, Newham and Southwark. The almost unlimited powers of the LDDC included the appropriation of publicly owned land,

approximately 5,000 acres and 55 miles of waterfront property which could then be sold off to private developers. Land prices leapt from £70,000 per acre in 1981 to £4m by 1987. Add to this the presence of the river, and the context is created for a battle over representations in which the complex historical and mythological connotations of the Thames and the City are played out in the conflict between dominant and subordinate cultures and economies.

If, in the modern period, the City signified the visual and tactile presence of modernity, a liberatory image of anonymity and freedom, in the postmodern period this has become a battleground of oppositions between the public and the private. The urban wasteland has been repositioned within the circuits of international finance capital and recoded as a site of consumption and the pursuit of leisure. Spatially, Docklands represents the principle of uneven development that is the hallmark of the geography of capitalism. The ideology of 'regeneration' (represented as a natural process of decay, death and rebirth) masks the economic and social relations that characteristically determine a history of neighbourhood decline and abandonment, followed by rediscovery and gentrification – a trajectory frequently initiated by the social migration of transitory groups of squatters, students and artists seeking affordable, temporary accommodation. This process can be traced across the maps of the modern and postmodern metropolises as localities and communities are included or excluded from the centres of wealth, decision-making and power.

The rationale behind Docklands as a redefined space has been the intended relocation of a proportion of the City's finance centre away from the overcrowded Square Mile to East London, with the inducement of rent reductions for office space from £70 to £20 per square foot. (Events are constantly overtaking my analysis. The recession has now more or less removed the disparity between the rents in Docklands and the Square Mile, where they have dropped by 30 per cent over the last year.) Besides the economic arguments, however, the developers have sought to construct a symbolic opposition between a myth of the waterfront – the Thames as the once-great artery of the capital, but allowed to silt up and become a sinister space of Dickensian dereliction and impoverishment – and a sparkling, utopian, vision of a 'Metropolitan water city of the twenty-first century'. In this scenario, a history of government neglect is turned into a pathology of despair, a narrative of outmoded industries and regressive labour relations, parasitical local authorities and fragmented communities, a spectre of the inner city as a virus within the social body. This 'official' labour history of Docklands constructs a recalcitrant and militant workforce resistant to the inevitable dynamic of modernization, specifically the containerization of cargoes. Mechanization removed the requirement for a large dockside workforce available up-river in East London and shifted the transfer of cargo to the purpose-built container port of Tilbury at the mouth of the Thames. Some containerization took place in the Royal Docks but was discontinued in 1981 because of the greater profits available from the sale of the land. The shipping lines were able to avoid having to employ a strongly unionized workforce drawn from London's docklands in favour of the individualistic and collaborative support of the transport companies. The fabricated and selective history of the docks represents a systematic refusal of class history and struggle; of multicultural communities and democratic organization around demands over housing, education, and the preservation of traditional and local skills and industrial experience. (Although, of course, it is important not simply to substitute one myth for another – there is also a history of anti-semitism, racism, sexual harassment – and their respective points of resistance.)

To take a ride through Docklands on the Light Railway is to experience the global postmodern as a building site. Here consortia and multinationals swallow up the generous offers of land that are made available in an enterprise zone and spew out varieties of architectural postmodernism and high-tech in paroxysms of construction that are as incoherent as they are unregulated. To cite just one example of the advantages available to potential developers: any proposals that fall within the zone benefit from a greatly simplified planning process, no rates payable until 1992, and the offsetting of building costs against income tax or corporation tax by 100 per cent for projects under construction by April 1992.

Olympia & York stepped in to rescue the financially ailing LDDC after the original American consortium (Crédit Suisse, First Boston, Morgan Stanley, and the Travelstead Group) had backed out of the Isle of Dogs schemes in 1987. Besides the monumental Canary Wharf development, Olympia & York are building 1,000 privately owned riverside apartments and a hotel across the water at Heron Quay; they own a third of the property group which has erected a vast shopping mall in the Royal Docks, and have bought into the Port East retail development at West India Dock. The magnanimity of their entrepreneurial activities (generally trumpeted as providing jobs, homes, etc.) is tempered by the conditions they made for their involvement. These included the initial land purchase at a fraction of the (then) market value, £500m from the government in capital allowances, and a further contribution of £550m to upgrade the transport facilities upon which Canary Wharf's financial viability depends. Much of this development centres around the 'politics of the view' – a battle over representations. This is not simply the desirability of the waterside, but the juxtaposition of private affluence with public deprivation – of luxury apartments facing high-rise council housing blocks. Recently the LDDC has provided the finance to screen off all the contradictory signifiers of uneven development through planting rows of trees, rebuilding picturesque dock walls, or renovating the façades of council estates which overlook expensive residential apartments.

Already the unoccupied low-rise development at Heron Quay is being demolished as the initial projections are redrawn and plans are revised upwards to accommodate an anticipated daily influx of 150,000 people by 1993, far in excess of the capacities of even the most ambitious transport scheme, a process described by the LDDC as 'second-wave regeneration'. The present deep recession and high interest rates have slowed development and forced a number of property companies out of business. Despite this, the proposed major extension to the Light Railway and the construction of the Docklands Highway are going ahead. The Highway will be a six-lane motorway scything through the East End from the City, a plan agreed secretly between the LDDC and the Department of

Transport, the first major road project since the war to be approved without any form of public inquiry. However, the building of the highway has become symptomatic of the general economic problems threatening the rate, if not the scale, of development, and symptomatic too of the crisis within the LDDC. The 1.1-mile Limehouse Link, now costed at over £300m, is set to become the most expensive stretch of road in the world. As the LDDC has been forced by the Treasury to cut costs, the schemes to be jettisoned are, predictably, those social projects that represented concessions to the needs and demands of the residential Docklands communities.

The publicity material produced by the LDDC and Olympia & York presents an image of harmony and coherence, a unity of places and functions not brutally differentiated into the respective spheres of work, home and leisure, but woven together by the meandering course of the river into a spectacular architectural myth of liberal *civitas*. Canary Wharf is indeed a fantasy of community: a city within the City populated by a migrant army of executive, managerial and office staff serving the productive signifiers of postmodernity – microelectronics, tele-communications and international capital – along with the relevant support structures and lifestyle accoutrements, from food to culture, an airport to shopping malls. These panoramic representations of the City characteristically adopt an elevated perspective that distances the viewer and creates an image of totality. We look from a distance across, or down upon, the river and adjacent buildings, each scene suffused with a gentle light which plays upon the towers and the water. Nothing is unharmonious or out of place – these are viewpoints that allow us to possess the City in imagination. Similarly, the estate agents' advertisements stress the desir-ability of a riverside lifestyle: windsurfing, cocktails in St Katherine's Yacht Club, a plethora of marinas, the proximity of other European capitals via the new Docklands airport, the jetfoil to the City or the West End, electronic surveillance and round-the-clock security systems – a familiar description of urban living cushioned by the privileges of wealth and power. As Michel de Certeau has argued, this voyeuristic gaze is in contrast to the everyday practices of inhabiting the street: the tactics of

lived space. The voyeuristic gaze implies mastery and objectification, a distancing relationship of knowledge and power that is founded upon repression and desire. The politics of these postmodern utopias is to legitimize the actual processes of redevelopment and gentrification over and above the everyday needs and experiences of the people inhabiting Europe's largest building site. Indeed, the experience of visiting Docklands is one of exclusion and alienation. The monumental scale overpowers the street; corporate architecture dominates and privatizes every vista; the mirror-glass façades disorientate the relations of people, place and space whilst the physical combinations of height and mass create climatic conditions that, on anything other than the brightest and calmest of days, threaten to sweep the unwary into the river. There is also the question of the spatial mobilities and experiences that are constructed by these forms of urban geography; the directives and constraints upon movement, access and enclosure which serve to remind that the spaces of the city, modern or postmodern, are gendered spaces reinforcing relations of power and subjectivity.

Against these powerful myths and realities of power, the Docklands Consultative Committee – comprising the democratically elected repre-sentatives of the various Dockland boroughs, neighbourhoods and community groups – produced in 1988 their own report and assessment of the claims and progress of government and developers. Not surprisingly, they found that the real beneficiaries were service-sector industries and employees, with only a minimal proportion of jobs being made available to local residents, mostly on govern-ment training schemes and temporary and part-time servicing contracts: cleaners, caterers, chauffeurs, domestic labour, etc. Similarly, the commitment to provide rented accommodation and affordable housing had fallen short of all the original estimates: in fact, to date, over 90 per cent of all residential property has been sold on the open market. With excessive land values, even after the slump, and the redirection of government funding from the public to the private sector, the local authorities are unable to construct alternative, cheap accommodation, or create employment opportunities: a cycle of high unemployment and urgent welfare requirements in the context of inadequate resourcing producing the very conditions that are then used to justify market intervention and private investment.

Against the destructive impact of international capital, complicit government, and a culture of enterprise, the collaborative activities of the Docklands communities, founded upon a history of political activism and the proliferation of points of resistance across the divisions of class, gender and ethnicity, have provided more than a token refusal of market forces, winning crucial victories over land and housing projects, and forcing the LDDC to establish a Community Development Unit. In 1983 a number of local organizations combined to produce The People's Plan for the Royal Docks, a campaign initiated by the Borough of Newham Docklands Forum aided by the Joint Docklands Action Group and the Docklands Community Poster Project. The focus for this activity was a report then being prepared on the feasibility of building a light airport in the Royal Docks. At each stage in the development of the report and throughout the subsequent political campaign, antagonistic and contradictory versions of 'official' discourse were produced – as posters, banners, leaflets, flyers and press reports. Although the eventual outcome was the opening of the airport, the enabling process of collaboratively challenging the dominant meanings provided a vehicle for the expression of cultural identities and community interests.

Occupying a crucial position in this process has been the Docklands Community Poster Project (DCPP), founded in 1982 out of the practice of two artists: Peter Dunn and Loraine Leeson. The discursive nature of the DCPP, and the co-ordination of research and production through collaboration with an advisory steering committee representing the varied interests of Docklands residents, is a recognition of the multi-faceted and mobile aspect of cultural politics today. The DCPP has been involved in the production of posters, publications and banners for specific issues and campaigns; the establishment of a visual archive of the history of Docklands; the generation of a travelling exhibition which situates the narrative of the docks within a broader social history and acts as a model for other sites subject to riverside redevelopment; and, most recently, the establish-

ment of a consultancy – Art For Change. The Project has also co-ordinated a number of highly theatrical street demonstrations, and between 1984 and 1986, it organized the People's Armada – flotillas of boats sailing up the Thames to the Houses of Parliament to protest against Docklands developments – events which attracted national press and media coverage. However, the central element of the DCPP has been an ongoing series of large-scale photomurals, initiated in 1981, charting the 'changing picture of Docklands'.

The photomurals are actually large billboards, each measuring 18 feet by 12 feet, containing photographic and drawn imagery in colour and black-and-white, that are located at key sites throughout the docklands boroughs. Each mural tells a story, an unfolding of collective history and lived experience, in selected moments of oppression and struggle as each successive scene represents a stage in a transformation from past memories, through present 'realities', towards a utopian resolution. The series The Changing Picture of Docklands consists of eight stages. Panels are changed at regular intervals until the original image representing the signs of urban dereliction – corrugated iron and graffiti-covered walls – has been transformed into a liberatory scenario of a schoolgirl leading a heterogeneous and multi-racial community of local residents into Our Future in Docklands. Intermediate stages visualize an economy of privatization and exploitation: towers of money, tenement blocks, the scrapheap of social planning, the controlling hand of capital. The final images, although problematic in their utopian aspirations, are an attempt visually to invert power relations and to recognize the necessity for alternative visions of social reality, besides emphasizing the city as representation – a field of meanings that can be contested and changed.

In a recent series on housing, the posters invert the strategies of corporate advertising. Advertising conventionally erases all traces of the labour history of the product prior to its entry into the circuits of distribution and consumption: its message is one of constant and immediate gratification. Like advertising, the official visual rhetoric of Docklands expresses the hallucinatory form of a wish, the voyeuristic gaze that involves subject and object in processes of fantasy and desire. The strategy of Dunn and Leeson in the various poster series is to confront the social and psychic construction of images through representing the specific histories, experiences and cultural practices that have formed individual and communal identities in this area of East London. The *Housing* series documents various moments and events in local history, connecting the home and the workplace in representations that make evident the gender politics that underlie the distinctions between the realms of the public and the private. Historically significant events in the labour history of Docklands are depicted using the dominant representational technologies of the period – social documentary photography to depict the Reformist movement of the late nineteenth century, and the techniques of the illustrated popular press to describe dock life in the 1920s. However, the apocalyptic imagery of industrialization, with its fascination with the sublime expressed as a tension between terror and awe, is here employed as a deconstructive device against the truth claims of the documentary tradition. The ideology of victimization and class passivity is challenged by examples of political and cultural production: the suffrage movement in East London and the organization of the Dockers' Union. The second half of the sequence takes the viewer through the labour disputes of the 1930s to the reconstruction of the postwar period, the slogan 'Homes for Heroes' reads as an ironic commentary as the widely deployed 'temporary' housing units (prefabs) are piled on top of each other to form a tower block. This symbol of postwar planning and social control bears the address Ronan Point, a reference to the collapse of a council housing block in 1968, an overdetermined moment in the crisis of welfarism and the bureaucratic state.

The visual rhetoric of the murals partly references the critical traditions of collage and montage within modernism, although Dunn and Leeson have deliberately constructed a deep spatial field as an analogue for the richness of social history. The montages are intended to produce narratives that accept the ultimate instability at the core of relations of representation, while also recognizing the necessity for points of stability. To subvert meaning, to place meanings *under erasure*, there have to be points of fixity against which ambiguities within the social formation, and the formation of subject identities, can be rearticulated to mean *otherwise*... The DCPP represents the ways in which culture as a 'whole way of life' can be made to articulate the wishes and desires of 'ordinary' people, and the quality of everyday experience. Located, for the most part, outside the sanctioned spaces of high art, and representing the collaborative efforts and ideals of communities instead of celebrating individualism, Dunn and Leeson are involved in a project 'to develop a critical means of celebrating solidarity, strength and the ability to survive and win over oppression'. Obviously there have been failures – the combination of multinational capital and government complicity is hard to resist – but there have also been victories, not the least of which has been the use-value of a cultural organization in producing a coherent identity and focus for social and political campaigning.

Today the iconography of Docklands development prepares the regenerated public space for the postmodern, postindustrial, upwardly mobile, credit-wealthy, information-rich subject. This new occupant of the City consumes and is consumed by the spectacle of an economic geography as satellite dishes and fibre-optic cables connect global marketeers to the universal data banks. As the architects and planners enfold the most technologically sophisticated products of the computer industries in the mantle of historicism and heritage, the confident beneficiaries of enterprise culture take their leisure at a waterside bar, stroll through the 'unique Georgian malls and arcades' of Docklands Shopping Village, or jet over to Paris from the Royal Docks City Airport.

Meanwhile, the bulldozed lots and boarded-up housing blocks testify to ghostly presences and forgotten histories. What East London – a neighbourhood of closely knit terraced housing, shops and pubs, riverside warehouses, schools and hospitals – used to mean was streets. Despite the contradictions and hardships concealed in memories of community (forms of nostalgia that have their own tendency to simulate 'heritage'), the street did provide a location for social gathering and for collective memory and shared forms of lived experience and resistance embedded in the very fabric of everyday life. It is, perhaps, too easy to slide into a melancholic discourse of loss, hopelessness and regret. Instead we can listen for the distinctive voices of challenge and resistance encoded in the rhythms of subcultural street life, the vibrancy of local and heterogeneous cultures redefining the relations of periphery to centre, the visible presence in the metropolitan world of alternative traditions of representation, the deconstructive strategies at work from street theatre to the combative rhetoric of the Poster Project's photomurals: all are elements contributing to an eclectic, postmodern culture of resistance.

In this uncertain moment of geocultural transformation, it is clearly hard to predict how this culture might grow and flourish. Although the recuperative and co-optive powers of global capital have effectively blunted the critical edge of the socialist tradition as a historical project – modernity's 'grand narratives' – it remains to be seen whether the vantage point of the *post* (as in postmodern, post-colonial, post-history, post-Marxism, etc.) leads to a productive reassessment of the values and means necessary for self-realization and cultural diversity. For the Left, the centrality of 'enterprise culture' as both economic dominant and symbolic order needs to be constantly questioned, from the marble halls and sanctioned spaces of official culture, to the shopping malls and urban geographies of cities and states as we approach the '*fin de millénium*'. We need a new 'politics of place', capable of recognizing and exploring the instabilities and contradictions present in all areas of social life and political formations, and we need practices that symbolize the bridging of individual experience, cultural difference and collective representations. Raymond Williams always recognized our responsibilities to the disenfranchised voices of past, present and future generations. For Williams, it was the commitment to reflect critically upon the complexity and diversity of social relationships and to discover their determination upon ourselves as subjects that provided the justification and promise of freedom. Never likely to be a convenient or dominant expression, 'its sound is usually unmistakable; the sound of that voice which, in speaking as itself is speaking, necessarily, for more than itself. Whether we find such voices or not, it is worth committing ourselves to the attempt.

POSTSCRIPT

Events continue to overtake my analysis. Currently Canary Wharf is set to become the largest financial collapse of the post-war period with Olympia & York having total debts in the region of $20b. Only ten of Canary Wharf's fifty floors are actually occupied and the most optimistic forecasts project this will only rise to 60 per cent of leased space by 1993. One response to this is a round of applause at the prospect of another dramatic example of the crisis of capitalism. However, like most such crises previously, there are few beneficiaries from the fall-out from multi-billion dollar fiascos other than the bailiffs and official receivers. Certainly the impact upon the local communities is likely to be even more demoralizing than the last decade of unregulated speculation and 'regeneration': the spectre of the ghetto is once again haunting the docklands of East London.

Whatever happens, Canary Wharf stands as a material legacy of the Thatcherite dream of remaking Britain through private development and corporate finance under the combined ideologies of enterprise, regeneration and heritage. Docklands is the most spectacular example (or the most grotesque) of the interweaving of global capital, global information networks, transnational media, and complicit government. It provides evidence of the dramatic (and disastrous) effects of multinational grids of investment and marketing upon the redefinition of social space and the relations between the spheres of the public and the private – of the antagonisms between the culture of the market and the life and history of the community.

56 LEBBEUS WOODS

'Everyday War'

from P. Lang (ed.), *Mortal City* (1995)

The statement "Architecture is war. War is architecture," has upset many people. War, it seems, occupies a particular place in human experience that peace does not. It is an echo of the awesome, the terrible, and the sublime that Edmund Burke, among others, reserved for special occasions. An equation of war and peace seems to offend this idea of war. Even more, it challenges the idea of the normal, for it seems that this category, which embraces, none too lovingly, the everyday and the ordinary, is something to be held outside of the sublime. Yet the offending statement proposes exactly the opposite – that war and peace are intertwined in direct and interdependent ways. It is as much to say that peace involves war, and the converse, that war contains a particular kind of peace.

Those who recall slogans of the Orwellian dystopia will detect a measure of "doublethink" in this corollary remark. And doublethink it is. Poor George simply was unable to enjoy the fact. He believed it was dictatorship that would most benefit from the exploitation of paradoxical thinking, simply because dictators choose to lie so much. The only "opposite" he could imagine was "truth" – objective truth, that is, which has never been more than the truth of those holding power at the moment. But lies are in no way paradoxical. Lies do not, as he exposed so clearly, express the condition of "holding two contradictory concepts in the mind at the same time, and believing both." Lies, or at least those made by the dictators he had in mind, are never believed by the liars. Rather, they are used to obscure an actuality. In this way, every lie confirms a truth. The liar is perhaps the greatest believer in truth, simply because he struggles so hard to conceal it. Rather, in doublethink, Orwell unintentionally introduced a more complex and subtle concept than mere lying. To hold two contradictory concepts in the mind at the same time, and to believe in both – that is closer to poesis than to crude propaganda. Indeed, it is the very essence of living.

In fact this concept best serves those individuals and societies (though there are none purely of this kind today) that are devoted to the liberation of the human mind. Living things, contrary to Aristotle's logical constrictions, not only are filled with contradictions, but thrive on them. The second law of his logic (and let no one today believe that this law is no longer absolutely in force) dictates that a cannot equal b, because a equals a; that is, persons and things have fixed "identities." But however expedient this belief may be in solving logical problems, it overlooks the fact that every person and thing is constantly changing, becoming "something else." For example, does not a person's body, which is considered integral with personhood when alive, become a mere thing when the person dies?

The attempt to salvage identity and the logic underlying it leads, in this case, to Descartes's duality, which is as arid and damaging a concept to the idea of liberation as has ever existed. A simpler example: a person is walking across the room. At any instance, is he "only" at a particular place in the room, one defined, say, by Cartesian coordinates? If so, how does he get to the next place – in other words, how does he cross the threshold of the limits of the increment? And where is he when he does? Between coordinates? Well, where, in such a system of logical description, is that? "Becoming" – whether in the form of simple motion or historial transformation – cannot be divided into discrete increments of identity, but flows as a continuum, so that at any one "point" (the concept of which is no more than a logical convenience) a thing is simultaneously what it "is" and what it is "becoming." Identities are transformational, sliding and shifting in an ongoing complex stream of becoming. In the continuum of living, a, it seems does equal b.

But perhaps this is where Orwell has a point – a does not actually "equal" b; nor does war "equal" peace anymore than ignorance "is" power. An equation inevitably opposes two entities, and this opposition, made to prove equivalency, invariably proves non-equivalency just as well. For no matter how thoroughly it is proven that two entities are equivalent, the question of their fixity at any "point" in the (re)solution returns the problem to its Aristotelian, noncontradictory form. Yet it is precisely a form of fixed identity that is lost (or perhaps it is better to say never gained) in the onrush of time and events. Things are not equivalent to one another, therefore maintaining separate and isolated existences; rather they are *part* of one another. Futhermore, they cannot be divided, except at the convenience of a particular verbal or mathematical language. It is in the context of this argument that war and architecture should be considered.

In particular, the ideas commonly described as "construction" and "destruction" need to be examined in the context of the paradoxicality inherent in experience. Few thoughtful people would fail to acknowledge that in order to build, something must be destroyed. Architects speak of ecological concerns today in a rather thoughtless way, believing that their buildings are energy efficient, use "natural" materials, do not pollute the environment, and so on. They are speaking in relative terms, of course, not only because the act of building and its consequences still create waste and pollution and remain highly "unnatural," but also because they introduce new forms of entropy into the existing environment. Not only do "natural" materials involve the mining of the earth and the felling of forests, resulting in the disruption of plant and animal life, the pollution of rivers and air, and the degradation of human existence, which is dependent upon these natural resources, but they produce buildings that begin to decay even as they are built. The effects of this decay, which continues regardless of all efforts at main-tenance, contribute to the spiral of human life, indeed of the whole environment, downward towards an ultimate heat death. Far-fetched as this may sound, it need not be understood only in millennial terms. Anyone who has studied the nature of buildings understands their relatively short life spans. These are measured not only in terms of the wearing-out or erosion of materials, but in their lack of durability in terms of use. The average life span of a building built in any city today is twenty-five years. After that, it will most likely be torn down, if not simply abandoned. In the former case, an entire industry has been created to deal with the dross of used-up buildings; in the latter, an industry of bureaucratic sloth. It is not too much, in light of these facts, to say that buildings are built in order to be destroyed, or at least partly so. Destruction is factored in, at the very least, to construction. They are inevitably and paradoxically intertwined.

But this is not the only paradox to be realized. Building is by its very nature an aggressive, even warlike act. For one thing, buildings are objects that disrupt existing landscapes. Older buildings, perhaps much-loved, must be torn down. Fields and farms are taken over. But perhaps, it can be argued, the reasons for building are benevolent ones. A certain type of building is "needed." But who needs this building? And who will *not* get a building they need in order that those who can command a building into existence get theirs? No society is rich enough to build all the buildings it "needs," if, that is, it takes into account the demands of all its people. A building will only be beautiful and useful for those who benefit directly and tangibly from its existence. As for others, blind to its benefits because they occur at the expense of those that might have been their own, the building is little more than an instrument of denial. But at great expense and on a monumental scale, it is also nothing less than an instrument of war.

This is no mere metaphor. Marshall Berman describes the war conducted by Robert Moses and the authority he controlled on many New York neighborhoods in the 1940s and 1950s. The wrecking machines that leveled houses and urban blocks were no less destructive to culture than if they had been the tanks and artillery of an attacking army. The finished highways and parks, purportedly benefits for the masses, were actually monuments to the victory of autocratic authority over the fragile lives of mere people. Even granting that every act of construction is not as wantonly indifferent to an existing fabric of culture as this, the fact that every building aggressively destroys such a fabric, even if only a little, cannot be denied. Nor can it be denied that the fabric destroyed is always that of an "inferior" class of culture (whether it be agrarian or simply old and in the way of progress), or of people (whether they be the wrong color or economic circumstances). The building of architecture is essentially political and, even though it is rarely presented as such, ideological. It is by nature warlike in the violence of its clearing of a site, the sheer energy of its construction, and the naked exercise of its power to change social and environmental conditions.

But there is a more profound paradox lying in wait between war and architecture. This is to be found in the destruction of an idea of normal that their correspondence brings. In short, the everyday is not innocent of the violence by which war is usually stigmatized, or elevated, depending on point of view; it merely conceals domestic violence and other forms of physical and emotional aggression under the label "abnormal." The "normal" actually consists of routines, procedures, protocols, manners, and customs that create the appearance of some fundamental relationships between the people of a culture and are asserted as being "true" by their universal nature. In fact, this commonality has nothing to do with the actual relationships between people, or the lack of them. Indeed, people hardly know, or even see, each other. What they have in common, besides the conventions that they are more or less coerced into observing, is exactly their lack of commonality – which is to say, their uniqueness and individuality. By disrupting or destroying the comforting, repetitious patterns of conventional thought and behavior, war in effect strips away the mask of the normal, revealing an actuality that is both liberating and terrifying.

To survive in war means to become extremely conscious of one's particular place and movements, of one's every decision, need, and action – in other words, to become much more independent and self-aware, and, in very tangible terms, self-responsible. This is liberating because consciousness is heightened; emotions and reasonings are necessarily more sharp, precise, and consequential. In taking personal responsibility for one's own survival, one feels in certain ways stronger and more in control. At the same time, the loss of a set pattern of predictability, of the sense that one is acting within a larger social whole in which responsibilities are shared, can be terrifying. The feeling of being alone, vulnerable, and isolated within the limitations of one's own individuality and abilities cannot help but produce anxiety and alienation. The loss of the normal is a loss of innocence that can never be recovered.

As a result of war, architecture, too, suffers this loss. Formerly it could be concerned simply with enhancing the normal, with giving routine behavior and cultural conventions aesthetic qualities that seem to elevate them beyond the power of their own monotonous claim to innocence. By exalting the everyday activities of people and the social institutions that direct them, architecture could dignify them, reinforcing their truth. It could suggest that life, even at its most ordinary, was worth living because it is part of a much larger world whose beauty, taken as a whole, is inaccessible, yet can still be glimpsed in moments of "good design." During war, "good design" can only refer to lost innocence. In the aftermath of war, it can only be reinstituted with a substantial degree of disingenuousness. What is called for then is neither something "good" nor "true," but rather something paradoxically human: an architecture of liberation and terror.

Alas, the poor architects – how roughly war treats them! It not only takes the moral high ground from under them, but it demands that they, like all those who are forced to come face to face with the harsher conditions of existence, painstakingly revise the superstructures of their practices and precepts – in short, their very conception of architecture, and of themselves. Everyday war is the worst. Where once they could believe they were creating beauty, raising humanity to a higher standard of living or even a higher level of existence – all of which might, on occasion, be true – they (or at least those with eyes to see) must become aware that in so doing they are soldiers in an army engaged in the conquest of space, and at the command of very particular individuals and institutions, whose ends, if they are to be realized, must be pursued ruthlessly, in other words, in the name of many. Whose idea of "beauty" is being served? Who benefits when the dust has settled? Whose idea of "higher" was monumentalized through immense expenditure of human lives and resources? What was lost in order that some could gain? The answers to these questions are not as clear as the individuals and institutions, the clients of architecture, would have the architects believe. Far from it, they can only be reached through an endless maze of questions and personal choices.

All of this is not to say that construction equals destruction, but only that they are, paradoxically, part of one another. "Architecture is war. War is architecture." – the first phrase would make no sense without the one that follows it. This does not imply that architects should not build, or that they should not lend themselves to wars between cultures, classes, and ideas, but only that they should not deceive themselves about what they are doing, and why. They should make their choices consciously, honestly, and with their eyes open, accepting personal responsibility for what they do. Today, this is seldom the case. If it were, the likelihood is that more well-intentioned architects would not choose to build what they do.

SECTION XII

POSSIBLE FUTURES

INTRODUCTION

The future of the city is not, of course, predictable. Given that the complexity of the city makes our understanding of it as it already exists a near impossibility, so any vision of the future is even more fraught with difficulty. This is not to say futurology is entirely impossible. Certain new trends, certain conditions, certain cultures can always be identified in contemporary cities that might be presumed, with reasonable confidence, to play an important part in urban futures. The trick, however, is to identify which of these many possible trends will prove to be the most significant.

One of the most common responses to this kind of conjecture is to turn to technology, and to try to see what the most advanced science of today (both realized and imagined) might lead towards. In 'Soft Cities' (1996), William Mitchell considers the technology of the computer and the cyberspatial world to which it gives access. Although first seen as part of computer games, cyberspace has become a global phenomenon of Internet link-ups and immediate, international data transfer. This is not just a medium of exchange, however. Cyberspace is a mental and conceptual space, a space that takes time to enter and use, a place of codes and instructions, a space of power. Cities will, Mitchell foresees, change accordingly – fewer face-to-face interactions will be accompanied by an accessing of 'places' in unknown geographic locations dispersed around the world. Above all, we will consider privacy over shelter, informational protection over social contact.

In a rather different treatment of technology, Corinne McLaughlin and Gordon Davidson, in an extract from *Builders of the Dawn* (1985, Book Publishing Company), consider the experimental city of Arcosanti in Arizona. Designed by architect Paolo Soleri as an 'arcology' (part 'architecture', part 'ecology') for 5,000 people, Arcosanti uses environmental design and technology to preserve water and energy, eliminate cars and promote farming and leisure activities. The culture of the embryonic city is also disseminated through visitors, temporary students and construction workers. Unlike Mitchell's cyberspace, this is a profoundly physical future, using architectural and landscape forms to generate much of its technological drive. Also contrary to cyberspace is the focus on social and cultural density, on immediate social exchanges, and a cultural response to the beauty of the natural landscape.

If Arcosanti represents the culture of density in marginal lands, then the 'Edge City' represents the lack of density of suburban developments. In his book of that name, (1991, Anchor), included in the list of suggested further reading, Joel Garreau argues that suburban growth has led to the formation of a new kind of a city: a city without a core, which is dispersed across a landscape of shopping malls, residential communities and business parks. Edge City residents live and work in no particular place, often driving large distances to connect up the distant places which they must visit. Edge City forms are, therefore, not skyscrapers, sidewalks and streets, but jogging paths, campuses, low-rise buildings, freeways and satellite dishes. But, however, we do not know how these new cities will eventually turn out, what forms they will adopt and cultures they will produce. For the moment, all we can do is note their existence and struggle to comprehend them.

Echoes of Garreau's observations are to be heard in 'Whatever Happened to Urbanism', in the architect Rem Koolhaas' *S,M,L,XL* (1995, Monacelli Press), another extract of which is also included in Section I, Forms and Spaces of the City. Koolhaas notes that just as urbanization has conquered the world, so urbanism as a profession has disappeared. The recent rediscovery of the classical city has also come at the very moment when it is of no relevance to the mega-city, the global city, of the twenty-first century. Koolhaas' vision of the future is one of despair for the urbanism profession, seeing cities as offering only individual buildings (our refuge in the parasitic security of architecture), unless, he suggests, we can learn to stage uncertainty and potentiality, to create indefinite forms and hybrids, to reinvent psychological space. Urbanism will then be not a profession or a thing, but a way of thinking.

In the essay by Anne Pendleton-Jullian, from *The Road That is Not a Road: The Open City, Ritoque, Chile* (1996, MIT), we find an even more extreme faith in the power of the imagination to help create and control our cultural futures. Pendleton-Jullian sees poetry, in particular, as something to be translated into all manner of imaginative activity, opening up different readings and realities. Referring to the experiments of both the French *poètes maudits* and the surrealist poets, she sees such creativity as involving at once image and words, discovery and adventure. What matters above all in such formulations is not the creative discipline itself (whether poetry, sculpture, painting or architectural design) but the essence of creativity itself. Herein, Pendleton-Jullian suggests, is the most productive of possible futures.

'ARCOSANTI: Urban "Arcology" — Ecologically Sound Architecture'

from Builders of the Dawn (1985)

For those who argue that small-scale rural communities can't possibly provide solutions to all the crises in the modern urban world, Arcosanti offers a model of a *city* for the future. Like the Socrates of Plato's *Republic*, Arcosanti members believe the city is a necessary instrument for the evolution of the human spirit. And Arcosanti is probably the most visually impressive and futuristic looking of all the communities. A visitor might feel as if s/he has come upon a set for a science fiction film. And yet Arcosanti also reminds one of something from the distant past – like the cities of ancient Aztecs, or perhaps even the old cathedrals of the Middle Ages, monuments that were never completed in their builders' lifetimes.

Arcosanti has been called "a minor wonder of the world ... probably the most important [urban architectural] experiment undertaken in our lifetime, by Douglas Davis of *Newsweek*." Both Arizona Senators Goldwater and DeConcini have recommended Arcosanti for federal funding, and Governor Babbitt declared a week in October as "Arcosanti Festival Week."

Arcosanti was designed by Paolo Soleri, an Italian laureate in architecture who studied with Frank Lloyd Wright. He calls his design for a city of five thousand people an "arcology" – the application of ecological principles to architecture. The word "arcology" points at the interdependence of the words "architecture" and "ecology" – the human abode within the planetary abode. Arcology is a positive response to the planetary problems of population, pollution, energy and natural resource depletion, food scarcity, and quality of life. It is estimated that Arcosanti will need 10% as much water as traditional development, due to greenhouses and recycling. Cars will be unnecessary in the city

itself, thus eliminating roads, traffic jams, gasoline lines, and pollution, but they can be available for use outside the city. Residents can walk to work, to friends' apartments, and to cultural events and entertainment. Arcosanti offers an alternative to current urban dilemmas and, at the same time, seeks to re-establish contact between people and an undisturbed landscape. Although Arcosanti will be a city for five thousand, only fourteen of its 860 acres will be used for the city structures, with the rest left as a nature preserve or for farming and recreation.

Arcosanti is located seventy miles north of Phoenix in the Arizona desert. Construction on the new city began in 1970 and is only about 2% complete today. There are currently about forty permanent residents. Although they often argue that they are a "construction site," not a "community," in fact a feeling of community is developing there through their shared work and common purpose. Up to 150 students come to Arcosanti for three-week workshops to learn this type of architecture. Over the years, more than three thousand people have participated in the building program, and sometimes a hundred people a day visit Arcosanti for a tour. Cement is used as the basic building material, as wood is not found in the ecology of the Arizona desert. Students learn formwork, silt-casting, concrete pouring, excavation, carpentry, metal work, welding, plumbing, and electrical work. Arcosanti appeals to people with a strong pioneering spirit, as its foundations have to be dug out of solid rock, often in the 110-degree heat of the Arizona desert. Donations, as well as income from workshops, Paolo Soleri's lectures, and the sale of bronze bells made in their foundry, support Arcosanti's building projects.

Arcosanti's "two suns arcology" (designed for the physical sun as well as the human sun) is an extraordinary energy-efficient design, integrating the effects of apse, greenhouse, chimney, and heat sink. The two apses (an apse is a quarter sphere) face south, providing shade during the hot summer when the sun is higher in the sky and letting in the warm rays of the sun during the winter when the sun is lower in the sky. The main housing structure, which has not been built yet, will be twenty-five stories high and resemble a modern skyscraper turned on its side, bent like a boomerang. Referred to as a double-bladed structure, it will face south for sun-warming in the winter and shade in the summer, like the smaller apses already built. Units in the structure are designed so that on one side they view the breathtaking wonder of the desert's mesas, and on the other side they view the pulsating inner-city life.

Below the main housing structure, a four and one-half acre terraced greenhouse is planned for growing food and trapping solar heat. The hot air will rise through giant chimneys to heat the housing structure in the winter. The top three levels of the greenhouse will use evaporation to cool the air, which will then sink through the chimneys to cool the lower parts of the housing structure during the summer. The concrete used in construction will act as a heat sink, collecting and storing solar heat, and then discharging it when the surrounding air is cooler. These elements will make it a passive solar heating and cooling system.

Work on the greenhouse began last year with horizontal semi-cylindrical retaining walls that define the terracing of the sloping greenhouse. The columns support the greenhouse membrane and are designed as large "tree pots" containing deciduous trees or vines that will shade the membrane in the summer months. In addition, Arcosanti has built two experimental greenhouses and cultivates more than forty acres of hay and green manure crops and five acres of organic vegetables and herbs. They've also been experimenting with vineyards and orchards, beekeeping, biological pest control strategies, and food processing-preservation techniques. Completed building projects include simple residences for forty members and workshop participants, a pottery apse, a foundry apse for making bells that are sold, a visitors' center with café, bakery, and gallery, a music center, a swimming pool, and two "vault" buildings for general usage. A new structure to be built, called the West Lab or Veleda Springs, will be a combination of residences, greenhouses, and a garment manufacturing business.

Many critics believe that clustering people so densely at Arcosanti into one huge structure will have disastrous effects. The dehumanizing high-rise apartments of our major cities make people feel like sardines, they claim. But to Arcosanti members, these were not designed from wholistic or ecological principles. They feel that contemporary cities are architecturally, economically, and socially fragmented, and are not whole.

"Arcosanti plans to prove that large-scale human development need not be inhuman," comments member Patrick Colvin. He also says they have plans to build more arcologies like Arcosanti and network them, so that people can move around and live at several of them during the year, thus creating "habitats for a mobile society."

Unlike most communities that concern themselves with trying to create the perfect system of governance, economics, relationships, etc., Arcosanti has a different approach. As member Russell Ferguson expresses it, "We need only concern ourselves with building, trusting that the urban effect will take over and the forms shape the inhabitants." It is an interesting experiment.

Arcology aims at creating a whole integrated environment, allowing its residents to identify with the whole city. "Arcosanti is an art object and is lovable," says designer Paolo Soleri. He believes that this will create a much better community spirit and alleviate the social alienation so common in our major cities. And he feels that the increasing complexity of cities enhances the evolution of the human spirit, unlike the standardized and sprawling suburbs, which appear counter to the nature of evolution, possibly lowering social intelligence. Life seems to cluster because people are social, Soleri feels. "The richness of crowding lies in its being the intrinsic facilitator of connectedness, i.e., coherence and sensitization."

Arcosanti members feel that the city structure must contract and miniaturize in order to support complex economic, social, and cultural activities and to give people a new perspective and renewed trust in society and in the future. "One direct pay-off of complexity-miniaturization is direct and instantaneous access to all the institutions of the town (living, learning, working, etc.) as well as to the open countryside as 'the foot' of the town," Soleri observes. "The democratic nature of such accessibility should be evident as it is critical."

Arcosanti proposes an economy of frugality as a more effective and equitable path into the future for a planet limited in resources. It is located on marginal land, not only to demonstrate the viability of a community on such land but also because of the beauty and inspiration that the land engenders. It intends to make use of its limited agricultural potential to partially satisfy the needs of its residents and keep them in touch with one fact of life: the food chain. Fruits and vegetables growing "at the foot of the town" are a tangible and educational fact, giving the town a greater resilience.

> If the sacred is where life is, then the whole habitat is sacred . . . The city is the producer and the market place, but the city is also the temple, the city is the palace, and the citizen is the sacred carrier of well-being and of grace to be. To build Arcosanti should be to build the church and the palace in one. This is why I hope . . . that the physical structure of Arcosanti is not just an instrument but also music.

Soleri says that the buildings at Arcosanti can glow with the light of matter transforming into spirit. When a visitor asked if anyone had seen them glow, she was met with a sudden, hushed reverence, as if entering sacred ground – a remarkable silence among such swaggering, brusque people. "No, no one has actually seen them glow . . . *yet*."

58 WILLIAM MITCHELL

'Soft Cities'

from *Architectural Design* (1995)

Each window on my computer screen is an electronic forest *avantureuse*, a digital Broceliande. When I choose to enter this microworld, I have to play strictly by its rules.

Any piece of software creates a space in which certain rules rigorously apply, but with video games, the rules are the whole point. Without them there would be no game, and hence no fun. *Vide*!

Lemmings! Minimunchkin-mannikins tramp across the pixel patterns, timed to the tinny beat of the endless electromuzak; cast as an old softy of a God, I must intervene to save these tiny numskulls from otherwise-certain self-destruction.

'Eat fire, bug-eyed scum!' Suddenly I'm lost in the testosterone-fuelled fascist funhouse of Strike Squad; I must blow away successively more fearsome insectoid gladiators or face gory extinction myself. Kill or be killed.

SimCity! I get to play earnest urban policy wonk, but don't have to face any angry community when (as wonks are apt to do) I screw up. Its creator wrote, 'Access to a "toy" city gave me a guinea pig on which I could try out my city planning experiments'. Good, clean fun until you notice the implicit political agenda. Its underlying structure is that of Jay Forester's mechanistic, conservative planning models from the 1950s and 60s. It presents you with a microworld magically free from racial divisions, labour unions, developers, or preservationists – one in which planners always win by building infrastructure (even when it carves up communities), lowering taxes, attracting industry, and creating jobs.

Realms of Arkania! 'Game features: more than 70 towns, villages, dungeons and ruins; twelve character archetypes; seven positive and seven negative character attributes; over 50 skills; twelve magic realms with over 80 spells; auto-combat option; and parties of up to six characters may be split and regrouped'.

F-15 Strike Eagle III! 'Surround yourself in a revolutionary new three-dimensional graphics system that provides you with a digitised map of downtown Baghdad complete with every bridge, the TV famous Air Ministry building and the "Baby Milk" factory! Cheat death in three explosive scenarios including Desert Storm, Korea and Central America!'

Metal & Lace: The Battle of the Robo Babes! 'Give your joystick a thrill . . . With the intense heat and action, you'll both end up with less than full body dress'.

Doom! The geography of this hack-and-slash masterpiece – the shareware sensation of 1994's gloomy winter – is three-dimensional, realistic, and extensive; you have a whole virtual city to get to know. The architecture is complex, the surfaces are textured, and the lighting and sound effects are eerily convincing. But the really big hook is that it's a fast-paced, cooperative, network game; physically separated game freaks can get 'together' in virtual spaces to explore them 'side-by-side'. You can see the computer-animated 'bodies' of your companions and they can see yours. You can communicate with each other as you blast thorny brown hominids with plasma rifles, or chainsaw flaming skulls that come flying at you out of nowhere. You get to gloat guiltily as your comrades-in-arms are mugged by the monsters that they meet and come to grisly, twitching ends. And in the 'death-match scenario', the instructions are to 'Kill everything that moves, including your buddies'.

Read the ads (bizarre as Borges, quirkier than

Calvino) in *Computer Gaming World* or *Electronic Gaming Monthly*. Imagine yourself a mips-driven Marco Polo, a cybersurfing Gulliver; visit a few video game microworlds and engage in the action. The petty-Faustian bargain that all software offers will soon become vividly apparent; enter a digitally constructed world, accept its constitution and its rules, and you buy into its ideology. Love it or leave it.

REAL ESTATE/CYBERSPACE

I was there at the almost-unnoticed Big Bang – the silent blast of bits that begat the universe of these microworlds. UCLA, the fall of 1969, and I was a very young assistant professor writing primitive CAD software and trying to imagine the role that designers might play in the emerging digital future; in a back room just down the hallways from the monster mainframe on which I worked, some Bolt Beranek and Newman engineers installed a considerably smaller machine that booted up to become the very first node of ARPANET – the computer network that was destined to evolve into the worldwide Internet.

From this inconspicuous origin point, network tentacles grew like kudzu to blanket the globe. Soon, cyberspace was busting out all over, and the whole loosely organised system became known as the Internet. During the late 80s and early 90s more and more networks connected to the Internet, and by 1993 it included nearly two million host computers in more than 130 countries. Then, in the first six months of 1994, more than a million additional machines were hooked up.

While the Internet community was evolving into something analogous to a ramshackle Roman Empire of the entire computer world, numerous smaller, independent colonies and confederations were also developing. Dial-in bulletin board systems such as the Sausalito-based Well – much like independent city-states – appeared in many locations to link home computers. Before very long, though, most of these erstwhile rivals found it necessary to join forces with the Internet as well. There would not have been a great deal to connect if computers had remained the large and expensive devices that they were when ARPANET began in 1969.

But as networks developed, so did inexpensive personal computers and mass-marketed software to run on them. The very first, the Altair, showed up in 1974, and it was followed in the early 80s by the first IBM PCs and Apple Macintoshes. Each one that rolled off the assembly line had its complement of RAM and a disk drive, and it expanded the potential domain of cyberspace by a few more megabytes of memory.

Somewhere along the line, our conception of what a computer really was began to change fundamentally. It turned out that these electronic boxes were not really big, fast, centralised calculating and data-sorting machines as ENIAC, UNIVAC, and their mainframe successors had led us to believe. No; they were primarily communication devices – not dumb ones like telephone handsets, that merely encoded and decoded electronic information, but smart ones that could organise, interpret, filter, and present vast amounts of information for us. Their real role was to construct cyberspace.

WILD WEST/ELECTRONIC FRONTIER

It was like the opening of the Western Frontier. Parallel, breakneck development of the Internet and of consumer computing devices and software quickly created an astonishing new condition; a vast, hitherto-unimagined territory began to open up for exploration. Early computers had been like isolated mountain valleys ruled by programmer-kings; the archaic digital world was a far-flung range in which narrow, unreliable trails provided only tenuous connections among the multitudinous tiny realms. An occasional floppy disk or tape would migrate from one to the other, bringing the makings of colonies and perhaps a few unnoticed viruses. But networking fundamentally changed things – as clipper ships and railroads changed the pre-industrial world – by linking these increasingly numerous individual fragments of cyberturf into one huge, expanding system.

By the 1990s, the digital electronics and telecommunications industries had configured themselves into an immense machine for the ongoing production of cyberspace. We are rapidly approaching a condition in which every last bit of computer memory in the world is electronically linked to every other. This vast grid is the new land beyond the horizon, the place that beckons the colonists, cowboys, con-artists, and would-be conquerors of the 21st century. And there are those who would be King.

It will be there forever. Because its electronic underpinnings are so modular, geographically dispersed, and redundant, cyberspace is essentially indestructible. You can't demolish it by cutting links with backhoes or sending commandos to blow up electronic installations, and you can't even nuke it. If big chunks of the network were to be wiped out, messages would automatically reroute themselves around the damaged parts. If some memory or processing power were to be lost it could quickly be replaced. Since copies of digital data are absolutely exact replicas of the originals, it doesn't matter if the originals get lost or destroyed. And since multiple copies of files and programs can be stored at widely scattered locations, eliminating them all with certainty is as hard as lopping Hydra heads.

Cyberspace is still tough territory to travel, though, and we are just beginning to glimpse what it may hold. 'In its present condition', Mich Kapor and John Perry Barlow noted in 1990, 'Cyberspace is a frontier region, populated by the few hardy technologists who can tolerate the austerity of its savage computer interfaces, incompatible communications protocol, proprietary barricades, cultural and legal ambiguities, and general lack of useful maps or metaphors'. And they warned: 'Certainly, the old concepts of property, expression, identity, movement, and context, based as they are on physical manifestation, do not apply succinctly in a world where there can be none'.

HUMAN LAWS/CODED CONDITIONALS

Out there on the electronic frontier, code is the Law. The rules governing any computer-constructed microworld – of a video game, of your personal computer desktop, of a word processor window, of an automated teller machine, or of a chat room on the network – are precisely and rigorously defined in the text of the program that constructs it on your screen. Just as Aristotle, in *The Politics*, contemplated alternative constitutions for city-states (those proposed by the theorists Plato, Phaleas, and

Hippodamos, and the actual Lacedaemonian, Cretan, and Carthaginian ones) so denizens of the digital world should pay the closest of critical attention to programmed polity. Is it just and humane? Does it protect our privacy, our property, and our freedom? Does it constrain us unnecessarily, or does it allow us to act as we may wish?

At a technical level, it's all a matter of the software's conditionals – those coded rules that specify if some condition holds then some action follows. Consider, for example, the familiar rituals of withdrawing some cash from an ATM. The software running the machine has some gatekeeper conditionals; if you have an account, and if you enter the correct PIN number (the one that matches up, in a database somewhere, with the information magnetically encoded on your ATM card), then you can enter the virtual bank. (Otherwise you are stopped at the door. You may have your card confiscated as well.) Next the program presents you with a menu of possible actions – just as a more traditional bank building might present you with an array of appropriately labelled teller windows or (on a larger scale) a directory pointing you to different rooms: if you indicate that you want to make a withdrawal, then it asks you to specify the amount; if you want to check your balance, then it prints out a slip with the amount; if you want to make a deposit, then yet another sequence of actions is initiated. Finally, the program applies a banker's rule; if the balance of your account is sufficient (determined by checking a database), then it physically dispenses the cash and appropriately debits the account.

To enter the space constructed by the ATM system's software you have to submit to a potentially humiliating public examination – worse than being given the once-over by some snotty and immovable receptionist. You are either embraced by the system (if you have the right credentials) or excluded and marginalised by it right there in the street. You cannot argue with it. You cannot ask it to exercise discretion. You cannot plead with it, cajole it or bribe it. The field of possible interactions is totally delimited by the formally stated rules.

So control of code is power. For citizens of cyberspace, computer code – arcane text in highly formalised language, typically accessible only to a few privileged high-priests – is the medium in which intentions are enacted and designs are realised, and it is becoming a crucial focus of political contest. Who shall write the software that increasingly structures our daily lives? What shall that software allow and proscribe? Who shall be privileged by it and who marginalised? How shall the writers of the rules be answerable?

PHYSICAL TRANSACTIONS/ ELECTRONIC EXCHANGES

Historically, cities have also provided places for specialised business and legal transactions. In *The Politics*, Aristotle proposed that a city should have both a 'free' square in which 'no mechanic or farmer or anyone else like that may be admitted unless summoned by the authorities' and a marketplace 'where buying and selling are done ... in a separate place, conveniently situated for all goods sent up from the sea and brought in from the country'. Ancient Rome had both its *fora civila* for civic assembly and its *fora venalia* for the sale of food. These Roman markets were further specialised by type of produce; the *holitorium* was for vegetables, the *boarium* for cattle, the *suarium* for pigs, and the *vinarium* for wine. Medieval marketplaces were places both for barter and exchange and for religious ritual. Modern cities have main streets, commercial districts, and shopping malls jammed with carefully differentiated retail stores in which the essential transaction takes place at the counter – the point of sale – where money and goods are physically exchanged.

But where electronic funds transfer can substitute for physical transfer of cash, and where direct delivery from the warehouse can replace carrying the goods home from the store, the counter can become a virtual one. Television home shopping networks first exploited this possibility – combining cable broadcast, telephone, and credit card technologies to transform the purchase of zirconium rings and exercise machines into public spectacle. Electronic 'shops' and 'malls' provided on computer networks (both Internet and the commercial dial-up services) quickly took the idea a step further; here customer and store clerk do not come face-to-face at the cash register, but interact on the network via a piece of software that structures the exchange of digital tokens – a credit care number to charge, and a specification of the required goods. (The exchange might then become so simple and standardised that the clerk can be replaced completely by a software surrogate.) As I needed books for reference in preparing this text, I simply looked up their titles and ISBN numbers in an on-line Library of Congress catalogue, automatically generated and submitted electronic mail purchase orders, and received what I requested by courier – dispatched from some place that I had never visited. The charges, of course, showed up on my credit card bill.

Immaterial goods such as insurance policies and commodity futures are most easily traded electronically. And the idea is readily extended to small, easily transported, high value speciality items – books, computer equipment, jewelery, and so on – the sorts of things that have traditionally been sold by mail order. But it makes less sense for grocery retailing and other businesses characterised by mass markets, high bulk, and low margins. Cyberspace cities, like their physical counterparts, have their particular advantages and disadvantages for traders, so they are likely to grow up around particular trade specialisations. Since you cannot literally lay down your cash, sign a cheque, or produce a credit card and flash an ID in cyberspace, payment methods are being reinvented for this new kind of marketplace. The Internet and similar networks were not initially designed to support commercial transactions, and were not secure enough for this purpose. Fortunately though, data encryption techniques can be used to provide authentication of the identities of trading partners, to allow secure exchange of sensitive information such as credit card numbers and bid amounts, and to affix digital 'signatures' and time stamps to legally binding documents. By the summer of 1994, industry standards for assuring security of Internet transactions were under development, and online shopping services were beginning to offer encryption-protected credit card payment. And the emergence of genuine digital cash – packages of encrypted data that behaved like real dollars, and could not be traced like credit card numbers – seemed increasingly likely.

In traditional cities, transaction of daily business was accomplished literally by handing things over; goods and cash crossed store counters, contracts were physically signed, and perpetrators of illegal transactions were sometimes caught in the act. But in virtual cities, transactions reduce to exchanges of bits.

STREET MAPS/HYPERPLANS

Ever since Ur, doorways and passageways have joined together the rooms of buildings, webs and grids of streets have connected buildings to each other, and roads have linked cities. These physical connections provided access to the places where people lived, worked, worshipped, and entertained themselves.

Since the winter of 1994, I have had a remarkable piece of software called Mosaic on the modest desktop machine that I am using to write this paragraph. Mosaic, and the network of World Wide Web servers to which it provides access, work together to construct a virtual rather than physical world of places and connections; the places are called 'pages' and they appear on my screen, and the connections – called hyperlinks – allow me to jump from page to page by clicking on highlighted text or icons.

A World Wide Web 'home page' invites me to step, like Alice through the looking glass, into the vast information flea-market of the Internet. The astonishing thing is that a WWW page displayed on my screen may originate from a server located anywhere in the Internet. In fact, as I move from page to page, I am logging into machines scattered around the world. But as I see it, I jump almost instantaneously from virtual place to virtual place by following the hyperlinks that programmers have established – much as I might trace a path from piazza to piazza in a great city along the roads and boulevards that a planner had provided. If I were to draw a diagram of these connections I would have a kind of street map of cyberspace. MUD crawling is another way to go. Software systems known as MUDs – Multi-User Dungeons – have burned up countless thousands of Internet log-in hours since the early 1980s. These structure on-line, interactive, role-playing games, often attracting vast numbers of participants scattered all over the net. Their particular hook is the striking way that they foreground issues of personal identity and self-representation; as initiates learn at old

MUDder's knees, the very first task is to construct an on-line persona for yourself by choosing a name and writing a description that others will see as they encounter you. It's like dressing up for a masked ball, and the irresistible thing is that you can experiment freely with shifts, slip-pages, and reversals in social and sexual roles and even try on entirely fantastic guises. How does it really feel to be a complete unknown?

Once you have created your MUD character, you can enter a virtual place populated with other characters and objects. This place has exits – hyperlinks connecting it to other such settings, and these in turn have their own exits; some MUDs are vast, allowing you to wander among thousands of settings – all with their own special characteristics – like Baudelaire strolling through the buzzing complexity of 19th-century Paris. You can examine the settings and objects that you encounter, and you can interact with the characters that you meet.

But as you quickly discover, the most interesting and provocative thing about a MUD is its constitution – the programmed-in rules specifying the sorts of interactions that can take place and shaping the culture that evolves. Many are based on popular fantasy narratives such as *Star Trek*, Frank Herbert's *Dune*, CS Lewis's *Chronicles of Narnia*, the Japanese animated television series *Speed Racer*, and even more doubtful products of the literary imagination; these are communities held together, as in many traditional societies, by shared myths. Some are set up as hack 'n slash combat games in which bad MUDders will try to 'kill' your character; these, of course, are violent, Darwinian places in which you have to be aggressive and constantly on your guard. Others, like many of the TinyMUDs, stress ideals of constructive social interaction, egalitarianism, and nonviolence – MUDderhood and apple pie. Yet others are organised like high-minded lyceums, with places for serious discussion of different scientific and technical topics. The MIT-based Cyberion City encourages young hackers – MUDders of invention – to write MUSE code that adds new settings to the environment and creates new characters and objects. And some are populated by out-of-control, crazy MUDders who will try to engage your character in TinySex – the one-handed keyboard equivalent of phone sex.

Early MUDs – much like text-based adventure video games such as Zork – relied entirely on typed descriptions of characters, objects, scenes, and actions. (James Joyce surely would have been impressed; city as text and text as city. Every journey constructs a narrative.) But greater bandwidth, faster computers, and fancier programming can shift them into pictorial and spatial formats.

ENCLOSURE/ENCRYPTION

In physically constructed cities, the enclosing surfaces of constituent spaces – walls, floors, ceilings, and roofs – provide not only shelter, but also privacy. Breaches in these surfaces – gates, doors, and windows – have mechanisms to control access and maintain privacy; you can lock your doors or leave them open, lower the window shades or raise them. Spatial divisions and access control devices are deployed to arrange spaces into hierarchies grading from completely public to utterly private. Sometimes you have to flip your ID to a bouncer, take off your shoes, pay admission, dress to a doorman's taste, slip a bribe, submit to a search, speak into a microphone and wait for the buzzer, smile at a receptionist, placate a watchdog, or act out some other ritual to cross a threshold into a more private space. Traditions and laws recognise these hierarchies, and generally take a dim view of illicit boundary crossing by trespassers, intruders, and peeping Toms.

Different societies have distinguished between public and private domains (and the activities appropriate to them) in differing ways, and cities have reflected these distinctions. According to Lewis Mumford, domestic privacy was 'a luxury of the well-to-do' up until the 17th century in the West. The rich were the people who could do pretty much what they wanted, as long as they didn't do it in the street and frighten the horses. Then, as privacy rights trickled down to the less advantaged classes, the modern 'private house' emerged, acquired increasingly rigorous protections of constitutional law and public policy, and eventually became the cellular unit of suburban tissue. Within the modern Western house itself – in contrast with some of its ancient and medieval predecessors – there is a carefully organised gradation from relatively public verandahs, entry halls, living rooms and parlours to more private, enclosed bedrooms and bathrooms where you can shut and lock the doors and draw down the shades against the outside world.

It doesn't rain in cyberspace, so shelter is not an issue, but privacy certainly is. So the construction technology for virtual cities – just like that of bricks-and-mortar ones – must provide for putting up boundaries and erecting access controls, and it must allow cyberspace architects and urban designers to organise virtual places into public-to-private hierarchies.

Fortunately, some of the necessary technology does exist. Most obviously, the rough equivalent of a locked gate or door, in cyberspace construction, is an authentication system. This controls access to virtual places (such as your electronic mail inbox) by asking for identification and a password from those who request entry. If you give the correct password, you're in. The trouble, of course, is that passwords – like keys – can be stolen and copied. And they can sometimes be guessed, systematically enumerated till one that works is found, or somehow extorted from the system manager who knows them all. So password-protection – as with putting a lock on a door – discourages illicit entry, but does not block the most determined break-in artists.

Just as you can put the valuables that you really want to protect in a sturdy vault or crypt, though, you can build the strongest of enclosures around digital information by encrypting it – scrambling it in a complex way so that it can only be decoded by somebody with the correct secret numerical key. The trick is not only to have a code that is difficult to crack, but also to manage keys so that they do not fall into the wrong hands, and the cleverest known way to do this is to use a technique called RSA public-key encryption. In this system, which derives its power from the fundamental properties of large prime numbers, each user has both a secret 'private' key and a 'public' key that can be distributed freely. If you want to send a secure message, you first obtain the intended recipient's public key, and use that to encode the information. Then the recipient decodes it using the private key.

Under pressure from cops and cold warriors, who anticipate being thwarted by impregnable fortresses in cyberspace, the US Federal Government has doggedly tried to restrict the availability of strong encryption software. But in June 1991, hacker folk-hero Philip Zimmerman released his soon-to-be-famous, RSA-based Pretty Good Privacy (PGP) encryption program. By May 1994, commercial versions had been licensed to over four million users, and MIT had released a free, non-commercial version that anybody could legally download from the Internet. From that moment, you could securely fence off your private turf in cyberspace.

Meanwhile, the Clinton Administration pushed its plans for the Clipper Chip – a device that would accomplish much the same thing as RSA, but would provide a built-in "trapdoor" for law-enforcement wiretapping and file decoding. The effect is a lot like that of leaving a spare set of your front door keys in a safe at FBI headquarters. Opinion about this divided along predictable lines. A spokesman for the Electronic Frontier Foundation protested, 'The idea that the Government holds the keys to all our locks, before anyone has even been accused of committing a crime, doesn't parse with the public'. But an FBI agent, interviewed in the *New York Times*, disagreed: 'OK, someone kidnaps one of your kids and they are holding this kid in this fortress up in the Bronx. Now, we have probable cause that your child is inside this fortress. We have a search warrant. But for some reason, we cannot get in there. They made it out of some new metal, or something, right? Nothing'll cut it, right? ... That's what the basis of this issue really is – we've got a situation now where a technology has become so sophisticated that the whole notion of a legal process is at stake here ... If we don't want that, then we have to look at Clipper'.

So the technological means to create private places in cyberspace are available, but the right to create these places remains a fiercely contested issue. Can you always keep your bits to yourself? Is your home page your castle?

59 REM KOOLHAAS

'Whatever Happened to Urbanism?'

from Rem Koolhaas and Bruce Mau, *S, M, L, XL* (1995)

This century has been a losing battle with the issue of quantity.

In spite of its early promise, its frequent bravery, urbanism has been unable to invent and implement at the scale demanded by its apocalyptic demographics. In 20 years, Lagos has grown from 2 to 7 to 12 to 15 million; Istanbul has doubled from 6 to 12. China prepares for even more staggering multiplications.

How to explain the paradox that urbanism, as a profession, has disappeared at the moment when urbanization everywhere – after decades of constant acceleration – is on its way to establishing a definitive, global "triumph" of the urban condition?

Modernism's alchemistic promise – to transform quantity into quality through abstraction and repetition – has been a failure, a hoax: magic that didn't work. Its ideas, aesthetics, strategies are finished. Together, all attempts to make a new beginning have only discredited the *idea* of a new beginning. A collective shame in the wake of this fiasco has left a massive crater in our understanding of modernity and modernization.

What makes this experience disconcerting and (for architects) humiliating is the city's defiant persistence and apparent vigor, in spite of the collective failure of all agencies that act on it or try to influence it – creatively, logistically, politically.

The professionals of the city are like chess players who lose to computers. A perverse automatic pilot constantly outwits all attempts at capturing the city, exhausts all ambitions of its definition, ridicules the most passionate assertions of its present failure and future impossibility, steers it implacably further on its flight forward. Each disaster foretold is somehow absorbed under the infinite blanketing of the urban.

Even as the apotheosis of urbanization is glaringly obvious and mathematically inevitable, a chain of rearguard, escapist actions and positions postpones the final moment of reckoning for the two professions formerly most implicated in making cities – architecture and urbanism. Pervasive urbanization has modified the urban condition itself beyond recognition. "The" city no longer exists. As the concept of city is distorted and stretched beyond precedent, each insistence on its primordial condition – in terms of images, rules, fabrication – irrevocably leads via nostalgia to irrelevance.

For urbanists, the belated rediscovery of the virtues of the classical city at the moment of their definitive impossibility may have been the point of no return, fatal moment of disconnection, disqualification. They are now specialists in phantom pain: doctors discussing the medical intricacies of an amputated limb.

The transition from a former position of power to a reduced station of relative humility is hard to perform. Dissatisfaction with the contemporary city has not led to the development of a credible alternative; it has, on the contrary, inspired only more refined ways of articulating dissatisfaction. A profession persists in its fantasies, its ideology, its pretension, its illusions of involvement and control, and is therefore incapable of conceiving new modesties, partial interventions, strategic realignments, compromised positions that might influence, redirect, succeed in limited terms, regroup, begin from scratch even, but will never reestablish control.

Because the generation of May '68 – the largest generation ever, caught in the "collective narcissism of a demographic bubble" – is now finally in power, it is tempting to think that it is responsible for the demise of urbanism – the state of affairs in which cities can no longer be made – paradoxically *because* it rediscovered and reinvented the city.

Sous le pavé, la plage (under the pavement, beach): initially, May '68 launched the idea of a new beginning for the city. Since then, we have been engaged in two parallel operations: documenting our overwhelming awe for the existing city, developing philosophies, projects, prototypes for a preserved *and* reconstituted city and, at the same time, laughing the professional field of urbanism out of existence, dismantling it in our contempt for those who planned (and made huge mistakes in planning) airports, New Towns, satellite cities, highways, high-rise buildings, infrastructures, and all the other fallout from modernization. After sabotaging urbanism, we have ridiculed it to the point where entire university departments are closed, offices bankrupted, bureaucracies fired or privatized. Our "sophistication" hides major symptoms of cowardice centered on the simple question of taking positions – maybe the most basic action in making the city. We are simultaneously dogmatic and evasive. Our amalgamated wisdom can be easily caricatured: according to Derrida we cannot be *Whole*, according to Baudrillard we cannot be *Real*, according to Virilio we cannot be *There*.

"Exiled to the Virtual World": plot for a horror movie. Our present relationship with the "crisis" of the city is deeply ambiguous: we still blame others for a situation for which both our incurable utopianism and our contempt are responsible. Through our hypocritical relationship with power – contemptuous yet covetous – we dismantled an entire discipline, cut ourselves off from the operational, and condemned whole populations to the impossibility of encoding civilizations on their territory – the subject of urbanism.

Now we are left with a world without urbanism, only architecture, ever more architecture. The neatness of architecture is its seduction; it defines, excludes, limits, separates from the "rest" – but it also consumes. It exploits and exhausts the potentials that can be generated finally only by urbanism, and that only the specific imagination of urbanism can invent and renew.

The death of urbanism – our refuge in the parasitic security of architecture – creates an immanent disaster: more and more substance is grafted on starving roots.

In our more permissive moments, we have surrendered to the aesthetics of chaos – "our" chaos. But in the technical sense chaos is what happens when nothing happens, not something that can be engineered or embraced; it is something that infiltrates; it cannot be fabricated. The only legitimate relationship that architects can have with the subject of chaos is to take their rightful place in the army of those devoted to resist it, and fail.

If there is to be a "new urbanism" it will not be based on the twin fantasies of order and omnipotence; it will be the staging of uncertainty; it will no longer be concerned with the arrangement of more or less permanent objects but with the irrigation of territories with potential; it will no longer aim for stable configurations but for the creation of enabling fields that accommodate processes that refuse to be crystallized into definitive form; it will no longer be about meticulous definition, the imposition of limits, but about expanding notions, denying boundaries, not about separating and identifying entities, but about discovering unnameable hybrids; it will no longer be obsessed with the city but with the manipulation of infrastructure for endless intensifications and diversifications, shortcuts and redistributions – the reinvention of psychological space. Since the urban is now pervasive, urbanism will never again be about the "new," only about the "more" and the "modified." It will not be about the civilized, but about underdevelopment.

Since it is out of control, the urban is about to become a major vector of the imagination. Redefined, urbanism will not only, or mostly, be a profession, but a way of thinking, an ideology: to accept what exists. We were making sand castles. Now we swim in the sea that swept them away.

To survive, urbanism will have to imagine a new newness. Liberated from its atavistic duties, urbanism redefined as a way of operating on the inevitable will attack architecture, invade its trenches, drive it from its bastions, undermine its certainties, explode its limits, ridicule its preoccupations with matter and substance, destroy its traditions, smoke out its practitioners.

The seeming failure of the urban offers an

exceptional opportunity, a pretext for Nietzschean frivolity. We have to imagine 1,001 other concepts of city; we have to take insane risks; we have to dare to be utterly uncritical; we have to swallow deeply and bestow forgiveness left and right. The certainty of failure has to be our laughing gas/oxygen; modernization our most potent drug. Since we are not responsible, we have to become irresponsible. In a landscape of increasing expediency and impermanence, urbanism no longer is or has to be the most solemn of our decisions; urbanism can lighten up, become a *Gay Science* – Lite Urbanism.

What if we simply declare that there *is* no crisis – redefine our relationship with the city not as its makers but as its mere subjects, as its supporters?

More than ever, the city is all we have.

60 ANNE PENDLETON-JULLIAN

'Extract'

from *The Road That is Not a Road — The Open City, Ritoque, Chile*

The central poetic preoccupation of the group, which was considered its "unconventional" approach, was the relationship of poetry to architecture, sculpture, and painting. It was to be a direct relationship, without mediating elements, and one which stated, and required, the bond of action to word. It was not poetry as a bias or sentiment but rather poetry as a way of acting and of doing creatively. The poetry around which the group united was that of the modern French poets – the *poétes maudits* – and of the surrealists: Baudelaire, Mallarmé, Rimbaud, Verlaine, Lautremont, Breton. This poetry involved a passionate quest in which poetry, no longer a commodity, transformed itself into poetic activity that aimed at recuperating the mind's original powers. The poet was an alchemist who employed the imagination to transform reality both mentally and physically, who embraced the mystery and adventure of creative activity as reality was opened up to a different reading, a different understanding, a different reality, ignited by the power of words to embrace multiplicity and plurality within the unifying body of poetic language.

The apperance of the poétes maudits can be attributed to a confluence of two key conditions: first, the deterioration of the social status of poetry and poets as wealthy patrons of the nineteenth century disappeared in the wake of the growth of the European bourgeoisie, who were neither entertained nor enlightened by poetry. Poetry lost its value as a commodity, as a means of employment. Concurrently, there was the birth of modernity and a heritage of a vision of the modern world formed by the German romantics Hölderlin and Novalis in which divine systems were no longer sufficient or relevant to existential meaning within modern reality.

Together, these provided the ground for the liberation of the poetic word from privileged claims and allowed it, in fact required it, to return to the source of language: to man. Man begins the return to himself for meaning and the responsibility that that implies. The "antagonism between the modern spirit and poetry begins as an agreement. With the same decision of philosophical thought, poetry tries to ground the poetic word on man himself. The poet does not see in his images the revelation of a secret power ... Poetic writing is the revelation of himself that man makes to himself."

For this revelation to form, it was imperative that the traditional subject matter supplied by previous poetries be dismissed along with the emotionalism attached to such subject matter. The effusive description of elements of the natural world and the tedious articulation of feelings or thoughts supplied by these poetries were no longer relevant as they obscured the perception of reality. In the modern French poetry, reality and "real things" are not described but are consciously put aside so that transparent contact with the profundity of reality may be discovered. It is a self-conscious program that begins from an intuitive suspicion that our relationship with reality is attached not only to surface perceptions but, even more important, to an enigmatic mental dimension that underlies physical reality in which all things are linked by "correspondences" between them. Not through similarities but through Charles Baudelaire's idea of the almost mystical familiarity and intimacy between things that are never the same, that can never be the same. They are connected while retaining a distance that can never be collapsed. Access to this unknown region is achieved by the imagination of the poet through its receptivity to the nonconcrete, through its use of language as a "forest of symbols" where mysterious and interconnected signs transport the mind and every sense.

Described as being "the original faculty of all human perception" by the English poet Coleridge, the imagination is capable of operating through reason, or of transcending reason, to perceive, select, judge, edit, transform. With the modern French poets, the making of poetic images combined with a necessary altering of language to discover as well as convey that which transcended reason. Words were not used to describe but to create images attached to the rhythm of language. For this reason, every new poem was to be a total recasting of its author's means of expression, disdaining all pre-established conventions and any innovations made prior to it in ways of thinking or ways of saying. Poetry was about discovery. It became adventurous. The fantastic and marvelous were stalked as the new poetry argued the reattachment of life to art: life consisting of the day to day, minute to minute, the mundane; and art being conceived through the word, specifically, because it is the element through which one engages both the world one acts within and the activity of the mind. It is a tool of perception and a tool of contemplation. Therefore, through the word, the poet is empowered to unite the processes of interpretation and transformation. "So we manage to have a synthetic attitude combining the need to transform the world radically and to interpret it as completely as possible."

To achieve this interpretation and transformation, it was necessary to alter language radically. Its very matter was operated on as words were forced to "lay bare their hidden life and reveal the mysterious trade they indulge in, independent of their meaning." It was asked to return to its origin, to discard the conventional relationship language had developed with meaning through arbitrary mental selection processes based on prescribed and learned value judgments. The "new" language was highly transparent and fluid, capable of joining sensation and thought by moving between the two with such facility as to totally obliterate the boundary between interior and exterior. It was to be, as Arthur Rimbaud described: "of the soul, for the soul, summarizing all, scents, sounds, colors, of thought hooking onto thought and pulling." The poem itself takes on an autonomous, ritualized quality as product of the imagination, and it becomes keyed to the participation of the reader as an experience in itself. Whereas poetry, the product, had previously been the ambition when it was marketable, it now becomes the by-product of poetic activity.

INDEX